Climate Change
Man and Environment

Dr. Goutam Kumar Saha passed B.Sc. (Honours in Zoology) from Burdwan University securing first class first position in the year 1981. In M.Sc. examination too, he stood first amongst the successful candidates in Zoology and received University gold medal and was awarded his Ph.D. degree in the year 1991. He has published 115 research papers in different journals of International repute and authored seven books/technical monographs to his credit. He has actively participated in 35 National and 14 International seminars and symposia and extensively visited different countries like Canada, Germany, France, Netherlands, Belgium, Italy, China, Thailand, Australia and Singapore in connection with his research work. He has successfully carried out a good number of research projects sponsored by DST, Govt. of India, DST, Govt. of W.B., UGC, Department of Forest, Govt. of W.B., MoEF, Govt. of India, W.B. Biodiversity Board, Central Zoo Authority, Peter Scott Foundation and World Pheasant Association (South Asia). He is best known for his pioneering work on Red Panda research in the Singhalila National Park, Darjeeling, India. His contribution in the field of wildlife research has been widely acclaimed by the international community and has been incorporated in the IUCN Action Plan 2002-2004. He started his professional carrier as a lecturer in Zoology at Post Graduate Department of Zoology, Darjeeling Govt. College, Darjeeling in the year 1990 and at present working as a Professor of Zoology, University of Calcutta. He is an honorary fellow of the Indian College of Allergy and Applied Immunology and The Zoological Society, Kolkata. Besides his excellence in the academic field, he is a good sportsman and represented the University as its blue and participated several National games.

Climate Change
Man and Environment

— *Editor* —

Goutam Kumar Saha

Professor of Zoology
University of Calcutta

2012

DAYA PUBLISHING HOUSE®

New Delhi – 110 002

© 2012 EDITOR
ISBN 9788170359845

Published by	:	**Daya Publishing House®** **A Division of** **Astral International Pvt. Ltd.** **ISO 9001:2008 Certified Company –** 4760-61/23, Ansari Road, Darya Ganj, New Delhi - 110 002 Phone: 23245578, 23244987 Fax: (011) 23260116 e-mail : dayabooks@vsnl.com website : www.dayabooks.com
Laser Typesetting	:	**Classic Computer Services** Delhi - 110 035
Printed at	:	**Chawla Offset Printers** Delhi - 110 052

PRINTED IN INDIA

Preamble

Diversity and perpetuation of life process is considered as a unique outcome of organic evolution influenced grossly by physical environment. Every biological form evolved under the backdrop of chemical and physical ambience of environment and is perfected by the process of natural selection. Undesirable change of global climate is an anthropogenic consequence, which has drawn a serious attention to the scientists and scholars of diverse disciplines. The dynamic equilibrium of ecological system has precariously being threatened by undesirable shift in the climatic parameters of planet Earth. The functional performance of intra and interspecific interaction has often been critical due to deterioration of environment. Till last decade, crisis of environmental pollution had been identified as a major detrimental issue to plant and animal life. Currently, the problem of climate change has definitely aggravated this problem of existence of life form on earth.

Climate change is a hard reality to human civilization affecting different facets of life and its environment. It is a fact that Earth's climate is undergoing an alteration at an unprecedented rate. It may be limited to a specific region, or may occur across the globe. Climate change can occur in different ways such as changes in regional and global temperature, changes in rainfall patterns, expansion and contraction of ice-sheets, sea level variation, plate tectonics, volcanism, ocean variability etc. However, the term is commonly used to refer specifically to climate change caused as a result of human activity. The observed temperature increase over the last fifty years has been mainly due to the increase in the concentrations of greenhouse gases like carbon dioxide, methane, water vapour and ozone to satisfy the selfish and one sided interest of the human being in the name of modern living. Changing weather patterns, retreating glaciers, polar ice melt, sea level rise, increased severity and occurrence of natural disasters like flood, super cyclone, drought, changes in hydrological cycles, crisis in freshwater and drinking water resources, uncertainity in cereal output, increasing vector borne diseases, changing migration pattern and flowering time, are just some of the consequences already being experienced by us. Fragile and delicate ecosystems like coral reefs and mangroves are extremely sensitive to the slightest variation in climate. Rapid and early melting of arctic glaciers is another disastrous consequence in the survival of many animal species of the region. Polar bears are going to become the first victim of this undesirable situation. Melting of floating ice, which provides them a jumping surface for prey capture and a suitable niche for reproduction has threatened the survival of this wonderful species.

In developing countries like India, climate change may contribute an additional burden on ecological and socioeconomic systems associated with rapid urbanization, industrialization and economic development. Since the Indian economy is mainly dependent on its natural resource base, India is extremely vulnerable to the impacts of climate change. Extreme temperatures and heat spells have already become evident over northern India, causing loss of animal and human life. Global warming may create favorable conditions for vectors transmitting infectious diseases like dengue, malaria, etc. Disease vectors such as mosquitoes, ticks, and flies may occur in greater numbers over longer periods even in colder regions. Diarrhoeal diseases and outbreaks of other infectious diseases like cholera and hepatitis have been reported to be influenced by climate-related factors. The effects, or impacts, of climate change may bear physical, ecological, social and/or economic dimensions. Climate change in future is expected to have considerable impacts on natural resource systems, which in turn, may lead to instability and conflict, resulting possible displacement of people and changes in habitation and mass migration leading to the creation of 'Environmental Refugees'. Recent reports indicate that changes in climatic pattern grossly affected the life cycle strategies and pollination efficiency of selected insect pollinators, resulting in a sharp decrement in crop yield. Needless to mention that, this situation would invite an unforeseen crisis in our national economy. In fact, the production loss due to climate change may drastically aggravate the problem of poverty, food insecurity and malnutrition and thus creating a long-term effect on the human society, environment and human civilization.

However, the ultimate and holistic impact of climate change in the biology of plants and animal is still a less studied area of environmental biology. Intellectual and scientific involvement of human beings would reveal the vast stretch of mystery of this unexplored area of biological study. Needless to mention, we have to carry out this responsibility with utmost sincerity and zeal at an enhanced pace to save the life process on this planet. Let us think together in a sustainable manner to ensure a healthy Earth, both in social and biological dimensions.

I express my deep sense of gratitude to the contributors for their wholehearted support. My special thanks are due to Professor Sajal Roy, Dr. Kamalesh Misra and Mr. Subhendu Mazumdar for editorial assistance. I am thankful to Mr. Anil Kumar Mittal of Daya Publishing House, New Delhi for publication of this book in time.

I dedicate this edited volume to the loving memory of my beloved friend Mr. Debasish Paul, affectionately called as "Bablu".

Rathayatra, 2011   ***Goutam Kumar Saha***

Contents

List of Contributors

Acharya, Krishnendu
Associate Professor, Department of Botany, University of Calcutta, Kolkata

Akhtar, Wasim
Institute of Agricultural Sciences, University of Calcutta, Kolkata

Basu, Rathindra Narayan
Former Vice Chancellor, University of Calcutta, Former Chairman, WB State Agriculture Commission, Kolkata

Bhattacharya, Sayan
Department of Environmental Sciences, University of Calcutta, Kolkata

Bhattacharyya, Chittaranjan
Former Professor, Department of Geology, University of Calcutta, Kolkata

Chowdhury, Ashim
Acharya P.C. Roy Professor, Institute of Agricultural Sciences, University of Calcutta, Kolkata

Garg, Ruchi
Institute of Vegetable Research, Varanasi

Ghose, Dipankar
Head, Eastern Himalayas Programme, WWF-India, New Delhi

Ghosh, Apurba Ratan
Director, Academic Staff College and Professor of Environmental Sciences, University of Burdwan, Burdwan

Ghosh, Asish
Former Director, Zoological Survey of India & Director, Centre for Environment and Development, Kolkata

Hati, Amiya Kumar
Former Director, Calcutta School of Tropical Medicine, Kolkata

Kaviraj, Anilava
Professor, Department of Zoology, University of Kalyani, Kalyani

Lahiri, Pulak
Former Sir Nil Ratan Sarkar Professor of Zoology, University of Calcutta, Kolkata

Mazumdar, Subhendu
Assistant Professor, Department of Zoology, Shibpur Dinabandhu College, Howrah

Misra, Tapan
Principal, Vidyasagar College, Kolkata

Mohan, Divya
WWF-India, New Delhi

Mukherjee, Asok Kumar
Professor, Department of Biotechnology, Bengal College of Engineering and Technology, Durgapur

Mukhopadhyay, Aniruddha
Head, Department of Environmental Sciences, University of Calcutta, Kolkata

Pradhan, Prakash
Department of Botany, University of Calcutta, Kolkata

Raghunathan, C.
Zoological Survey of India, Andaman and Nicobar Regional Centre, Portblair

Ramakrishna
Former Director, Zoological Survey of India, Kolkata

Ray, Krishnangshu
Director, Calcutta School of Tropical Medicine, Kolkata

Roy, Somenath
Professor, Department of Human Physiology with Community Health, Vidyasagar University, Midnapore

Roy, Subrata
Former Dean, Faculty of Science and Professor of Zoology, University of Burdwan, Burdwan

Saha, Goutam Kumar
Professor of Zoology, University of Calcutta, Kolkata

Saha, Sujay
Institute of Vegetable Research, Varanasi

Saha, Tapan
Senior Scientist, Institute of Environment Studies and Wetland Management, Kolkata

Sanyal, Asok Kanti
Director in Charge, Zoological Survey of India, Kolkata, Member Indian Antarctic Expedition Team

Sanyal, Pranabesh
Former Director, Sundarban Biosphere Reserve & Visiting Professor, School of Oceanographic Studies, Jadavpur University, Kolkata

Sarkar, Banwarilal
Deputy Director, Department of Microbiology, National Institute of Cholera and Enteric Diseases, Kolkata

Satyakumar, S.
Professor and Head, Department of Endangered Species Management, Member Indian Expedition Team to Antarctica, Wildlife Institute of India, Dehradun

Sinha, Shirish
Head, Climate Change Programme, WWF-India, New Delhi

Sivakumar, K.
Wildlife Institute of India, Dehradun

Climate Change: Man and Environment (2012)   *Pages 1–6*
Editor: **Goutam Kumar Saha**
Published by: **DAYA PUBLISHING HOUSE, NEW DELHI**

1

Climate Change: Cause and Concern

☆ *Asish Ghosh**

Introduction

'Climate change' is considered as a major critical issue facing the planet Earth. The causes and effects of climate changes have now been fairly well established; the group of gases *viz.*, carbon dioxide, methane and nitrous oxides besides others has become infamous as 'Green House Gases' (GHG) leading to global warming and change in climatic conditions. Obviously, the global warming potential of each of these GHG's varies, as does their lifespan in the upper atmosphere.

Of all the GHG's, carbon remains the major target, released due to human activities, largely through thermal power generation and automobile fuel emissions, besides massive rate of deforestation, carbon has been accumulating since the days of industrial revolution, even after the major part of carbon emission being absorbed by green plants and the deep sea.

Climate Change and Global Warming can largely be attributed to the human activities are now well established (IPCC, 2007). The question that was raised in Kyoto, Japan in 1997 about sharing the responsibility of reducing emission still remains unresolved in 2010.

International Scenario

The UN Framework Convention on Climate Change (UNFCCC, 1992) laid the foundation for an international consensus in 1992 and the Kyoto Protocol laid out

* Centre for Environment and Development, Kolkata. E-mail: cedkolkata@yahoo.com

the road map in 1997. Countries identified as 'Historic Emitters' were grouped together and given the task of reducing CO_2 by 5.2 per cent below 1992 level between 2008 and 2012. USA didn't accept it; countries took long years to ratify the Protocol till 2005. But the reports at the end of 2009 in Copenhagen meet of Conference of the Parties (CoP15) showed the target of reduction remains unfulfilled promises, though it was legally binding. Collectively, developed countries emitted 17 per cent more carbon when compared to 1990 level. In the major Economics Forum Declaration at L' Aquila, Italy in 2009, President of USA stated "climate change is one of the defining challenges.........the scenario is clear and conclusive and impacts can no longer be ignored." While admitting, 'Every nation on this planet is at risk and just as no one nation is responsible for climate change, no one nation can address it alone.' But major developed countries are responsible for more than 75 per cent of carbon pollution. The target set at the conference is to reduce carbon emission by 80 per cent by 2050 by the developed countries but global target will be to cut emission by 50 per cent. This is targeted to limit global warming 'to no more than 2°C' to comply with the findings of scientists across the world. Developed countries demand that reductions must be 'measurable, reportable and verifiable'. The politics of climate change, not withstanding the continuing crisis of global warming related phenomena created international pressure of the global community to act; the much awaited Copenhagen Meet in December, 2009 belied the hope. "Kyoto Protocol" clearly defining the responsibilities and other issues seems to take a back seat.

National Scenario

India, along with China, is now recognized as one of the fastest developing economies. India's economic growth is dependent on carbon based energy sector, amongst others and therefore has to commit to cut CO_2 emission by 20–25 per cent. India has announced National Action Plan on Climate Change (NAPCC) in 2008 with eight mission–based targets:

1. National Solar Mission
2. National Mission on Enhanced Energy Efficiency
3. National Mission on Sustainable Habitat
4. National Water Mission
5. National Mission on Sustaining Himalayan Ecosystem
6. National Mission for Sustainable Agriculture
7. National Mission for a Green India
8. National Mission on Climate Change for Strategic Knowledge

These eight missions involving both mitigation and adaptation measures remain India's position. The target of Solar Energy Mission with a goal of 20,000 MW of solar power capacity being installed by 2020 is considered most ambitious. Agriculture and water remain critical issues for India's billion plus population and deserved separate missions; so is the vital role played by Himalayan ecosystem.

The increase in green cover has been reported by 'Forest Survey of India' in its latest report but demand of development specially in the mining and industrial sector including thermal power plant continues to take a heavy toll on India's forest. So the mission for green cover is a welcome feature.

India's Energy Efficiency Efforts seem to start working by early 2010. The Bureau of Energy Efficiency had launched 'Bachat Lamp Yojna' to replace 400 million incandescent light bulbs with CFL bulbs and prevent 40 million tonnes of carbon from entering the atmosphere annually. Supported by Clean Development Mechanism (CDM) scheme of United Nation, it will entitle Indian to win World's largest Carbon Credit Project. The investors in the project can sell the Carbon Reduction Unit (CRU) in the international market where buyers like manufacturers and power producers buy the Units to meet their own GHG emission reduction targets (currently it sells at 10–12 Euros). So the private investors in India offering the CFL at minimum price (15 per cent) can recover the investment by 6th–7th year and continue to earn significant project thereafter.

Impacts of Climate Change

Impacts of climate change have been discussed globally over last nearly two decades. Sea level rise, crisis in freshwater resources, uncertain cereal output, increasing vector borne diseases, changing migration pattern and flowering time are but some of the significant impacts predicted. The impacts will also lead to mass migration of human population now named 'Environmental Refugees' – a new category not yet included in UN Commission of Human Refugees.

Impacts on Biodiversity

Response of biodiversity to climate change may be varied. Climate change may seriously affect ecosystem and living organism as can be seen from the history of post climate shift. While some population may flourish, others may face crisis for survival. Some species, scientists believe, will be unable to adjust with the shift in temperature and the changing precipitation rate and face mass extinction. One such study shows ability of the well known pest, the pine beetles completing life cycle at an altitude that was previously hostile to these insects. Such extension of vertical distribution range can create new problems in the management of pine forests in the temperate region. On the other side, it is observed while some trees can bloom earlier with warmer temperature but pollinators do not hatch early, thereby causing disruption in the propagation of the species. Such mismatches can occur between predators and prey, herbivorous insects and host plants and pollinators and flowering plants (Ghosh, 2007). Population decline due to changing climate such as decline of Krill population in the Antarctic seas may have significant impact on the food chain, which includes Penguin, Seals, Whales and Albatross. Scientists have observed northward shift of habitats in a wide variety of species to an average of 16 km per year; these include diverse species of birds, butterflies and alpine herbs in the Palaearctic and Nearitic zone. As many as 677 species recorded a shift to earlier spring cycle due to global warming (Parmesan and Yohe, 2003).

In a recent study 'Surface Air Temperature Variability' over India during 1901 – 2007 and its association with ENSO' conducted recently at Indian Institute for Tropical Meteorology (IITM), Pune, showed that temperature data collected from 388 weather station indicate 0.51°C rise in India during last 100 years but since 1970, minimum temperature rose by 0.20°C per decade faster than maximum temperature 0.17°C. Rising night temperature may affect rice grain in the field, leading to drop in the yield. At ecosystem level, possible impacts on biodiversity due to climate change has been focused (Ghosh, 2007; 2008). The relationship between climate change and biodiversity is shown in Figure 1.1.

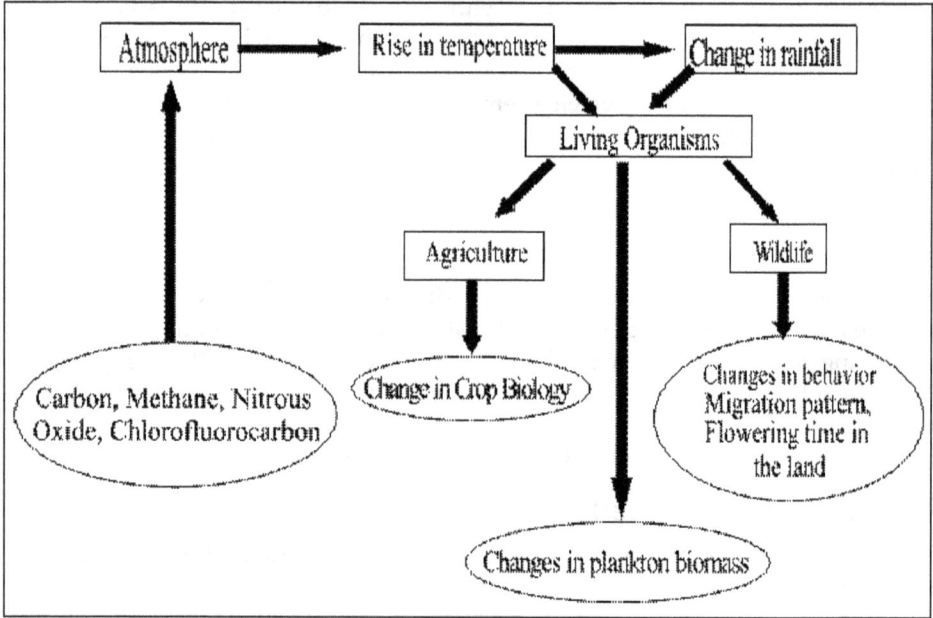

Figure 1.1: Schematic diagram showing climate change and its relationship with biodiversity.

Polar ecosystem witnessed an increase in temperature 10 times faster than the observed surface mean temperature–Penguin, Polar Bear and other animals adapted to typical polar habitat are apt to suffer.

The possible impact on agroecosystem besides the loss of food grain may include phonological changes amongst pest, predator and parasitic species. Soil condition, when becoming drier will not only affect crop but also affect the soil micro–arthropods, changes in flowering time may lead to early harvesting resulting in massive reduction in yield.

Deserts, both hot and cold are projected to be either hotter or drier or both. Organisms at the apex of heat tolerance limit are likely to perish under maximum threshold level. Increased rate of wild fire could change the species composition. Desert flora in the Rann of Kutch is comprised of 700 species of flowering plants with nearly 50 per cent species being endemic; more than 300 species of vertebrates inhabit in the area. More than 4500 species of fauna live in Thar Desert (ZSI, 2002).

Forest ecosystem, with small changes in temperature and precipitation can have adverse impact on forest growth. Many forest dwelling large mammals, half of the large primates and nearly 10 per cent of all tree species are already said to be under risk of extinction. Forest fauna, in many cases, depend on fruits and flowers of specific flora and as such will have consequences on the rate of reproduction and survival. Wildlife species in designated areas also become vulnerable unless they are able to shift at higher altitude.

Mountain ecosystem across the world has specific and typical biota stratify at different altitudinal level. Current researches show retreat and sometimes disappearance of alpine species that become trapped on mountain summit. In Alps, some plant species have been migrating upward by one to four meters per decade and some other plants previously resolved from the top of the mountains have disappeared. A recent study from Institute of Tropical Meteorology, Pune, India shows that among regions, the hardest hit area is western Himalaya encomprising portions of Jammu and Kashmir, Himachal Pradesh and Uttarakhand. Here the mean temperature rise in the last century was 0.80°C, but more recently it has become 0.46°C per decade. This rapid warming would not affect glaciers but the unique biota which include more than 3000 flowering plants and 5700 faunal species representing 6.4 per cent of Indian fauna as evidenced from the study conducted at IITM, Pune (unpublished).

Freshwater ecosystem, fed by both melting of ice and annual rainfall, can undergo serious changes in terms of faster melting of glaciers and inadequate formation of ice in succeeding years as also from uncertain rainfall pattern. The warming of rivers, altered mixing regime, altered flow regime and greater frequencies of flood and drought can all affect growth, reproduction and distribution of aquatic biota including food fishes. Migratory birds dependent on lake's and stream's water resource and quality are also likely to suffer. It is to be mentioned that freshwater flora in India includes 87 species of pteridophytes and 690 species of angiosperms.

Marine and coastal ecosystem perhaps can be considered most vulnerable with potential sea level rise due to climate change. Coastal flooding, storm surges, cyclones and phenomena like Tsunami, are likely to affect fragile coral reef region and breeding ground of marine fishes. Changes in the phytoplankton biomass is predicted due to expected increase in atmospheric CO_2 which in turn will give rise to a three fold increase in surface water CO_2 contamination. These will affect seawater pH and enzymatic activities of phytoplanktons (Riebesell *et al.*, 2000; Riebesell *et al.*, 2001).

Remarks

The phenomenon of climate change and global warming is happening now at a pace faster than any time in the past. The economic cost of climate change is calculated at 2.5 times more for inaction rather than action, both adaptive and mitigative. Of the impact change, one of the most worrying is on biodiversity. Human societies still depend on biodiversity for its food security, fodder, fuelwood, medicinals, textile, leather and hides, oilseeds, beverages to a variable degree. Changes in biodiversity are likely to originate largely from changes in land use followed by climate change (Chapin *et al.*, 2000; Sala *et al.*, 2000). Consequences of changing climate and changing biodiversity are being researched but it is believed 'Human alteration of global environment has triggered the sixth major extinction event in the history of life'.

These undoubtedly will have 'profound consequences for services that human derive from ecosystem'.

While climate change may endanger the lives and livelihood of millions of people across the world, the surviving population will perhaps face the maximum crisis due to disruption of ecosystem services. May be a new international body comparable to IPCC should assess changes in biodiversity and their consequence as an integral component of the assessment of the societal impacts of global change (Chapin *et al.*, 2000).

References

Chapin, F.S., Zavaleta, E.S., Eviner, V.T. *et al.* 2000. Consequences of changing biodiversity. *Nature*, 405: 235–242.

Ghosh, A. K. 2007. *Biodiversity Conservation.* APH Publishing Corporation, New Delhi. Pp. 304.

Ghosh, A. K. 2008. *A Comprehensive Handbook on Biodiversity.* TERI–The Energy and Resources Institute. TERI Press, New Delhi. Pp. 128.

IPCC, 2007. Fourth Assessment Report. Intergovernmental Panel on Climate Change, Bali, Indonesia.

Parmesan, C. and Yohe, G. 2003. A globally coherent fingerprint of climate change impacts across natural systems. *Nature*, 421: 37–42.

Riebesell, U., Zondervan, I., Rost, B. *et al.* 2000. Reduced calcification in marine plankton in response to increased atmospheric CO_2. *Nature*, 405: 364–367.

Riebesell, U., Zondervan, I., B. Rost, B. *et al.* 2001. Effects of increasing atmospheric CO_2 on phytoplankton communities and the biological carbon pump. *Global Change Newsletter*, IGBP, 47: 12–15.

Sala, O. E., Chapin, F.S., Armesto, J.J. *et al.* 2000. Global Biodiversity Scenarios for the year 2100. *Science*, 287: 1770–1774.

United Nations Framework Convention on Climate Change (UNFCCC). 1992. *Article 1.* New York: United Nations.

ZSI, 2002. Ecosystem of India. (Alfred, J.R.B., Das, A.K., and Sanyal, A.K. eds.), Envis Centre, Zoological Survey of India, Kolkata.

Climate Change: Man and Environment (2012)   Pages 7–23
Editor: **Goutam Kumar Saha**
Published by: **DAYA PUBLISHING HOUSE, NEW DELHI**

2

Climate Change: Impacts, Adaptation and Mitigation

☆ *Sayan Bhattacharya and Aniruddha Mukhopadhyay**

Introduction

In the past few decades, scientists have assembled a growing body of evidence showing the extent of change of the earth's climate and the role of human activities behind it. United Nations Framework Convention on Climate Change defines climate change as "a change of climate which is attributed directly or indirectly to human activity that alters the composition of the global atmosphere and which is in addition to natural climate variability observed over comparable time periods."Without urgent and concerted action, it will damage fragile ecosystems, impede development efforts, increase risks to public health, frustrate poverty alleviation programs, and force large-scale migration from water or food–scarce regions.

On the other hand, global warming is the increase in the average measured temperature of the Earth's near–surface air and oceans since the mid–20th century, and its projected continuation. The observed temperature increase over the last 50 years has been due to the increase in the atmosphere of greenhouse gas concentrations like water vapour, carbon dioxide (CO_2), methane and ozone. Greenhouse gases are gaseous constituents of the atmosphere, both natural and anthropogenic; those absorb and emit radiation at specific wavelengths within the spectrum of thermal infrared radiation emitted by the earth surface, the atmosphere and clouds. Most of the short wave radiation of the sun received by the earth is absorbed by the land and oceans and some by the atmosphere. The rest is reflected back to the space. After receiving the radiation the earth surface becomes warm and emits long wave radiation. Green

* Department of Environmental Science, University of Calcutta. E-mail: am_cuenvs@yahoo.co.in

house gases trap this long wave radiation and re–emit some part of it which increases the atmospheric temperature. Concentration of green house gases in the atmosphere has been rising since the industrial revolution in the 18th century. Measurements over the past 130 years show that atmospheric temperatures have already risen considerably and have been the highest in the last few years (Bhattacharya and Mukhopadhyay,2009). Global average surface temperature of the earth has increased by about 0.7°C over the 20th century. The 1980s and 1990s were the warmest decades since accurate records began in the mid of late 1800s (Garvey, 2009). The latest warning from a group of scientists from NASA is "2005 was the warmest year on record since the end of last major ice age nearly 12000 years ago" (Nair, 2009)

'Global warming' is quite different from 'climate change', which is a more complex phenomenon. Climate change can occur in different ways such as changes in regional and global temperature, changes in rainfall patterns, expansion and contraction of ice–sheets, sea level variation, plate tectonics, volcanism, ocean variability etc; temperature is not the only changing factor in case of climate change.

Effects of Climate Change

Melting of Ice and Sea Level Rise

One of the many warning signals of global warming is the change in the ice caps due to melting, thinning, shrinking, retreating, freezing and disappearing (Nair, 2009). Satellite images have shown that the parts of our planet that are covered in snow each winter have decreased by 10 per cent (Garvey, 2009). The snows of Kilimanjaro have melted more than 80 per cent over the last 100 years and will vanish by 2020 (Garvey, 2009; Nair, 2009). Settlements in mountain regions are at enhanced risk of floods caused by melting glaciers. Spiti River, a high altitude Himalayan river located in the western Himalayan region, the annual snowmelt run–off, glacier melt run–off and total stream flow increase linearly with increase in temperature, but the most prominent effect of temperature increase has been noticed on glacier melt run–off (Singh and Kumar, 1997).

The ice covering parts of Greenland is already melting and may cause up to 7 meters of sea level rise (Garvey, 2009). From 1980–2000AD, the average annual temperature of the Arctic region has increased by 4°C, and as a result, the thickness of the ice sheets has decreased by 40 per cent. The melting ice also can wash away the algae of the arctic region and can affect the food chain. The melting of ice, in turn, is changing the hydrological cycles and is also affecting the ocean currents. The flow rate of Atlantic circulation has dropped by 6 million tons of water per second from 1957 to 1958 (Garvey, 2009).

In the past century, sea level rise has occurred at a mean rate of 1.8 mm per year (Douglas, 1997) and the IPCC (2007) report suggested that sea levels would rise by between 190 mm (7.5 inches) and 590 mm by the end of this century (Maslin,2004). One meter sea level rise can flood up to 75 per cent of the dry lands of Maldives in the Indian ocean and Marshall islands of the pacific, and can make these place uninhabitable (IPCC, 2007). Many island nations of the Pacific, notably the Polynesian islands of Tuvalu, which based on more severe flooding events in recent years, was thought to be "sinking" due to sea level rise. Sea level rise could also displace many shore–based populations; it is estimated that a sea level rise of just 200 mm could create 740,000 homeless people in Nigeria (Paehler,2007). Many small islands have poorly developed infrastructure and limited resources, and often small island populations are dependent on marine resources to meet their protein needs. So, climate change could completely change the social and economic conditions of the islands.

The IPCC report of 2007 estimated that accelerated melting of the Himalayan ice caps and the resulting rise in sea levels can increase the frequency of floods in the short–term during the rainy season and can greatly magnify the impact of tidal storm during the cyclone season. A sea-level rise of just 400 mm in the Bay of Bengal would put 11 per cent of the Bangladesh's coastal land under water, and can create 7 to 10 million "climate refugees". Recent report suggests that 45 cm rise in sea level (likely by the end of the 21st century), combined with other forms of anthropogenic stress on the Sundarbans, could destroy 75 per cent of the Sundarbans mangroves (UNESCO, 2007). Already, Lohachara Island and New Moore Island/South Talpatti Island have disappeared under the sea, and Ghoramara Island is half submerged (Douglas, 1997). Low island mangroves are most at risk, as demonstrated in Bermuda, where sea level rise has exceeded peat accumulation rates periodically, resulting in landward die back of mangrove stands. Climate change may reduce global mangrove area by 10–15 per cent (Ellison, 1993). Direct climate change impacts on mangrove ecosystems are likely to be less significant than the effects of associated sea level rise. Rise in temperature and the direct effects of increased CO_2 levels are likely to increase mangrove productivity, change the timing of flowering and fruiting, and expand the ranges of mangrove species into higher latitudes. Changes in precipitation and subsequent changes in aridity may affect the distribution of mangroves (Alongi, 2002). Sea level rise will also destroy about 40–50 per cent of the world's coastal wetlands by 2080 (Nair, 2009). A World Bank report concluded that human activities in the deltas can cause the areas to sink much faster than any predicted rise to sea level, thus can increase their vulnerability to storms and floods (Maslin, 2004). Sea level rise in the Nile delta can change the water quality, can affect many freshwater fishes, can increase the salinity of the groundwater and also can inundate the fertile agricultural lands. Rising sea levels will threaten coastal aquifers. Many of India's coastal aquifers are already experiencing salinity ingress including Saurashtra coast in Gujarat and Minjur aquifer in Tamil Nadu. Increasing frequency and intensity of droughts in the catchment area will lead to more serious and frequent salt–water intrusion in the estuary and thus can deteriorate surface and groundwater quality (Xu, 2003; Thanh *et al.*, 2004). Some scientists suggest that climate change may alter the physical characteristics of aquifers themselves. Higher CO_2 concentrations in the atmosphere may change carbonate dissolution and can promote the formation of crust, which in turn may negatively affect infiltration properties of topsoil. Due to melting of ice sheets, there will be an increase in the frequency of floods in the rivers of India originating from Himalayas.

Effects on Agriculture and Biodiversity

Biodiversity in all its components (*e.g.* genes, species, ecosystems) increases resilience to changing environmental conditions and stresses. Climate change could affect a number of physical and biological processes in which the health and composition of terrestrial ecosystems depend. Genetically diverse populations and species–rich ecosystems have greater potential to adapt to climate change. Leemans and Eickhout (2004) showed that 1°C warming alters more than 10 per cent of all ecosystems. At 2°C and 3°C rise, 16 per cent and 22 per cent of all terrestrial ecosystems can change significantly. IPCC report lists the species which will be under threat as a result of climate change and global warming which includes the mountain gorilla in Africa, amphibians that only live in the cloud forests of the neotropics, forest birds of Tanzania, Bengal tiger and other species in the Sundarban wetlands, rainfall sensitive plants found only in the Cape Floral Kingdom of South Africa, polar bears, penguins etc. Coral reefs, mangroves, coastal wetlands, mountain ecosystems found in the upper 200–300 meters of mountainous areas, prairie wetlands, and permafrost ecosystems are some of the natural habitats which are threatened due to global warming (IPCC, 2007). These species or ecosystems are unable to migrate in response to climate change because of their particular geographical locations. Several studies revealed that due to increased temperature in recent decades, certain species began breeding and migrating earlier than expected (Root *et al.*, 2002). Other studies found that the geographical range of numerous species had shifted poleward or moved to a higher elevation – indicating that some plants and animals are occupying areas that were previously too cold for survival. They conclude that a rapid temperature rise in combination with other environmental pressures "could easily disrupt the connectedness among species" and possibly lead to numerous extinctions. Malcolm *et al.* (2006) assessed the potential effects of climate change on terrestrial biodiversity on a global scale, rather than considering only individual species. They reported that doubling of present CO_2 concentration and resulting temperature rises could potentially eliminate 56,000 plant and 3,700 endemic vertebrate species in the 25 hotspot regions. Areas particularly vulnerable to climate change include the Cape Floristic region of South Africa, Caribbean regions, Indo–Burma, the Mediterranean Basin, southwest Australia, and the tropical Andes as the species in these regions have restricted migration options due to geographical limitations. If climatic conditions shift quickly enough, slower moving species may be left behind, especially when human activities have destroyed and fragmented existing habitat. Due to climate change, habitats can be shifted, and eventually habitats can be lost from many areas. However, species with short generation time, such as microbes and insects, may adapt more successfully to climate change than those species with long generation time (Pamela *et al.*, 2009). Here are some examples of habitat loss, change in productivity and species extinction in response to climate change and global warming as revealed recently:

☆ A vast region of Amazon Rain Forest, which holds a high proportion of global biodiversity, is on the brink of being turned into desert, with catastrophic consequences for the world's climate. The forest can't withstand more than two consecutive years of drought without breaking down.

☆ Crop productivity is projected to increase slightly at mid to high latitudes for local mean temperature increase of up to 1–3°C depending on the crop, and then decrease beyond that in some regions. At lower latitudes, especially seasonally dry and tropical regions, crop productivity is projected to decrease for even small local temperature increases (1–2°C), which would increase the risk of hunger. The poorest countries would suffer most, with reductions in crop yields in most tropical and sub–tropical regions due to decreased water availability and new or changed insect pest incidence. In Africa and Latin America many rain fed crops are near their maximum temperature tolerance, so that yields are likely to fall sharply for even small climate changes; falls in agricultural productivity of up to 30 per cent over the 21st century are projected. Production of rice, maize and wheat in the past few decades has declined in many parts of Asia due to water shortage, increasing frequency of El Nino and reduction in the number of rainy days. All of these factors are related with climate change (IPCC, 2007). In a study at the International Rice Research Institute, the yield of rice was observed to decrease by 10 per cent for every 1°C increase in growing–season minimum temperature (Peng *et al.*, 2004).

The climatic change could affect agriculture in several ways:

- *Productivity*, in terms of quantity and quality of crops.
- *Agricultural practices*, through changes of irrigation conditions and agricultural inputs such as herbicides, insecticides and fertilizers.
- *Environmental effects, i.e.* frequency and intensity of soil drainage (leading to nitrogen leaching), soil erosion and reduction of crop diversity.
- *Rural space*, through the loss and gain of cultivated lands, land speculation, land renunciation, and hydraulic amenities.
- *Adaptation, i.e.* organisms may become more or less competitive, as well as humans may develop urgency to develop more competitive organisms, such as flood resistant or salt resistant varieties of rice.

☆ Due to global warming, spring arrives earlier in Europe as a result of which birds are being forced to change their migration patterns. They arrive in northern Europe earlier in time for the start of spring. The familiar bird species of Britain will be driven hundred of miles further north by the end of the century.

☆ Amphibian populations are declining due to global warming in the designated wilderness areas and national parks (Pounds and Crump, 1994). Pounds *et al.* (2006) reported that global climate change has made conditions more favorable for a new disease (chytridiomycosis), thus indirectly leading to extinctions and declines of amphibians. Because of the permeable skin, biphasic lifestyles and unshelled eggs, amphibians are extremely sensitive to small changes in temperature and moisture. Amphibians in temperate regions may be even more susceptible to increased temperature. In eastern Australia, Ingram (1990) found a correlation between draught and massive declines of stream–dwelling rain forest amphibians. Local changes in the environment can also decrease immune function of the amphibians and lead to pathogen outbreaks and elevated mortality. Gervasi and Foufopoulos

(2007) found that amphibian immune responses become increasingly weaker and leukocyte counts were increasingly lower with higher desiccation.

☆ Due to the loss of sea ice habitat, polar bear is under threat. Stirling *et al.* (1999) observed that due to loss of ice sheets, Hudson Bay polar bears are coming ashore for several months of fasting in progressively poorer condition. Hudson Bay polar bears prey primarily on ringed seals (*Phoca hispida*), the population of which is in decline due to a loss of these stable ice flows. Due to nutrient deficiency, polar bears will use stored fat as an energy source, which can remobilize the Persistent Organic Pollutants stored in their tissues and potentially resulting in the dual stresses of starvation and chemical toxicity (Macdonald *et al.*, 2005).

☆ The pH is an important water quality indicator as fishes and other organisms are sensitive to pH. Ocean surface pH has already decreased by 0.1 pH units in colder waters and almost 0.09 pH units in warmer waters. If atmospheric CO_2 concentrations continue to increase, another 0.3 pH unit decrease of oceanic surface waters may occur (Haugen, 1997). Furthermore, as temperature increases, an increased proportion of the water molecules dissociate to H^+ and OH^-, decreasing water pH, which will affect the ocean biodiversity.

☆ Extreme environmental conditions, such as elevated water temperature, low dissolved oxygen or salinity and pH, can have deleterious effects on fishes (Moyle and Cech, 2004). Suboptimal environmental conditions can decrease foraging, growth, and fecundity, alters metamorphosis, and affects endocrine homeostasis and migratory behavior (Barton and Barton, 1987). The progressive acidification of oceans is expected to have negative impacts on marine shell–forming organisms (*e.g.*, corals) and their dependent species.

☆ Corals are vulnerable to thermal stress and have low adaptive capacity. Increases in sea surface temperature of about 1–3°C are projected to result in more frequent coral bleaching events and widespread mortality, unless there is thermal adaptation or acclimatization by corals. Most of the pigmentation within corals is within the symbiotic algal cells–the zooxanthellae. Corals losing their zooxanthellae cause coral bleaching. Thermal bleaching occurs when the coral is exposed to prolonged above–normal (or below–normal) temperatures, resulting in additional energy demands on the coral, depleted reserves, and reduced biomass (Muller–Parker and D'Elia, 1997). Bleaching causes a decrease in the growth rate of corals, and the time taken for a coral to recover from a bleaching event may be several years or decades. If the frequency of bleaching increases, then the capacity for coral reefs to recover is diminished (Done, 1999).

☆ Due to warm habitats, in polar regions there is a possible shift in spawning times, alteration of bioenergetics and changes in transport of larvae in the populations of north sea cod, haddock, herring, and sardines (Alheit and Hagen, 1997; Ottersen *et al.*, 2001). In temperate regions, the distribution of pacific salmons (*Oncorhynchus* sp.) shifts northwards, their population decreases and the size is also changed (McFarlane *et al.*, 2000; Ishida *et al.*, 2001).

☆ The bioavailability and toxicity of Persistent Organic Pollutants and pesticides in wildlife may increase in response to rising temperatures and salinity. An underlying mechanism of this toxicity is that temperature can alter the toxicokinetics of chemical pollutants in organisms; or the increasing temperature can alter homeostasis and other key physiological mechanisms, thereby can magnify the adverse effects of contaminants (Pamela *et al.*, 2009).

☆ Species that are especially sensitive to climate change may be used as indicator species ('bio–indicators') for assessing the climate sensitivity of whole ecosystems. A number of herbaceous plant species, butterflies and birds are identified as suitable bio–indicators for climate change. Natural ecoclimatic transitions or ecotones may be especially suitable for monitoring effects of climate change, because they are likely to be especially sensitive to climate change (De Wilde and Van der Maessen, 1997).

Effects on World's Water Resources

A warmer climate will accelerate the hydrological cycle; can alter the intensity and timing of rainfall. Warm air can hold more moisture and can increase evaporation of surface moisture, which in turn can intensify rainfall and snowfall events. So, intensity of flood will also increase. If there is deficiency of moisture in the soil, solar radiation will increase the temperature, which could contribute to longer and more severe droughts (Trenberth *et al.*, 2005). In a number of studies, it has been proved that global warming and decline in rainfall may reduce net recharge and can affect groundwater levels. Decrease in winter precipitation would reduce the total seasonal precipitation being received during December–February, and can impose greater water stress. Intense rain for few days will result increased frequency of floods and the monsoon rain would also be lost as direct run–off, thus can decrease the groundwater recharging potential (Mall *et al.*, 2006). Increased rainfall amounts and intensities will lead to greater rates of soil erosion.

Changed freshwater inflows into the ocean will lead to changes in turbidity, salinity, stratification, and nutrient availability, all of which affect estuarine and coastal ecosystems (Justic *et al.*, 2005). The status of present precipitation and a rise in temperature by 4°C caused a decrease in run–off by 2–8 per cent, depending upon the areas considered and the model used (Sharma *et al.*, 2000). The river basins of central India, located in a comparatively drier region, are more sensitive to climate change (Mehrotra, 1999).

Climate change will affect the Indian river basins significantly. River basins of Mahi, Pennar, Sabarmati and Tapti shall face water shortage conditions. On the other hand, river basins belonging to Godavari, Brahmani and Mahanadi shall not face water shortages, but severity of flood shall increase in these areas (Gosain and Rao,2004). A change in field–level climate may alter the need and timing of irrigation. In India, roughly 52 per cent of irrigation is dependent on groundwater use. So, it can be an alarming situation with decline in groundwater and increase in irrigation requirements due to climate change (Mall *et al.,* 2006).

By the mid of the 21st century, annual average river runoff and water availability will increase by 10–40 per cent at high latitudes and in some wet tropical areas, and decrease by 10–30 per cent over some dry regions at mid–latitudes and in the dry tropics, some of which are presently water–stressed zones (Trenberth *et al.,* 2005). There will be an increase in drought affected areas and the frequency of heavy precipitation will increase, which, in turn, will increase flood risk.

Effects on Human Society and Health

IPCC reported that increased global temperature will increase the death rate. Higher temperatures will likely increase the concentration of ozone in troposphere that will contribute to cardiovascular and pulmonary illness (IPCC, 2007). Global warming may create favorable conditions for vectors conveying infectious diseases like dengue, malaria, etc. Disease vectors such as mosquitoes, ticks, and flies may occur in greater numbers over longer periods during the year. As northern Europe becomes warmer, ticks, which carry encephalitis and lyme disease and sand flies, transmit visceral leishmaniasis are likely to move in. Endemic morbidity and mortality due to diarrhoeal disease primarily associated with floods and droughts are expected to rise in east, south and southeast Asia due to projected changes in the hydrological cycle associated with global warming. Diarrhoeal diseases and outbreaks of other infectious diseases (*e.g.,* cholera, hepatitis, malaria, dengue fever) have been reported to be influenced by climate–related factors. There is a strong correlation between increase in sea surface temperature and sea level and the annual severity of cholera epidemics in Bangladesh (Maslin, 2004). Warmer sea–surface temperatures of south and southeast Asia would support higher phytoplankton blooms, which are excellent habitats for survival and spread of infectious bacterial diseases such as cholera (Pascaul *et al.,* 2002).

In general, increased warmth and moisture will enhance transmission of diseases (Maslin, 2004). Mathematical models proved that by 2080, the potential exposure of people by mosquitoes could increase by 2–4 per cent (Maslin, 2004). An empirical model projected that the population at risk of dengue fever will be larger in India and China (Hales *et al.,* 2002).

Several studies have confirmed and quantified the effects of high temperatures on common forms of food poisoning, such as salmonellosis (IPCC, 2007). The methylation of mercury and its subsequent uptake by fish and human beings will increase due to climate change (Booth and Zeller, 2005; McMichael *et al.,* 2006).

A large number of deaths due to heat waves–mainly among the poor, aged people and labourers such as rural daily wage earners, agricultural workers and rickshaw pullers have been reported in Andhra Pradesh, Orissa and elsewhere in India during the past five years (Lal, 2002). Exposure to heat can be a significant risk factor for cerebral infarction and cerebral ischemia during the summer months (Honda *et al.,* 1995). The combination of air pollutants and allergen production linked to climate change may enhance allergic disease and asthma incidences in vulnerable individuals, especially children, infants, and asthmatics (Epstein, 2007).

Economics of Climate Change

The valuation of costs and benefits of climate change is difficult because some climate change impacts are difficult for analysis, *e.g.*, ecosystems and human health. It is also impossible to know the preferences of future generations, which affects the valuation of costs and benefits (Le'tard *et al.*, 2004). Economic estimates of the impact of climate change are typically based on "damage functions" that relate GDP losses in response to temperature increase. The estimates of GDP costs in response to the impacts of climate change usually grouped as market impacts and non-market impacts. Market impacts include effects on climate–sensitive sectors (such as agriculture, forestry, fisheries, and tourism), damage to coastal areas from sea–level rise, changes in energy consumption and changes in water resources. Non-market impacts cover effects on health, leisure activities (sports, recreation, and outdoor activities), ecosystems and human settlements. There will be shifts in international trade, in capital and migration flows, and in the prices of commodities, other goods and services, and assets in response to climate change.

Tourism is a major economic sector in many small islands. The impacts of climate change on tourism resources in small islands will have significant effects, both direct and indirect (Bigano *et al.*, 2005). For example, the sustainability of island tourism resorts in Malaysia is expected to be compromised by rising sea level, beach erosion and saline contamination of coastal wells (a major source of water supply for island resorts (Tan and Teh, 2001).

A cross–sectional analysis of per capita income and temperature suggests that people are poor because of the climate (Nordhaus, 2006). So, global warming could cause economies to shrink or grow slower. As poverty implies higher impacts, this would drag the economy down further. The Indira Gandhi Institute of Development Research has reported that, according to IPCC report, climate–related factors could decrease India's GDP by up to 9 per cent.

Climate can be a factor in an area's comparative advantage for economic production and growth. Not only can climate affect an area's own economic patterns, it can also affect the competitive position of its markets and competitors, and thus affect prospects for local employment and individual livelihoods. For example, in 2003 heat wave in Western Europe affected settlements and economic services in a variety of ways. Economically, it created stress on health, water supplies, and food storage and energy systems. The IPCC provides evidence that global insured and uninsured property losses currently amount to over $40 billion per annum compared to just $4 billion per annum (all in real terms) some 50 years ago. Infrastructures such as bridges, roads, pipelines or transmission networks, industry can experience substantial economic losses. Where the climate warms due to climate change, less temperature will be needed for industrial, commercial and residential buildings, and cooling demands will increase, which in turn, will increase the energy demands. According to assessment reports, the value of damages caused by climate change would cost about 4.8 trillion US$, about half the GDP of the whole European Union.

If we try to prevent the global temperature increase above 1.5°C, it would cost around 37 trillion US$ (IPCC, 2007).

Nicholas Stern (2006) released a 700–page report for the British government, which discusses the effect of global warming on the world economy. It proposed that one per cent of global gross domestic product (GDP) per annum will require avoiding the worst effects of climate change, otherwise there will be up to 20 per cent decrease in GDP in response to the adverse changes.

Sustainable development has become part of all climate change policy discussions at the global level. It is 'the development that meets the needs of the present without compromising the ability of future generations to meet their own needs' (WCED, 1987). Sustainable development is a multidimensional concept including economic, social and environmental issues. Environmental conservation for sustainability of natural resources is not a luxury but a necessity when considering long-term economic growth and development, particularly in the least developed countries. Linking the concept of sustainable development to climate change provides an opportunity to explore long-term societal responses to global environmental change.

If the mean summer temperature rises by 1°C in Japan, consumption of summer products such as air–conditioners, beer, soft drinks, clothing and electricity are projected to increase about 5 per cent, which have indirect negative effects on human health and economy of the country (Harasawa and Nishioka, 2003).

In Asia, the pressure on land in the 21st century will increase, due to the increasing food grain demand, the growing economic development, as well as climate change. Encroachment into forest zones for cultivation, grazing, fuel wood and other purposes will change the ecology of the forest zones. This, in turn, can affect the process of carbon sequestration, as forests generally act as carbon sink.

Future climate change is expected to have considerable impacts on natural resource systems, which, in turn, can lead to instability and conflict, often followed by displacement of people and changes in occupancy and migration patterns (Barnett, 2003). The production losses due to climate change may drastically increase the problem of poverty, food insecurity and malnutrition in several developing countries in Asia (Wang *et al.*, 2006).

Adaptation Strategies

In the context of climate change, mitigation is defined as "implementing policies to reduce greenhouse gas emissions and enhance sinks", while adaptation to climate change is defined as, "[an] adjustment in natural or human systems in response to actual or expected climatic stimuli or their effects, which moderates harm or exploits beneficial opportunities" (IPCC, 2007). The capacity of adaptation varies considerably among regions, countries, and socioeconomic groups and will vary over time. Countries with limited economic resources, low levels of technology, poor information and skills, poor infrastructure, unstable or weak institutions, and inequitable empowerment and access to resources have little capacity to adapt and are highly vulnerable. Enhancement of adaptive capacity represents a practical means of coping with changes and uncertainties in climate. In this way, enhancement of adaptive capacity reduces vulnerabilities and promotes sustainable development. The adaptive capacity of a nation is likely to be greater when the nation has a stable and prosperous economy, technological strength and a good resource base. Climate change adaptation will be needed in a variety of ecosystems, including agro–ecosystems (crops, livestock, grasslands), forests and woodlands, inland waters and coastal and marine ecosystems. Adaptation practices require extensive high quality data and information on climate, and on agricultural, environmental and social systems affected by climate, with a view to carrying out realistic vulnerability assessments and looking towards the near future. Some of the adaptation strategies against climate change are discussed below:

1. Crops and vegetables that can tolerate abiotic stresses should be used. The selection of appropriate rice varieties that have different abilities to tolerate high temperature, salinity, drought and floods deserves consideration for adaptation to climate change.

2. Land cover assessment and monitoring of its dynamics are essential for sustainable management of natural resources, assessing the vulnerability of ecosystems and food security planning. Crop rotations, agro–forestry, crop–livestock associations, crop–fish systems and the use of hedges, vegetative buffer strips, controlling insect outbreaks and other farm landscaping practices can be beneficial in this matter.

3. Trees and shrubs in farming systems can play a significant role in mitigating the impacts of climate change and the resulting threats to food security. In addition to benefits such as the provision of wood and non wood forest products, soil fertility maintainace and the conservation of biological diversity, trees and forests can improve the microclimate by buffering winds, regulating the water table, providing shade to crops and animals, and stabilizing coastal areas (*e.g.* through mangrove rehabilitation and reforestation). They thus contribute to sustainable agricultural production and food security.

4. In coastal areas, impacts of climate change can be minimized by:
   - Developing county–scale maps depicting which areas will require shore protection and which areas will be allowed to adapt naturally.
   - Analyzing the environmental consequences of shore protection and promoting sustainable shore protection techniques.
   - Improving early warning systems and flood hazard mapping for storms and protecting water supplies from saltwater intrusion.
   - Improving water use efficiency, planning for alternative water sources (such as treated wastewater or desalinated seawater) and making changes to water allocation.
   - Water availability, cost and competition from other sectors must be considered.

5. Grain storage, emergency feeding stations, adjusting clothing and activity levels and increase in fluid intake can directly or indirectly help to combat

with climate change. The most important and cost–effective adaptation measure is to rebuild public health infrastructure properly.

6. For protecting biodiversity, high genetic diversity and population size should be maintained and heterogeneity and redundancy of populations should be increased. Controlled burning should be applied to reduce fuel load and potential for catastrophic wildfires. Migration corridors should be protected and enhanced to allow species to migrate properly as the climate changes.

7. Risk assessment systems should include a detail historical climate data archive, advanced climate data analysis procedure and monitoring tools using systematic meteorological observations.

8. Environment monitoring programme, climate prediction and weather forecasting, market signals related to climate variability and change, response capacity, information dissemination and public awareness – all these areas should be revised, re–examined and improved to ensure better protection.

Mitigation of Climate Change

The scientific debate of the last two or three decades on global warming has brought climate forcefully to the attention of governments and included the topic in international political debates. But, in a heterogeneous world, an understanding of different regional conditions and priorities are essential for implementing climate change mitigation policies. Some of the mitigation and adaptation strategies of climate change are discussed below:

☆ Energy supply and distribution efficiency should be increased to control climate change. Renewable sources of energy like hydropower, solar energy, wind energy, biogas and geothermal energy should be used to minimize pollution as these sources reduce the amount of greenhouse gases emitted. They are currently small overall contributors to global heat and electricity supply, but are increasing most rapidly.

☆ Carbon dioxide capture and storage (CCS) is a process consisting of the separation of CO_2 from industrial and energy related sources, transport to a storage location and long-term isolation from the atmosphere. Carbon sequestration is a geo–engineering technique for long-term storage of carbon dioxide or other forms of carbon to mitigate global warming. Conservation and sequestration result in higher carbon stocks, but can lead to higher future carbon emissions if these ecosystems are severely disturbed by either natural or direct/indirect human–induced disturbances (Maslin, 2004).

☆ The Intergovernmental Panel on Climate Change (IPCC) estimates that the cutting down of forests is now contributing close to 20 per cent of the overall greenhouse gases entering the atmosphere. The UK government's Stern Review on the economics of climate change argued that curbing deforestation was a "highly cost–effective way of reducing greenhouse gas emissions". Afforestation and reforestation, like forest protection, may also have beneficial hydrological effects. The use of forestry products for bio–energy can replace fossil fuel use (Maslin, 2004; IPCC, 2007).

☆ Smit *et al.* (2001) reported that activities which can help to deal with the problem of climate change are equivalent to the sustainable development policies like improving access to natural resources, poverty reduction, equal distribution of resources, improvement of education and information, improvement of infrastructures, institutional capacity and efficiency.

☆ In agricultural sectors, following measures should be taken:

1. Improved crop and grazing land management to increase soil carbon storage.
2. Restoration of cultivated peaty soils and degraded lands.
3. Improved rice cultivation techniques and livestock and manure management to reduce CH_4 emissions.
4. Improved nitrogen fertilizer application techniques to reduce N_2O emissions.
5. Production of bio–energy crops may reduce nutrient leaching and soil erosion and generate additional environmental services such as soil carbon accumulation, improved soil fertility, and the removal of cadmium and other heavy metals from soils or wastes.
6. In water–scarce regions, water supply may take place (partly) by desalinization of saline water.
7. Methane and nitrous oxide emissions can be reduced, such as those from livestock enteric fermentation, rice paddies, nitrogen fertilizer use and animal wastes.

☆ In transport sectors, modern energy efficient technologies, such as plug–in hybrid electric vehicles, and development of new technologies, such as hydrogen cars, may reduce the consumption of petroleum and emissions of carbon dioxide.

Remarks

Climate change has the potential to create a long term effect on the human society as well as on ecosystems. Humans and human civilizations have developed at a time in the earth's history when climate, in a geological sense, has been relatively stable, and that stability has been a major factor in the evolution and development of our society. Now the stability has been affected and we should recognize the harmful effects of anthropogenic activities, which have the potential to change our climate. Let us think together in a sustainable manner to ensure a healthy earth, both in social and in biological dimensions.

References

Alheit, J. and Hagen, E. 1997. Long-term climate forcing of European herring and sardine populations. *Fish.Ocean.*, 6:130–139.

Alongi, D.M. 2002. Present state and future of the world's mangrove forests. *Environmental Conservation,* 29: 331–349.

Barnett, J. 2003. Security and climate change. *Global Environ. Change,* 13:7–17.

Barton, M. and Barton, A.C. 1987. Effects of salinity on oxygen consumption of *Cyprinodon variegatus.Copeia,* 230–232.

Bhattacharya, S. and Mukhopadhyay, A. 2009. Biological Perspective of Global Warming. *Global Warming: A Threat to the World.* Kolkata. Pp. 82–96.

Bigano, A., Hamilton, J.M. and Tol, R.S.J. 2005. The impact of climate change on domestic and international tourism: a simulation study. Working Paper FNU–58, Hamburg University and Centre for Marine and Atmospheric Science, Hamburg.

Booth, S. and Zeller, D. 2005. Mercury, food webs, and marine mammals: implications of diet and climate change for human health. *Environ.Health Persp.,* 113:521–526.

De Wilde, J.J.F.E. and Van der Maessen, L.J.G. 1997. Diversity and Distribution of Caesalpinioid Species in the Lopé Reserve in Gabon: A Search for Patterns and Their Explanation. NWO Proposal. Wageningen Agricultural University, Wageningen, The Netherlands.

Done, T.J. 1999. Coral communities adaptability to environmental change at the scales of regions, reefs and reef zones. *American Zoologist,* 39: 66–79.

Douglas, B.C. 1997. Global Sea Rise: A Redetermination. *Surveys in Geophysics,* 18:279–292.

Ellison, J.C. 1993. Mangrove retreat with rising sea–level, Bermuda. *Estuarine, Coastal and Shelf Science,* 37:75–87.

Epstein, P.R. 2007. Climate change and human health. *New. Engl. J.Med.,*353(14):1433–1436.

Gervasi, S.S. and Foufopoulos, J. 2007. Costs of plasticity: responses to desiccation decrease post–metamorphic immune function in a pond–breeding amphibian. *Functional Ecology,* 22(1): 100–108.

Garvey J. 2009. *The Ethics of Climate Change– right or wrong in a warming world.* Continuum International Publishing Group, New York.

Gosain, A.K. and Rao, S. 2004. Impact of Climate Change on Water Sector. In. *Climate Change and India – Vulnerability Assessment and Adaptation* (Shukla, P. R. *et al.* eds.). Universities Press, Hyderabad. Pp. 159–192.

Hales, S., de Wet, N., Maindonald, J. *et al.* 2002. Potential effect of population and climate changes on global distribution of dengue fever: an empirical model. *Lancet,* 360: 830–834

Harasawa, H. and Nishioka, S. (eds.) 2003: *Climate Change on Japan.* KokonShoin Publications, Tokyo.

Haugen, P.M. 1997. Impacts on the marine environment from direct and indirect ocean storage of CO_2. *Waste Manage.,*17:323–327.

Honda, Y., Ono, M., Uchiyama, I. and Sasaki, A.1995. Relationship between daily high temperature and mortality in Kyushu, Japan. *Nippon Koshu Eisei Zasshi.*, 42:260–268.

Ingram, G.J. 1990. The mystery of the disappearing frogs. *Wildlife Australia,* 27(3): 6–7.

IPCC. 2007. *Climate Change 2007: Impacts, adaptation and vulnerability.* Cambridge University Press, Cambridge, UK.

Ishida, Y., Hariu, T., Yamashiro, J., *et al.*, 2001. Archeological evidence of Pacific salmon distribution in northern Japan and implications for future global warming. *Prog. Oceanog.*, 49:539–550.

Justic, D., Rabalais, N.N. and Turner, R.E. 2005. Coupling between climate variability and coastal eutrophication: evidence and outlook for the northern Gulf of Mexico. *J. Sea Res.*, 54: 25–35.

Lal, M. 2002. Global climate change: India's monsoon and its variability. Final Report under Country Studies Vulnerability and Adaptation Work Assignment with Stratus Consulting's Contract of the U.S. Environmental Protection Agency. Pp. 58.

Leemans, R. and Eickhout, B. 2004. Another reason for concern: regional and global impacts on ecosystems for different levels of climate change. *Global Environ. Change*, 14: 219–228.

Létard, V., Flandre, H. and Lepeltier, S. 2004. La France et les Français face à la canicule: les leçons d'une crise. Report No. 195 (2003–2004) to the Sénat, Government of France. Pp. 391.

Macdonald, R.W., Harner T. and Fyfe J. 2005. Recent climate change in the Arctic and its impact on contaminant pathways and interpretation of temporal trend data. *Sci. Total Environ.*, 342(1–3): 5–86.

Malcolm, J., Liu, C., Neilson, R.P., *et al.*, 2006. Global Warming and Extinctions of Endemic Species from Biodiversity Hotspots. *Conservation Biology*, 20 (2): 538–548.

Mall, R.K., Gupta, A., Singh, R., *et al.*, 2006. Water resources and climate change: An Indian perspective. *Current Science*, 90(12): 1610–1626.

Maslin, M. 2004. *Global Warming: A Very Short Introduction.* Oxford University Press, USA.

McFarlane, G.A., King, J.R. and Beamish, R.J. 2000. Have there been recent changes in climate? Ask the fish. *Prog. Oceanogr.*, 47: 147–169.

McMichael, A.J., Woodruff, R.E. and Hales, S. 2006. Climate change and human health: present and future risks. *Lancet*, 367:859–869.

Mehrotra, R. 1999. Sensitivity of Runoff, Soil Moisture and Reservoir Design to Climate Change in Central Indian River Basins. *Climate Change*, 42:725–757.

Moyle, P.B. and Cech, J.J. Jr. 2004. *Fishes:An Introduction to Ichthyology (5th Ed.).* Prentice Hall, Upper Saddle River, NJ. Pp. 726.

Muller–Parker, G. and D'Elia, C.F. 1997. Interactions between corals and their symbiotic algae. In: *Life and Death of Coral Reefs*. Birkeland, C. (ed.) Chapman and Hall, New York. Pp. 96–112.

Nair, G.P.S. 2009. *Earth in Peril*. Publication Division. Ministry of Information and Broadcasting, Govt. of India.

Nordhaus, W.D. 2006. Geography and Macroeconomics: New Data and New Findings. *Proc. Natl. Acad. Sci. USA.*, 103(10): 3510–3517.

Ottersen, G., Planque, B., Belgrano, A., *et al.*, 2001. Ecological effects of the North Atlantic Oscillation. *Oceanogr.*, 49: 539–550.

Pamela, D. and Noyes, P.D. 2009. The toxicology of climate change: Environmental contaminants in a warming world. *Environment International*, 35: 971– 986.

Pascual, M., Bouma, M.J. and Dobson, A.P. 2002. Cholera and climate: revisiting the quantitative evidence. *Microbes Infect.*, 4: 237–245.

Paehler, K.H. 2007. Nigeria in the Dilemma of Climate Change. *Country Reports*, Konrad–Adenauer–Stiftung, Nigeria.
Available at: http://www.kas.de/nigeria/en/publications/11468

Peng, S., Huang, J., Sheehy, J.E. *et al.* 2004. Rice yields decline with higher night temperature from global warming. *Proc. Natl. Acad. Sci. USA.*, 101: 9971–9975.

Pounds J.A. and Crump M.L. 1994. Amphibian declines and climate disturbance: The case of the golden toad and the harlequin frog. *Conservation Biology*, 8(1): 72–85.

Pounds, J.A., Bustamante, M.R., Coloma, L.A., *et al.*, 2006. Widespread amphibian extinctions from epidemic disease driven by global warming. *Nature*, 439: 161–167.

Root, T.L., Price, J.T., Hall, R.K. *et al.*, 2002. Fingerprints of global warming on wild animals and plants. *Nature*, 421: 57–60.

Sharma, K.P., Moore, B. and Vorosmarty, C.J. 2000. Sensitivity of the Himalayan Hydrology to Land–use and Climatic Changes. *Climate Change*, 47: 117–139.

Smit, B. and Pilifosova, O. 2001. Adaptation to Climate Change in the Context of Sustainable Development and Equity. In: *Climate Change 2001: Impacts, Adaptation and Vulnerability*. (McCarthy J.J. *et al.* eds.). Cambridge University Press, Cambridge, U.K.

Singh, P. and Kumar, N. 1997. Impact Assessment of Climate Change on the hydrological response of a snow and glacier melt runoff dominated Himalayan River. *J. Hydrol.*, 193: 316–350.

Stern, N. 2006. *The Economics of Climate Change: The Stern Review*. HM Treasury, London.

Stirling, I., Lunn, N.J. and Iacozza, J. 1999. Long-term trends in the population ecology of polar bears in western Hudson Bay in relation to climatic change. *Arctic*, 52(3): 294–306.

Tan, W.H. and Teh, T.S. 2001. Sustainability of islands and tourism resorts: a case study of Perhentian Islands. *Malaysian Jr. Tropical Geography*, 32: 51–68.

Thanh, T.D., Saito, Y., Huy, D.V. *et al.*, 2004. Regimes of human and climate impacts on coastal changes inVietnam. *Reg. Environ. Change*, 4: 49–62.

Trenberth, K.E., Fasullo, J. and Smith, L. 2005. Trend and variability in column–integrated atmosphere water vapor. *Climate Dynamics*, 24: 741–758.

Wang, X., Chen, F. and Dong, Z. 2006. The relative role of climatic and human factors in desertification in semiarid China. *Global Environ. Change*, 16: 48–57.

WCED. 1987. *Our Common Future*. Report of the World Commission on Environment and Development (WCED), Oxford University Press, New York.

Xu, C.X. 2003. *China National Offshore and Coastal Wetlands Conservation Action Plan*. China Ocean Press, Beijing. Pp.116.

Climate Change: Man and Environment (2012)   *Pages* **24–55**
Editor: **Goutam Kumar Saha**
Published by: **DAYA PUBLISHING HOUSE, NEW DELHI**

3

Impact of Climate Change on Geological System

☆ *Chittaranjan Bhattacharyya**

Introduction

By climate we normally mean that it is simply related to the behaviour of the atmosphere. But it should be mentioned that climate of a region can be changed by the changes in other parts of the Earth's surface over long periods of time. "Thus, if the pattern of circulation of the ocean changes so as to warm some parts of the ocean surface and cool others, patterns of atmospheric moisture content, temperature, pressure and wind direction and strength will also change; if biological productivity changes so as to change the atmospheric content of carbon dioxide or methane, which are greenhouse gases, the surface temperature of the Earth will change. If ice sheets grow they will cool parts of the Earth by a latent heat effect, change patterns of wind circulation around them, increase the Earth's reflectivity and so cool the Earth, lower the ocean level and change the ocean circulation pattern" (Duff, 1994). These have led to think of the atmosphere, oceans and ice sheets and the biosphere as parts of the climate system, which is "strongly a coupled system, meaning that change in any one part will lead to changes in the other components" (Duff, 1994).

Geological system or Earth system is constituted broadly of core (innermost shell, 2900–6371 km depth), mantle (intermediate shell, ≈35–2900 km depth), crust (outermost solid cover (up to av. ≈35 km depth), hydrosphere, biosphere and atmosphere. Besides the atmosphere, hydrosphere, biosphere and mountain ranges are components of both climate system and part of the Earth system. Both intrinsic causes (*e.g.,* volcanism, tectonism) and extraterrestrial causes (*e.g.,* solar radiation,

* Department of Geology, University of Calcutta, Kolkata. E-mail: *crbc.cu.geology@gmail.com*

bolide impacts, variation of the axial tilt of the Earth, eccentricity of the orbit and precession of the equinoxes) bring about changes in the climate system. Man also should be considered as a part of the geologic system because man is a natural animal (cf., palaeolife) and has become an important agent of denudation, landscape and climatic changes. In this paper our purpose is to present a brief outline of the impact of climates on the surface geologic system during the geological history of the Earth. It is obvious that the surface and near–surface domains of the Earth will experience the impact of climate change.

The Earth was formed from the solar dusts (nebula) around 4600 m.y. ago (Peck and Valley, 2009). It is inferred that during the first ~ 200 m.y. the Earth probably was a 'magma ocean', a boiling ball of molten rock material. During the beginning of the Earth's history the surface was hot and unsuitable for formation of oceans and continents. The primitive atmosphere, formed by the gases coming out (degassing) of the boiling 'magma ocean' is thought to have been composed of hydrogen, helium, CO, CO_2, H_2O, NO_x, H_2S and NH_3. Hydrogen being the lightest gas followed by helium escaped to the space. In the history of evolution of atmosphere and hydrosphere the 'degassing' hypothesis (atmospheric gases derived from within the Earth itself mainly by volcanic eruptions throughout later part of the Earth history) has been very popular. Recent thinking attributes much of the secondary atmosphere to extraterrestrial sources (Farquhar, 2009). Anyway, under the very hot climatic condition of the primitive Earth only igneous rocks were solidified from magma. Currently the oldest known rocks are from the Acasta Gneisses of the Great Slave Lake area in northwest territories of Canada, dated 4030–4000 m.y. (previously 3960 m.y.) based on U–Pb isotopes (Bowering *et al.*, 1989; Bowering and Williams, 1999). Possibly the primary rocks, older than this have disappeared by remelting. The formation of sedimentary rocks was possible only when the hot climate became cooler, water basin was formed and the climate became suitable for weathering. The discovery of detrital zircon crystals of 4400–4200 m.y. age from sedimentary rocks (now metamorphosed) of Narryer Gneiss complex, Western Australia indicates that the protocontinental crustal source rock and water–bearing basin/ocean formed before this time (*i.e.* 4400 m.y. ago). The weathering of zircon–bearing crustal source rock gave rise to zircon–bearing sediments for deposition and formation of early sedimentary rock in the ocean basin, which indicates that the Earth was relatively cool to form water in the ocean basin during the above time interval (Mojzsis *et al.*, 2001; Wilde *et al.*, 2001; Peck and Valley, 2009).

The oxygen isotope evidence indicates liquid or liquid–steam hydrosphere ≤ 250°C during 4400–4200 m.y. ago (Peck and Valley, 2009). These oxygen isotope studies of Archean chemical sediments further suggest that the Archean ocean temperature as late as ~ 3500 m.y. ago was likely to be up to ca. 40°C higher than today (Peck and Valley, 2009).

The Earth has witnessed climatic changes throughout its history. The causes of climate change have been described by many authors (*e.g.* Frakes, 1979; Duff, 1994; Merritts *et al.*, 1998; Keller, 1999; Lal, 2006; Jansen *et al.*, 2007). The climate changes have left their signature in the past rock record, particularly sedimentary rocks, the formation of which is due to surface processes and pressure–temperature–humidity conditions in the geological past and hence sedimentary rocks have a high potentiality for preserving the signatures of the climate under which they were formed. The interpretation of the impact of climate on the geological system is based on the Hutton's Law of Uniformitarianism or, expressed in a different way, as 'Present is the Key to the Past'.

Impact of Climate Change on Igneous Process

Among the three groups of rocks (igneous, sedimentary and metamorphic), only igneous rocks could be formed by crystallization from the magma in the primitive Earth and when the climatic condition became suitable (lowering of temperature, accumulation of water in the basins, precipitation) for weathering the sedimentary rocks were formed. Metamorphic rocks were formed by transformation of both igneous and sedimentary rocks in the solid state at higher P–T conditions prevailing at depths of the crust (surface sedimentary and igneous rocks might have been subducted to depth by tectonic movement and were metamorphosed there). The metamorphic rocks at depth and those igneous rocks which crystallized in magma chamber at various depths of crust are not subjected to change by climate, the domain of action of which is restricted to the surface or near–surface regions. However, such rocks when exposed to the surface either by erosion of overlying burden of rocks or tectonic movements, are affected by weathering process which is largely climate–controlled. Magma, formed at deeper part of the earth's crust (average depth ~35 km) or upper part of the underlying mantle, may crystallize at depth with gradual cooling over hundreds and thousands of years forming a coarse grained plutonic igneous rock (*e.g.* the rock gabbro crystallizes from a basaltic magma at depth, *i.e.* under plutonic condition). If the magma, instead of solidification at depth, finds its way upward through fractures in the crust, it will reach to the surface of the Earth either as volcanic eruptions or as fissure–fillings forming volcanic rocks. Due to the sudden change of climate, *i.e.* exposure of the high temperature magma (1200°C) to the surface temperature (and pressure) of the earth it (now called lava) will rapidly crystallize, as a result of which a fine grained rock known as volcanic or extrusive rock will form instead of a coarse grained rock crystallized at depth. Thus a basaltic magma will crystallize a coarse grained rock (due to slow cooling), gabbro, at depth and a fine grained rock (due to rapid cooling), basalt, on the surface of the earth.

If the basaltic magma extruded into the glacial region, it will consolidate so quickly that there will be very little or no time for crystallization of minerals (which have definite atomic arrangement) and hence the lava will solidify as dark coloured glassy rock, called tachylite. If the basaltic lava enters into the sea, it commonly crystallizes rapidly forming pillow–like structure forming pillow basalt.

The basaltic magma, before reaching the surface, may partly crystallize at depth slowly to minerals of coarse sizes; if such a partly crystallized magma, suddenly extrudes the surface, it will rapidly consolidate to fine grained basalt containing some earlier crystallized (at depth) phenocryst (coarse crystals of minerals) giving rise to porphyritic basalt. During solidification on the surface the gases trapped in the basaltic lava will exert pressure to come out forming vescicular structure on the exposed surface of the basalt.

Impact of Climate on Metamorphic Process

Metamorphism of crustal rocks takes place at depths at higher P–T conditions and hence metamorphosed rocks are not apparently affected by climate of the surface and near–subsurface domains until such rocks are exposed by erosion and tectonic activity. The excessive rainfall in a region may have some impact on metamorphic reactions taking place over a long period of time at crustal depth. The minute amount of water present in pore spaces of rocks has a profound effect in bringing about metamorphism. If the water concentration in the fluid phase of the rock at depth increases, a particular metamorphic reaction will take place at higher temperature in non-carbonate rocks than when the rock in question contains lower concentration of water. However, in case of carbonate rocks (limestone undergoing metamorphism to marble) the higher concentration of water than CO_2 in the fluid phase will cause the reaction $CaCO_3$ (calcite) + SiO_2 (quartz) → $CaSiO_3$ (wollastonite) + CO_2 to take place at lower temperature.

In the desert area the thunderstorm along a definite line on dune sand occasionally melts the quartz grains which solidify quickly (quenching) to silica glass, called Lechatelierite. Such silica glass occurs in a tubular form along a particular direction on sand dune, which may be attributed to the process of an ultrametamorphism in which extremely high temperature of thunderstorm, a natural climatic phenomenon, has completely melted the crystalline quartz followed by quenching within a few seconds.

Impact of Climate on Weathering of Rocks

Mechanism of Weathering

Weathering is a climate–controlled process by which rocks and its constituent mineral grains, the most important constituent of geologic system, are broken into smaller and smaller pieces and the constituent mineral grains are separated from one another and some minerals are decomposed to other minerals. Some of these loose materials may remain practically in the same location on the partially altered bed rock while the others are transported by gravity, wind and surface run–off, the later also includes dissolved ions–to lakes, streams, rivers and ultimately ocean for deposition and/or precipitation as sediments forming eventually sedimentary rocks.

There are two processes of weathering, physical and chemical. In physical weathering solid rocks are fragmented by physical processes. For example, the exposed rocks in the day time, when the temperature is higher, undergo expansion and in the night, when temperature is much lower, undergo contraction; continuity of this expansion and contraction for several years, eventually break down the rocks and mineral grains into smaller pieces. The other processes of physical weathering include (*i*) the separation of sheets of homogeneous and unlayered rocks, such as granite, due to release of load pressure on the underlying unweathered rock by the removal of the overlying weathered materials, (*ii*) frost wedging in which water, percolated into fractures of rocks, freezes to ice whose greater volume (9 per cent greater than that of liquid water) exerts a force great enough to split rocks, (*iii*) salt wedging in arid climate of desert where salt (NaCl), due to evaporation of water below the rock surface, exerts pressure giving rise to flaking of rock, (*iv*) biological disintegration of rocks or rock particles by growth pressure of tree roots in weak zones of rocks and activity of worms and (*v*) disintegration of fragmented materials by saltation and abrasion in river beds or beneath glaciers.

Chemical weathering, in contrast, causes chemical alteration of minerals present in rocks, which further helps disintegration. In reality, both physical and chemical weatherings occur together and both contribute to mass movement. High temperature and heavy rainfall accelerate chemical weathering, cold and arid (for example, desert) climates impede the process. Chemical weathering is most active in warm, humid areas where rainfall supports plant growth that yields humic and organic acids to facilitate chemical reactions. The agents for decomposition at the surface are water, oxygen, carbon dioxide, heat, acids, alkalies, plants and animal life and some of the soluble materials produced by decomposition of the rocks themselves (Bateman, 1956).

The chemical reaction between minerals in rocks and air and water, facilitated by higher temperature, results in chemical weathering. The SO_2, NO_2 and CO_2 in the atmosphere react with H_2O (occasionally O_2) to form different acids as follows (Duff, 1994; Murck *et al.*, 1996; Blatt, 1997; Merritts *et al.*, 1998; Press and Siever, 1999; Johnsson, 2000):

$$2SO_2 + 2H_2O + O_2 \rightarrow 2H_2SO_4$$

$$4NO_2 + 2H_2O + O_2 \rightarrow 4HNO_3$$

$$CO_2 + H_2O \rightarrow H^+ + (HCO_3)^- \rightarrow H_2CO_3$$

The acidic water thus formed attacks rocks, particularly limestone or marble :

$$CaCO_3 + 2H_2O + H_2SO_4 \quad \rightarrow \quad CaSO_4.2H_2O$$
(calcite in limestone/marble)   (gypsum)

$$CaCO_3 + H_2CO_3 \rightarrow Ca^{2+} + 2(HCO_3)^-$$

Some of the reactions, grossly oversimplified, that take place by chemical weathering of rock minerals are

$$KAlSi_3O_8 \quad + H_2O \quad \rightarrow \quad Al_2Si_2O_5(OH)_4$$
(K–feldspar)   (kaolinite clay)

$$4FeSiO_3 \quad + O_2 \quad \rightarrow 2Fe_2O_3 + 4SiO_2 \text{ (dissolved silica)}$$
(iron pyroxene)

$$2NaAlSi_3O_8 + 2H_2CO_3 + 9H_2O \rightarrow 2Na^+ + 2(HCO_3)^- + 4H_4SiO_4 + Al_2Si_2O_5(OH)_4$$
(Na–feldspar) (carbonic acid)   (bicarbonate ion) (silicic acid) (kaolinite clay)

$$4FeS_2 \quad + 15O_2 + 8H_2O \quad \rightarrow \quad 2FeO_3 \quad + 8H_2SO_4$$
(pyrite)   (hematite)

$$4FeO \quad + 2H_2O + O_2 \rightarrow 4FeO.OH$$
(iron oxide)   (goethite)

$$2FeOOH \quad \rightarrow \quad Fe_2O_3 \quad + H_2O$$
(goethite)   (hematite)

Carbonic acid (H_2CO_3), formed by the reaction of CO_2 with rain water in the atmosphere, is a powerful reactant in the weathering of carbonate and silicate rocks. In addition to the atmospheric contribution of H_2CO_3 in rain water, the organic matter in the soil at the surface can increase the acidity of reacting waters through the introduction of organic acids. The immense chemical weathering in tropical and subtropical humid climates is facilitated by high precipitation and temperature coupled with high organic acid concentration.

The rounded form of the granitic rocks and dolerites is the result of both physical and chemical weathering. Both the rocks are rather impermeable and hence, acidic water percolates through the fractures ultimately bringing about what is known as spheroidal weathering. The gypsum formed on marble by reaction, as shown above, is easily washed away by rainwater, thus altering the polished marble to a rough surface (Blatt, 1997).

Impact of Climate on formation of Soil, Sedimentary Rocks and Mineral Deposits

Soil

Soil, more particularly fertile soil, is one of the most important precious natural resources and is a medium for growing food, timber and wood. A hand specimen of soil contains billions of microbes that decompose plant, animal and mineral matter and generate nutrients. Soils are the result of interaction among the lithosphere, atmosphere, hydrosphere and biosphere. The broad principles of rock weathering which give rise also to soil have been described earlier. The soil is defined as an internally organized, natural body of weathered minerals and organic constituents. The soilsphere or pedosphere varies in thickness from 1m to 200m. The depth of the pedosphere is greater where rainfall and temperature are high and weathering process extends deeply in the crust. The soils are arranged in horizons such as O horizon (top), A, B, C, R etc. horizons, the vertical sequence of which is called soil profile. The characteristics of soil horizons vary with climate (moist, dry, wet). Clays are important constituents of soils. Clays are formed by chemical weathering of feldspars and some other silicate minerals of rocks. The dying plants and animals contribute organic matter, the decay of which releases CO_2 and nutrients. Most of this CO_2 recirculates to the atmosphere (thus adding to the greenhouse gas of the atmosphere). The weathering of rocks releases sodium, potassium, magnesium, calcium, chloride and other ions to the soil. The nature of the soil is controlled not only by the composition of the bed rock but also by the climate–dependent weathering process. The nitrogen from atmosphere enters the soil and back through a series of biological transformations (Duff, 1994; Mason and Moore, 1985; Blatt, 1997; Merritts *et al.*, 1998).

Sedimentary Rocks

The basic requirements for formation of sedimentary rocks are the availability of loose sediments by weathering of preexisting rocks, transport by gravity, wind and water to the basin of deposition such as lakes, rivers and ultimately sea forming clastic sedimentary rocks within the broad range of surface pressure–temperature conditions. The root cause is weathering, both physical and chemical, which is dependent on climate – the higher temperature accelerating the process. The precipitation of some sediments from aqueous solution in lakes, rivers and abundantly in sea water is controlled by some factors such as pressure, temperature, Ph and Eh (oxidation-reduction potential) and concentration (Mason and Moore,1985); such rocks are called chemical sediments or non-clastic sediments (*e.g.*, chert and most limestones of the sea bed). Some examples of impact of paleoclimate on sedimentation are briefly described below. The discovery of detrital or clastic mineral zircon (age: 4400–4200 m.y. old) in metasedimentary complex of Australia, as already mentioned, indicates that weathering of primary igneous rocks forming the primitive crust commenced probably around this time interval or slightly younger and ocean basin, presumably having shallow water depth, was formed in the early period of the earth. The record of increasing amount of sedimentary rocks from 3900 Ma to 2500 Ma (Figure 3.1) indicates that climatic conditions for weathering and sedimentation were favourable. However, lower free oxygen concentration in the then atmosphere and hydrosphere did not allow formation of much Red beds (red colour is due to oxidation of iron) and evaporites like gypsum and anhydrite rocks and oxidation of pyrite (FeS_2). The common presence of warm humid climate with a general drying out of global climate caused increasing formation of arkosic sedimentary rocks until about 2000 m.y. (Figure 3.1). The increase of free oxygen in the atmosphere between 2500 and 2000 m.y. (Farquhar, 2009) and/or luxuriant development of algae in the vast ocean supplied oxygen for the oxidation of dissolved ferrous iron so as to precipitate Fe_2O_3 (hematite) giving rise to banded iron formation (BIF) mostly during Archean–Proterozoic boundary, 2500–2200 Ma (Cloud, 1968, 1973, 1976; Simonson and Kaufman, 2009). The warm humid climate from 2500–570 Ma (Proterozoic) and more oxidizing condition caused abundance of Red beds and arkoses and also precipitation of warm-water carbonate rocks, limestone and dolomite.

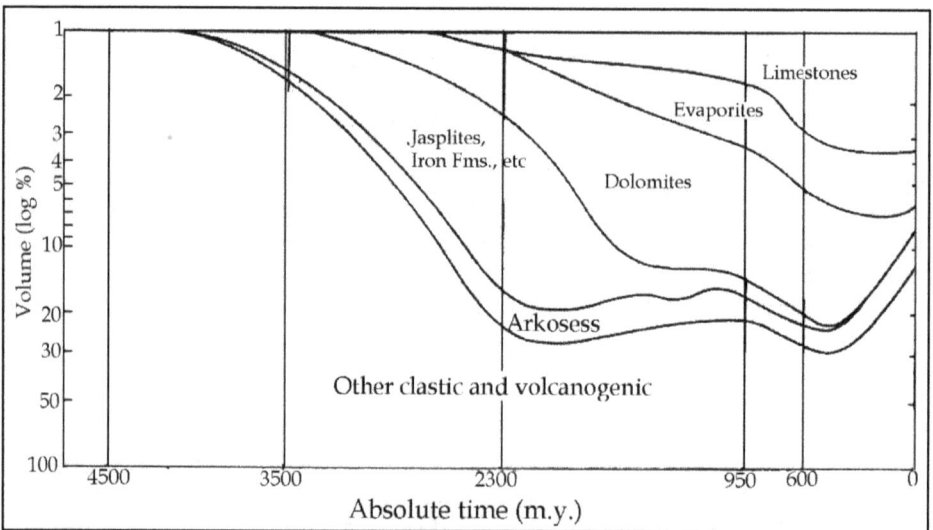

Figure 3.1: Variation of amount of different sedimentary rocks during the geological history of the Earth. Source: Frakes (1979).

The substantially warm climate of Cambrian to Devonian, similar to present day or slightly higher temperature favoured sedimentation of evaporites in Cambrian-Ordovician time in Australia, Siberia, the Aden–India, the Spanish Sahara, Canada and USA and of halites (NaCl) in Eurasia, and anhydrite ($CaSO_4$) in Americas. In Silurian evaporites occur in latitude <25° while Devonian evaporites in Asia extend to 60°. The high humidity and warm climate in Devonian, restricted to relatively lower latitudes, caused formation of coals (amphibians appeared at this time). In early Carboniferous the same climate favoured growth of organic reefs in low paleomagnetic latitudes (*e.g.*, 'Horseshoe Atoll' of the Midland Basin, Texas). In the upper Carboniferous to Lower Permian glaciations in the southern hemisphere caused formation of glacial boulder bed or till at the base of Permian strata of the Gondwanaland. With increasing warming and humidity from the basal beds of Permian to the upper part the development of luxuriant *Glossopteris* flora gave rise to rich coal deposits of the Lower Gondwana rocks. *Gossopteris* became almost extinct in the Triassic when the climate changed to arid condition giving rise to the appearance of *Dicroidium* flora, which, in turn, was replaced by *Ptilophyllum* flora in the Jurassic Gondwana due to the reappearance of warm humid climate. The warm tropical climate may have favoured the development of robust ectotherm labyrinthodont amphibians and reptiles in the continental Permian rocks in different parts of the Earth.

The fossil vertebrates (for example, *Lystrosaurus* in South Africa, Europe, Gondwana rocks of India (Bandyopadhyay *et al.*, 2002), Asia, North America, recorded from Triassic and Jurassic rocks indicate moist, tropical to subtropical conditions. The Lower Triassic oxidizing arid climate favoured deposition of Red beds (between 50° latitudes, encompassing much of Europe, former USSR, North America and Northern Brazil).

The distribution of hermatypic hexacorals, *Dictyophyllum* and *Sagenopteris*, the land–living giant reptiles (as far north as 50°N), marine reptiles, *Plesiosaurus*, and marine crocodiles indicate warm climate. The Late Jurassic is known as the age of giant dinosaurs in the reptilian faunas of the world. Oxygen isotope data from Jurassic belemnites at 75°S latitude just off Antarctica indicate that sea temperature reached 14°C (roughly 7°C higher than average temperature today) as suggested by Merritts *et al.* (1998). The warm oxidizing climate of Jurassic favoured formation of evaporite sediments. The Cretaceous climate was warmer than now which is indicated, among other evidence, by abundance of oxygen isotope paleothermometric data over the full range of latitudes from equator to poles (Figure 3.2), which promoted formation of evaporites, coals, reefs and bauxites. The Cretaceous climate corresponds to tropical to subtropical conditions and beyond 70°S latitude the climate was warm to cool-temperate. The striking feature of Cretaceous is the enormous accumulation of heat within the atmosphere–ocean system (Frakes, 1979).

Sedimentary Mineral Deposits

The warm climatic condition facilitates the evaporation of lake water or sea water to give rise to direct precipitation of salts yielding evaporite deposits such as sodium carbonate, sodium sulphate and borax. Marine evaporites form when a portion

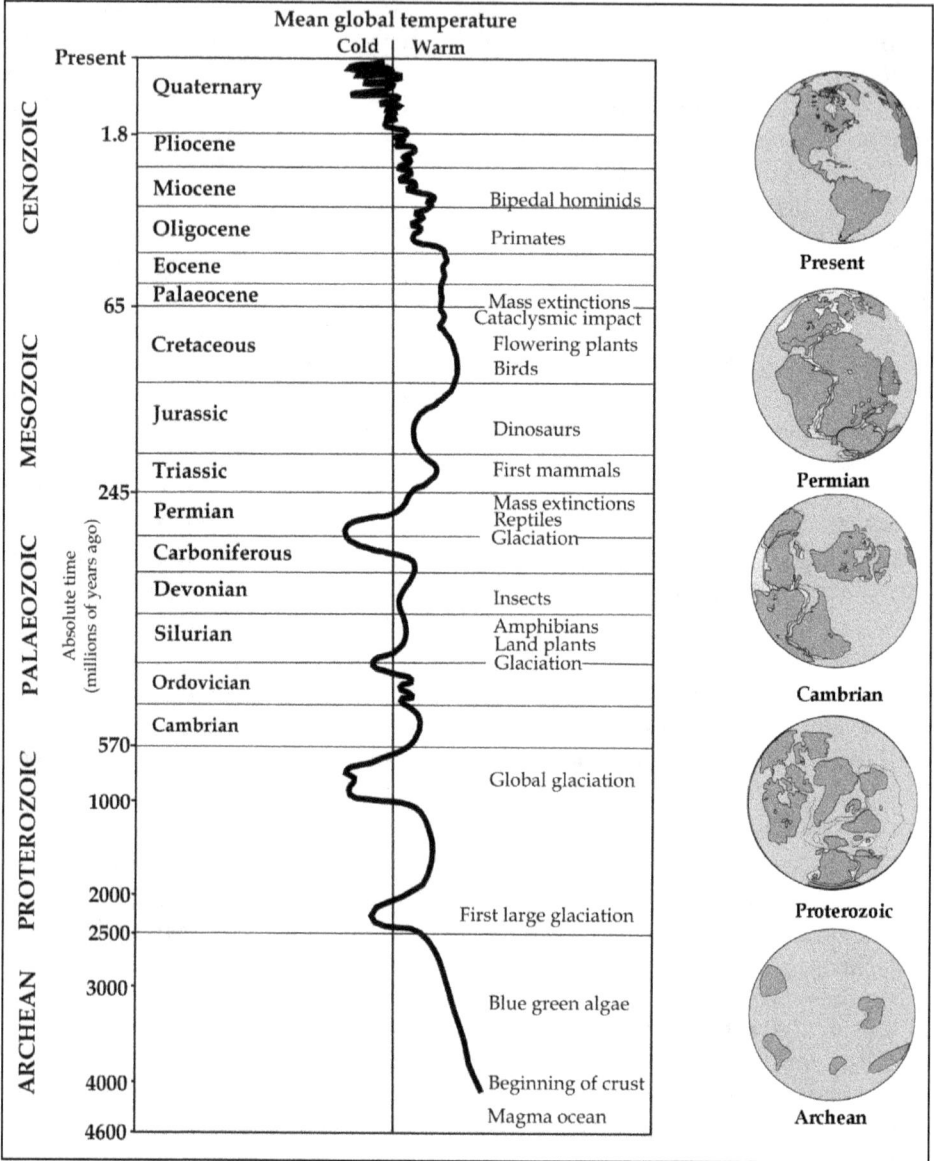

Figure 3.2: Temperature variation around the mean global temperature from about 4000 m.y. ago to present. Source: Merritts et al. (1998).

of the shallow sea becomes cut off from the main part of the ocean by a reef or other barriers. The most important salts that precipitate from sea water are gypsum ($CaSO_4$.$2H_2O$), halite or common salt (NaCl).

The sedimentary mineral deposits formed by precipitation from sea water solution are phosphates, iron deposits (e.g. banded iron formation) and manganese nodules. It may be mentioned in this context that average content of iron in river water is about 1.0 ppm whereas in seawater the amount is exceedingly small, about 0.008 ppm. Thus weakly acidic iron–bearing solution flowing into the sea from neighbouring land areas must precipitate most of their iron in the weakly alkaline marine water (Mason and Moore, 1985).

The concentration of heavy minerals takes place by mechanical process (for example, winnowing action of waves and currents of the river or sea water).The deposition of such heavy mineral concentrates in suitable places of the river or sea water gives rise to placer deposits (example: placer gold, platinum, diamond, monazite etc.).

Residual Mineral Deposits

Due to weathering of rocks soluble parts may be removed and the insoluble parts may be concentrated over the bed rock to form residual mineral deposits. In temperate climate, silica of silicate rocks is not extensively removed but forms clays with other constituents. Thus residual clays or clayey soils are common products in all temperate climates. Tropical and subtropical climates, characterized by alternate wet and dry seasons, warm weather and warm surface waters throughout the year and generally luxuriant vegetation with profuse supply of bacteria and organic compounds promote rock decay, more leaching, more thorough decomposition of silicate rocks and removal of silica in solution. The wet season of the tropics allows formation of Al_2O_3 and Fe_2O_3 and dry season allows removal of SiO_2 in solution. Thus, instead of clay (hydrous aluminium silicate), hydrous aluminium oxide (bauxite) is formed as a residual mineral deposit in the tropical–subtropical climates.

The conditions necessary for formation of bauxite as a residual deposit are (i) a favourable parent rock (high in aluminium silicate minerals and low in iron and free quartz or absence of quartz, *e.g.* nepheline syenite) with easily soluble minerals whose leaching will leave a residual enrichment of aluminium and/or iron (for laterite deposit), (ii) humid tropical or subtropical climate, (iii) effective rock porosity permitting free circulation of water, (iv) high rainfall with intermittent dry spells, (v) good drainage, (vi) low to moderate topographic relief, (vii) availability of reagents to bring about breakdown of the silicates and solution of silica, (viii) presence of vegetation including bacteria, (ix) prolonged stability, and (x) preservation (Bateman, 1956; Evans, 1993). In the temperate regions of Arkansas, Georgia and France the large bauxite deposits were formed during warmer Tertiary climates (Bateman, 1956). Most of the commercial deposits of bauxite in the world are found between Middle Cretaceous and Middle Eocene indicating that optimum conditions for their formation, say temperature, existed during this time (Figure 3.2). However, some Palaeozoic and Late Tertiary to Recent bauxites are also known. Apart from bauxite, valuable deposits of iron ore, manganese, clays, nickel, phosphate, kyanite, barite, ochers, tin, gold and few others are known as residual deposits.

Supergene Mineral Deposits

Climate plays an important role in enriching the ore content of an ore body. The key process in this enrichment is weathering. The warm humid climate of the tropical areas accelerates the process as already mentioned. Weathering may expose a low grade ore body of depth to the surface of the earth by removal of overlying economically unimportant weathered rock material. During further weathering, the surface water, while percolating through the low grade ore body, reacts to yield solvents which dissolve valuable materials. Thus many of the valuable materials are leached out from the upper oxidized zone and trickle downward to the water table. The zone above the water table is called the oxidized zone. In this zone of oxidation the leaching solutions may loose a part or all of their metallic content and give rise to oxidized ore deposit which may be viable for commercial exploitation. However, when the downtrickling solutions penetrate the water table, below which is the zone of reducing environment, the metallic content may be precipitated in the form of secondary sulfides giving rise to a zone of secondary or supergene sulphide enrichment of the sulphide protore (the primary unaltered ore body). Below the zone of supergene enrichment of ore body lies the lower grade unaltered protore. Copper, zinc and silver sulphides are soluble in the upper part of the ore body in the oxidized zone and are more susceptible to supergene enrichment. Apart from climate, the other factors which accelerate supergene enrichment are water table, rate of erosion, time, chemical, structural and textural properties of the enclosing rocks and physiographic development of the area. Oxidation of the protore at the surface/near–surface zone is the primary requirement for supergene enrichment. Supergene ore bodies occur in many of the non-glaciated land areas of the world. For more details the reader is referred to the books of Bateman (1956) and Evans (1993).

Impact of Climate on Desertification and Desert Landscape

The areas of little rainfall are eventually converted into deserts. They are primarily located between 15° and 30°N and S of the equator. This zone is dry due to global circulation pattern of atmosphere (Keller, 1999). The deserts of Africa, Australia and North America lie within this zone. However, some of the semiarid lands and deserts (North America and Central Asia) are dry lands because they are situated far in land in continents or in the "rain shadow" of mountain ranges that create obstruction to the rainfall and storms and some deserts (Chile and Peru) are due to intercept of air by a cold offshore ocean current. Human–induced land degradation, in addition to natural disastrous drought, together caused deaths of 250,000 people and several million cattle, sheep, goats and camels in the Sahel region of west Africa in the early 1970s (Keller, 1999). The impacts of desertification are (i) lowering of groundwater table, (ii) salinization of soil and near–surface soil water and (iii) reduction in aerial extent of surface water in streams, ponds and lakes.

In the desert area, the physical weathering is predominant partly due to scanty rainfall and mainly due to scorching heat at day time and severe cold at night, eventually disintegrating the rocks due to expansion and contraction during several years. The high wind velocity in the desert area causes enormous amount of dusts to darken the sky and these dusts are carried over great distances. These wind–blown sands not only abrade each other but also the rock surfaces. As a result the sand grains become finer and gradually rounded due to innumerable impacts. The sand grains (1–2 mm in diameter) become well–rounded like 'millet–seed' while the finer grains (0.006–0.2 mm in diameter) become generally angular. Due to winnowing action of the wind the transported particles become sorted. Thus the wind–blown dune sands are relatively clean and composed of uniform grains commonly with frosted surface. Ripple mark, formed by gentle wind, is a common feature of dune sands. The finer particles are blown out and the larger fragments, such as pebbles and gravels are left behind and concentrated on the wind–swept surfaces of the original rock waste (Duff, 1994).

The term dune is used for wind–blown sand deposits which frequently form mounds or ridges. The dunes show three geomorphological types: (i) Transverse dunes, (ii) Crescent–shaped dunes, called *barchans* (Figure 3.3) and (iii) Longitudinal dunes (sand deposited in rows of long linear dunes the axes of which are more or less equally spaced and parallel to regional wind direction (Duff, 1994). In *barchan* the crescent wings point downward and the curving bow faces the wind (Gilluly *et al.*, 1959). If the wind directions vary, the barchan dunes are unstable and instead the dunes are stung out in long chains at an angle to the winds and such dunes are called *seif* (meaning in Arabic, 'sword'). The height of some *seifs* in Iran is more than 700 feet above the base and 3/4ths of a mile wide. The length of individual *seif* ridges may be as much as 60 miles and the length of groups of *seifs* extend more than 200 miles in western Egypt (Gilluly *et al.*, 1959). Apart from the above geomorphological types of sand dunes the wind erosion produces three other types of landscape (Duff, 1994) in the desert surface such as (i) the rocky desert (the *hammada* or *hamada* of the Sahara, (ii) the stony desert composed mainly of gravels or pebbles (the *reg* of the Algerian Sahara, the *serir* of Libya and Egypt or the *gibber plain* of Australia) and (iii) the sandy desert (the *erg* of the Sahara).

In the desert area the sand blasts erode away the inclined softer bed rocks which may alternate with harder bed rocks, thus forming ridge–and–furrow structure or

Figure 3.3: In desert area wind–blown dune sands have changed the landscape to what is called 'barchans' forming a colony in Mauritania, West Africa. Source: Duff (1994).

passages between deeply undercut ridges, which are called *yardangs* in Asiatic deserts. The wind–blown sand may form also smoothed or pitted structure on the exposed bed rock of the desert floor. For example, massive granitoids become smoothed or pitted, gneisses develop rib and flute structures, particularly when the gneissic foliation plane is parallel to the wind flow direction, whereas compact limestone becomes polished by wind–blown sand. If there is seasonal change of wind direction or the pebbles in the desert are turned over by wind flow, two or more facets may be cut by wind blast and each pair of facets meet in a sharp edge. Such wind–faceted polished pebbles are called *ventifacts* or *dreikanter* (Gilluly *et al.*, 1959; Duff, 1994).

Climate Change and Mass Extinction

The signature of oldest life, till now claimed to have been recorded (a yeast–like spherical microorganism *Isuasphaera*, a dubiofossil) is from a metamorphic rock ~ 3800 m.y. old from Isua Geenstone Belt, Greenland (Taylor *et al.*, 2009). Since then both evolution and extinction and occasionally mass extinction of mainly animals throughout geologic history of the earth have taken place and have been recorded from the rock formations of different ages. By mass extinction we mean significantly higher rate (than normal) of death of life within a short time frame and over wide geographical areas. The major Mass Extinction events have been recorded from the rocks of the following geological times (Jones, 2006):

(1) Late Precambrian; (2) Early Precambrian; (3) Late Cambrian; (4) End Ordovician (death between 70 and 80 per cent of all species and 22 and 33 per cent of all families); (5) Late Devonian (between 70 and 80 per cent of all species and 20 per cent of all families); (6) End Permian (between 70 and 95 per cent of all species and between 50 and 60 per cent of all families); (7) End Triassic (80 per cent species and 20 per cent families); (8) Late Cretaceous; (9) End Cretaceous (between 46 and 76 per cent of all species and 14 per cent families); (10) End Paleocene; (11) End Eocene; (12) Pleistocene; (13) Holocene (since 17th century 116 species of birds (1 per cent of total number), have been extinct and 1029 species (11 per cent) are threatened or endangered).

Causes of Mass Extinction: Role of Climate Change

The causes of mass extinction can be classified into two groups: (1) Intrinsic, (earth–bound causes) and (2) Extrinsic (extraterrestrial causes). The intrinsic causes include volcanism, sea–level change and associated anoxia and climate. The extrinsic causes include bolide impacts and associated effects. Whatever the causes might be, the ultimate effect is climate/environment change leading to mass death of life during a span of short time interval repeatedly in the geological history of the earth. It is interesting to note that there are differences in the levels of extinction between the animal and plant kingdom. This is partly due to the greater ability of plants to withstand major ecological trauma (Willis and McElwain, 2005).

Intrinsic Causes

A good correlation between flood basalt magmatism and mass extinction events during the Palaeozoic to Upper Cretaceous has been shown by Rampino and Stothers (1988). Flood basalt volcanism would emit huge amount of dusts and several gases including greenhouse gases and other toxic materials causing darkening of the sky which obstructs solar radiation causing 'volcanic winter', stop of photosynthesis, intoxication of the atmosphere, global warming due to accumulation of CO_2 and other greenhouse gases and ultimately acid rain; wildfires may also be generated affecting the atmospheric composition. Intermittent basaltic volcanism simultaneously in different places of the Earth during over a few million years, for example, Palaeozoic to Upper Cretaceous basaltic volcanism as demonstrated by Rampino (1987), might cause climatic changes bringing about mass extinction. In India, wide–scale flood basalt volcanism of Deccan Traps in a major part of Central and Western regions took place about 65 m.y. ago (Upper Cretaceous), thus coinciding with the Cretaceous – Tertiary (K/T) mass extinction event.

Ocean Anoixic Events (OAE) are suggested by some to be correlatable with many mass extinction events (Leary and Rampino, 1990; Wilde *et al.*, 1990). The oxygen-poor condition may develop at a depth of the ocean due to high rate of organic carbon deposition in the sediments there and during such change mass extinction of biota might have taken place. The geological evidence of such OAE have been recorded from Palaeozoic and Mesozoic marine sediments from different parts of the world (Leary and Rampino, 1990).

A number of factors may cause OAEs, such as dissociation of methane from deep sea hydrates causing increase of greenhouse gases CO_2 and CH_4, changes in ocean circulation patterns and water temperatures, periods of marine transgressions, increase in nutrient supply and productivity, changes in salinity of bottom water and influx of organic matter (Leary and Rampino, 1990). The OAEs have also been connected by some authors to global tectonic pulsations such as rapid sea–floor spreading and increased volcanism giving rise to increase of atmospheric CO_2 and warm climate and warm stagnant bottom waters (Leary and Rampino, 1990). Ocean turn–over events giving rise to mixing of anoxic water throughout the ocean–water column may have been caused also by bolide impacts. A model of OAE development by increased influx of trace metals (*e.g.* Fe, Co, Mn, Mg, Mo, Sc) has been suggested (Leary and Rampino, 1990). According to this model, the increased trace elements would bring about high phytoplankton productivity and eventual death and extinction when the toxic levels had reached the levels of upper limit of tolerance by the marine animals (Leary and Rampino, 1990). The different sources of the trace metals detected at bioevent boundaries are extraterrestrial, volcanism, global wildfires, acid rain and upwelling of anoxic deeper water. The correlation among mass extinctions, large–body impacts, OAEs and flood–basalt volcanism led Leary and Rampino (1990) to propose multiple–related causes for mass extinctions.

Extrinsic Causes

Bolide impacts include impact of meteorite and also showers of comet. The theory of multiple large–sized meteorite impacts causing physical and chemical stresses imposed on biota leading to stepwise extinctions over a period of few million years (Rampino and Stothers, 1984; Hut *et al.*, 1987) has received much attention and momentum after the discovery of Chicxulub meteorite impact crater (diameter 180 km) in the northern Yucatan Peninsula, New Mexico, about 65 m.y. ago (K/T boundary) (Murck *et al.*, 1996). "Discovery of anomalous iridium and other trace metals in generally chondritic (chondrite, the most abundant type of meteorite) ratios in a thin band of clay, coinciding with the major mass extinction of nannoplankton and foraminifera at more than 75 sites worldwide and subsequent discovery of microspherules, shocked quartz and feldspar grains and soot in the boundary clay layer provided evidence for the impact scenario" (Caldeira *et al.*, 1990). The source of iridium in the clay bed at the basal part of Tertiary sedimentary rocks is thought to be the meteorite, the impact of which in Upper Cretaceous time (~ 65 m.y.) created pulverization of the impacted rock and this pulverized masses were transported for deposition in the water basin to form iridium–rich clay bed in the younger Tertiary time (K/T boundary). According to some authors the iridium anomaly may also originate by repeated volcanism. The atmospheric changes caused by prolonged volcanism may be created also by meteorite impacts.

The specific causes of some of the important mass extinction events suggested by different authors are summarized in Table 3.1.

Impact of Glacial Climate

The earth has witnessed glacial climates in varying extent repeatedly throughout its history. The earliest large glaciations took place in Middle Precambrian and mostly around 2300 m.y. ago in several continents. The second large glaciations between the range of 900 to 600 m.y. ago (Precambrian) occurred in continents which were then located at low latitudes. In Palaeozoic time, glaciations have been recorded during four periods. Of these, the Middle Carboniferous to Early Permian glaciation was the most intense and of great duration. Among all the glaciations, the Late Precambrian, Late Paleozoic and Late Cenozoic glaciations were probably the most extreme, judging from the areas affected (Frakes, 1979). During the Pleistocene epoch (~ 1.8 m.y. ago) glaciers occupied as much as 30 per cent of the land area of the earth. This is referred to as Ice Age. Around A.D. 1400 the cold period is known as Little Ice Age. Today glacial ice occupies 10 per cent of the land area (Keller, 1999).

Three causes have been suggested for the development of glacial climate: (1) the amount of solar radiation reaching earth's surface has regularly changed cyclically (called Milankovitch cycles) and results from regular changes in earth's orbit and orientation of earth's axis of rotation, (2) decline of CO_2 in the atmosphere (CO_2 in glacial time, ~190 ppm and during interglacial time, ~280 ppm, as recorded from the ice core samples of Antarctica) and (3) changes in the circulation patterns of the oceans and atmosphere (Keller, 1999). If there was no greenhouse effect, the earth would be approximately 33°C cooler than it is now and all surface water would be frozen (Keller, 1999).

Some of the impacts of glacial climate on the geologic system are summarized below:

1. The landform is variously changed by glacier movement. For example, it may convert V–shaped valley in the hilly region, which is characteristic for river erosion, into U–shaped valley in the landscape.

Table 3.1: Impact of climatic and other causes of some mass extinctions. The types of organisms affected most are also mentioned. Organisms in italics are more severely affected by mass extinction event (*Source*: Jones, 2006)

Period and Biota Affected	*Climatic and Other Causes*
End Ordovician: (acritarchs, corals, brachiopods, stenolaemate bryozoans, multiloids, ostracods, trilobites, echinoderms, graptolites, and chitinozoans, reef communities, and also trace fossils).	Global cooling up to 8°C associated with glaciation, rapidly followed by global warming accompanied by marine transgression and oceanic anoxia, a 'superplume event'.
Late Devonian: (cyanophytes, phytoplankton, calcareous algae, plants, stromatoporoids, rugose and tabulate corals, brachiopods, *mollusks,* ostracods, trilobites, echinoids, crinoids, fish and amphibians, reef– building organisms).	Global cooling and associated sea–level fall, global warming to as high as 34°C and associated sea–level rise and oceanic anoxia, or impact of meteorites in combination with (?) consequent – global cooling.
End Permian: (phytoplankton, plants, foraminifers, radiolarians, sponges, *rugose* and *tabulate corals,* articulate *brachiopods*, stenolaemate bryozoans, goniatite ammonoids, trilobites, insects echinoderms and tetrapods, *crinoids*, and trace fossils).	Global climatic change (evidenced from paleopedological studies from Karoo basin, Africa, Eastern Australia, and Antarctica), sea–level change, superanoxia, volcanic eruptions yielding CO_2 (Siberian trap), dissociation of methane from deep–sea hydrates causing increase of greenhouse gases CO_2 and CH_4, or some combination thereof, and meteorite impacts. The Permian–Triassic (P/T) Mass Extinction, about 250 m.y. ago, is the most severe event in the history of life which Erwin uttered as "world went to hell" (McKinney and Tolliver, 1995).
End Triassic: (plants, sponges, corals, *brachiopods,* stenolaemate bryozoans, *bivalves,* gastaropods, insects, echinoderms, amphibians and reptiles, locally trace fossils, *ceratite ammonoids, conularids, nothosaurs, placodonts,* labyrinthodonts, the mammal–like *reptiles,* the *conodonts,* the *procolophonids,* the *prolacertiforms,* and the *rhyncosaurs*).	Global climatic change, sea–level change, volcanic eruption (Atlantic magmatic province flood basalt eruption), dissociation of methane from deep–sea hydrates, impact of extraterrestrial bodies (*e.g.,* Manicaugan Impact Crater, Canada, 206–213 Ma). In the Gondwanaland the change of climate to more arid condition is indicated by replacement of older *Dicroidium* flora by a new, essentially cosmopolitan, conifer–benettitalean flora.
Late Cretaceous: (dinoflagellates, calcareous nannoplankton, foramianifers, sponges, caprinid rudists, cephalopods, echinoids, bony fish and ichthyosaurs).	Global warming, sea–level rise, oceanic anoxic event, volcanic eruptions with associated emissions of large quantities of CO_2 causing greenhouse warming (*e.g.,* Ontong–Java plateau, Carribean–Columbian plateau, Madagascar Ridge totaling some 1000,000 km² (Kerr, 1998).

Contd...

Table 3.1–*Contd...*

Period and Biota Affected	Climatic and Other Causes
End Cretaceous: (foraminifers, diatoms, cyclostomate and some gymnolaemate bryozoans, inoceramid and rudist bivalves, ammonites, belemnites, and marine reptiles/mosaurs and plesiosaurs, the plants, dinosaurs and pterosaurs).	Catastrophic impact of extraterrestrial bolide (*e.g.,* Chicxulub of the Yucatan Peninsula in Mexico).Apart from the immediate effect (wildfires on land, tsunamis at sea) the effect of such impact is far reaching, eg., the release of huge amount of dusts in the atmosphere causing blackout of sunshine for months and years consequently causing global cooling, stop of photosynthesis and a collapse of food production in sea and land. The long term effect would be global warming by greenhouse gases.
End Eocene: (plankton and open water bony fish in the sea, and mammals on land, at least in Europe).	Most often attributed to global cooling, impacts of extraterrestrial bodies have also been suggested (*e.g.,* impact crater at Chesapeake Bay, Eastern USA).
Pleistocene: (large land mammals, mammoths, mastodons and wholly rhinoceroses).	Global warming and environmental change associated mainly with the Pleistocene Ice Age, hunting activity by human, or some combination thereof. The change from glacial to interglacial climate might have caused extinction of some animals.
Holocene: (birds, mammals), (ongoing mass extinction).	Climatic and associated environmental changes, hunting activity by human, habitat destruction or anthropogenic global warming ongoing for thousands of years.

2. Glacier movement erodes the basement and lateral wall rocks which may be deposited on pre–glacial river valleys forming till deposits (highly unsorted sediments whose sizes vary enormously from boulders to very much fine clays). A glacier boulder bed, known as Talchir boulder bed, occurs at the base (Upper Carboniferous or Lower Permian) of the coal-bearing Gondwana rocks of Peninsular India. The characteristic feature of the fine grained sandstones deposited in river basin during this glacial climate is the presence of unaltered feldspars which undergoes alteration with increasing temperature and moist climate in the overlying coal–bearing Barakar sandstones in the basal Gondwana rocks of India. Varve sediments (composed of thicker layers of coarse materials alternating with thinner layers of fine grains) are deposited in fluvioglacial lakes by alternate summer and winter climates.

3. The plant fossils (*Glossopteris* flora) recorded from the basal part of the Gondwana rocks, which experienced glacial climatic condition, are narrow

in contrast to the much coarser luxuriant flora present in the overlying Barakar and Ranigunge rock formations in India.

4. Ice avalanche in Pune, "triggered by a great earthquake in 1970 killed about 20,000 people while burying several villages in debris" (Keller, 1999). When glaciers enter into sea (icebergs), the movement of the icebergs into shipping lanes may cause hazard to navigation (for example, Titanic disaster of April 15, 1912, claiming 1501 lives). The mixing of volcanic eruptions on a mountain peak with the glacial ice on the slope may result in volcanic mudflows which may destroy property and human lives.

5. Glaciation gives rise to two main types of glacial landscapes: (i) dome–shaped glaciers which submerge the topography and (ii) glaciers which flow in channels. The other types of landscapes formed by glaciers, which submerge the topography, occur as (a) Ice–sheets of continental size, for example, Antarctic and Greenland Ice–sheets with volume of 27×10^6 km^3 and 2.6×10^6 km^3 respectively, (b) Ice–caps, which are smaller ice–domes, for example, Vatnajökoll in Iceland and Penny and Barnes ice–caps in Baffin island and (c) Summit Ice–caps, which are small glacier domes covering some flat–topped mountains, for example, Kibo summit of Kilimanjaro in Tanzania. The glaciers which flow in channels include (a) Valley glaciers, which are long compared with their width and (b) Cirque glaciers which are relatively small ice bodies occupying cirque hollows high on mountain sides, the width and length of the glaciers having similar magnitude (Duff, 1994).

6. Glacial climate cause perennially frozen ground called permafrost. In polar and high mountain regions permafrost is present where average annual temperature is less than $-2°C$. The permafrost depth may be several hundred meters. In the summer the surface of permafrost undergoes melting to a water–saturated layer up to 1 to 2 m thick. During strong winter cooling and contraction, polygonal cracks are developed in permafrost, which is called *tundra polygons* (Duff, 1994). Such polygons may be several hundred meters in diameter. The continuous permafrost areas are Greenland and northernmost Asia and North America. The discontinuous permafrost occurs towards south of the above regions and percentage of unfrozen ground increases towards lower latitudes. About 99 per cent area of Alaska is covered by continuous and discontinuous permafrost; its thickness varies from 400 m in the north to 0.3 m at the southern margin of the frozen ground (Keller, 1999).

7. Glacial climate causes change in the shell chemistry of some organisms. For example, the oxygen isotopic composition of the CaCO$_3$ shell of foraminifers from the ocean of glacial climate is enriched in ^{18}O than ^{16}O in comparison with the ocean water whereas in the foraminifers from the warm interglacial ocean this ratio ($^{18}O/^{16}O$) is lower. This temperature dependence of oxygen isotope composition of calcite (CaCO$_3$) precipitating in the sea water was discovered by H. Urey and has been a powerful tool for palaeotemperature determination (Duff, 1994). In drill core samples of

Neogloboquadrina pachyderma from Norwegian sea, it has been observed that one species of this foraminifer shows right coiling in ice–free surface waters and the another species shows left coiling where sea ice is common (Duff, 1994).

8. The lava erupted through subglacial volcanic vent will melt the overlying glacial ice. As a result, due to very rapid cooling of the lava, glassy or extremely fine–grained solidified rock and hyaloclasites will be formed. If the lava is of basaltic composition, basaltic glass, pillow basalt and pillow breccias will be formed by chilling. The typical shape of the sub–glacial volcano is flat–topped, steep–sided and this type is called *tuya* (named after Tuya Butte in British Columbia). Such sub–glacial volcanoes can cause dangerous floods due to melting of ice and 'lahars' or mud flows formed by mixing of the ice–melt with ashes given off by the volcano. Such volcanoes most commonly form today in Iceland and Antarctica.

Impact of Pleistocene Climate

Pleistocene and Holocene (or Recent) *epochs* are the final two epochs of the Cenozoic Era (Figure 3.2) and represent about 2 m.y. or so; the widely accepted age of beginning of Pleistocene is 1.8 m.y. During Pleistocene, over 40 million km³ of snow and ice accumulated on about one–third of the land surface of the globe. The Pleistocene ice age is suggested by some geologists to have ended between 11000 and 12000 years ago. However, historical records and C–14 dating of old terminal moraines suggest that cold rushes have recurred periodically into the Holocene. The period between AD 1500 and 1900, witnessed cooler and drier conditions (temperatures were often 2° to 4° F cooler than today) and this spell of interval is called the Little Ice Age. The impacts of Pleistocene glaciations are summarized below (Levin, 1988).

1. In the northern hemisphere the climate zones were expanded southward and the USA and northern Europe (high latitude) developed Arctic conditions. On the other hand rainfall increased in lower latitudes causing generally beneficial effects on plant and animal life. The presently arid regions in north and east Africa, even as late as the beginning of the Holocene, were well–watered, fertile and populated by nomadic tribes.

2. It is estimated that the sea level may have fallen at least 75m during maximum ice coverage (Levin, 1988). A land bridge stretched from Alaska to Siberia and the British Isles was joined to Europe.

3. Glacial landforms developed extensively and the great weight of the ice depressed the crust in large parts of the glaciated area to a level of about 200 to 500m below the preglacial position; and with the removal of the glacial ice sheet, downwarped areas began to return to their former positions.

4. The old drainage channels were changed and new channels were created by the movement of the great continental glaciers.

5. The great lakes of North America are shown to have been formed by movement of glaciers to low land to scour them deeper and subsequent melting.

6. Glacier movement transported fertile topsoil from over the bedrock in many areas forming productive farm lands away from the original location of the topsoil.

Effects of Global Warming

The scientists are now more or less convinced that population explosion and human–induced increases in greenhouse gases have a high potential for significant warming of the planet. Figure 3.4 shows the pattern of global temperature change over the 140 years (1850–1990 AD) in the Northern hemisphere, Southern hemisphere and global. The IPCC (Jansen *et al.*, 2007) has estimated that over the next 100 years the average global temperatures would rise 1.5 to 6 °C.

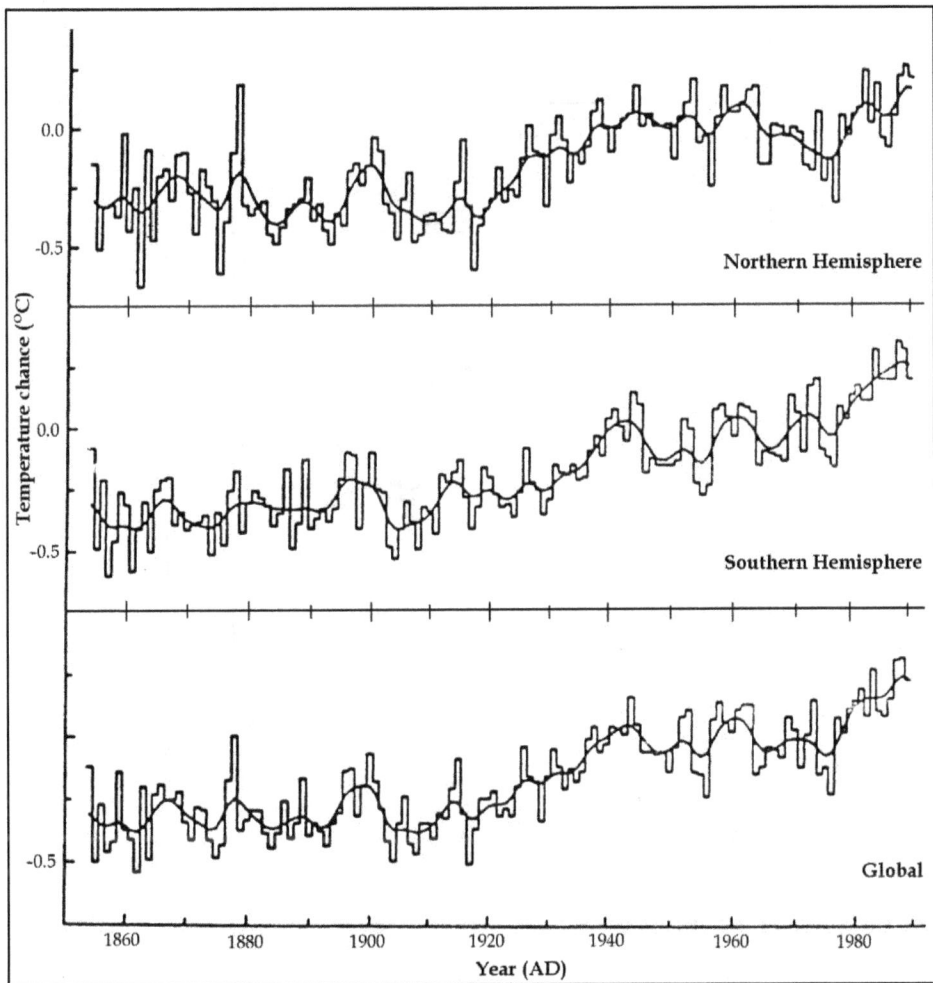

Figure 3.4: The pattern of global temperature change over 140 years (1850 – 1990). Source: Duff (1994).

Sea–level Change

Among the various causes of sea–level changes (*e.g.*, tectonic movement, extensive sea bottom volcanism) global warming, leading to expansion of sea water and melting of glaciers and ice–sheets, is most important for sea–level rise. It cannot be denied that there has been a rapid retreat of most of the world's glaciers, for example, glaciers in the north Atlantic area(Arctic Sea ice) have retreated during the 1920s to the mid 1960s and also since 1980 mainly due to temperature increase(Barry and Chorley, 2003). During the last 100 years the estimated rise of sea–level was 10 to 25 cm or more. This is due to thermal expansion of ocean waters (0.3–0.7 mm/yr), glacier and small ice–cap melting (0.2–0.4 mm/yr) and thinning of Greenland ice sheet (0.0 to 0.1 mm/yr).According to some models the sea–level may rise between 5 and 35m if CH_4, N_2O and O_3 maintain their present rate of increase and if all fossil fuels are burnt.

The pattern of sea–level change on Earth since Cambrian is shown in Figure 3.5. This shows that the maximum sea–level rise took place in the Late Cretaceous to

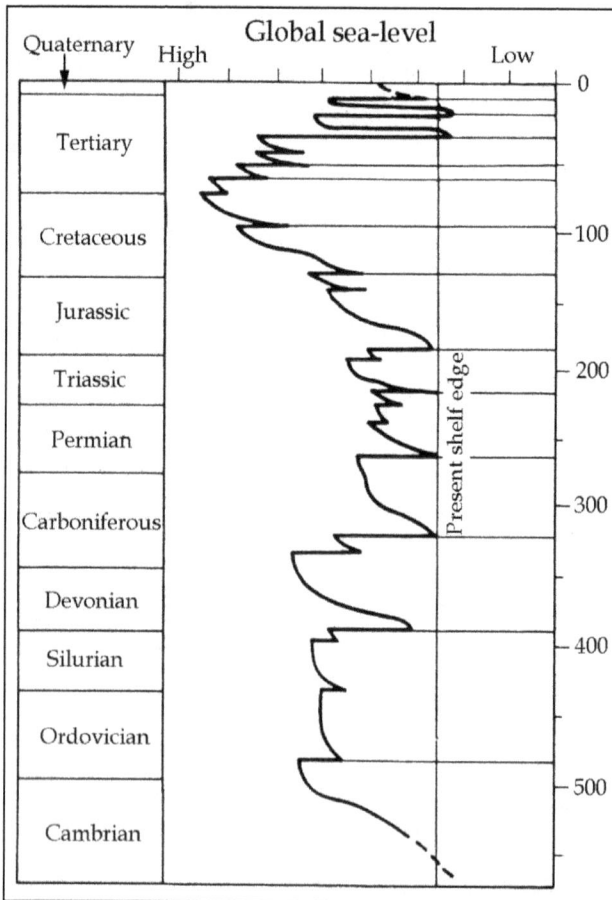

Figure 3.5: The pattern of global sea–level change since the Cambrian. Note that the sea–level rise was maximum in the Late Cretaceous when the global temperature was also the highest. Source: Duff (1994).

Early Tertiary boundary period and subsequently there has been an irregular decline of sea–level. One school of thought holds the view that global warming may cause also sea–level to fall due to snow accumulation at high rate at high latitudes (Barry and Chorley, 2003). There are still uncertainties regarding sea–level rise because of our inadequate knowledge on the behaviour of large ice sheets, especially Antarctica.

In spite of great uncertainty the magnitude of sea–level rise, even the lowest estimate rise, however, will flood many coastal areas including Bangladesh, Maldives off the coast of India, Netherlands; and the islands in the Pacific will be obliterated. The estimated highest sea–level rise would cause disaster for most coastal cities where live half of the world population. These people will be forced to migrate towards interior of the land which may cause socio–political and other problems.

Glacial Retreat

Glacial retreat has been accelerated in some areas, especially in Alaska and central Asia. It has been estimated that by AD 2050 "the present glacial mass may disappear with critical and irreversible long-term consequences for water resources in Alpine countries" (Barry and Chorley, 2003). In spite of fluctuating retreats of Arctic sea ice since 1920s to 2001 there is a tendency for retreat of Arctic sea ice while such tendency is not definite in case of Antarctic ice extent (Barry and Chorley, 2003).

In different mountain chains of the earth, snows and glaciers, accumulated in ridges and furrows, above the snow line, have long started melting due to global warming. Melting of major portions of the ice blocks may lead to long lasting off–season floods. With increasing temperature the glaciers will gradually diminish in volume and eventually the reduced supply of water would be a threat to forest cover of the mountains as well as the agricultural production of the adjoining planes. The glacial lakes on the mountain chains, due to mounting huge pressure of water derived by melting of glacier, will outburst the unstable bank causing devastating glacial lake outburst flood (GLOF), which may form multiple avalanches, landslides, destruction of forests of the mountains and floods in the river basins. The ICIMOD (International Centre for Integrated Mountain Development) has reported 15003 glaciers and 8790 glacial lakes in the Himalayan range (> 2400 km in length) (Das, 2009). The total amount of snow and glacial ice in the Himalayas is estimated to be next to Antarctic and Arctic circles (Das, 2009). If this huge reserve of glacier rock undergoes a major amount of melting by global warming it will be a catastrophic event with far–reaching consequences in the geological system of the Indian subcontinent and the neighboring countries.

Glacial melt water is a resource on earth and is used for irrigation and water supply to urban areas in some parts of the world, *e.g.*, the city of La Paz (Capital of Bolivia), one of the highest cities in the world (altitude ranging from 3–4 km), is dependent largely on glacial melt water in dry years for its water supply (Keller, 1999). Water from run–off streams is used in Europe and other places for generating electric power. Hence as a result of increasing global warming there is every possibility for water crisis in such areas.

Climate Changes: Floods and Desertification

The increase in temperature will promote more evaporation and hence more rainfall. The frequency of both droughts and rain storms is expected to increase. The global warming will cause expansion of arid (annual precipitation ~ 25 cm/year) and semiarid (annual precipitation ~ 25–30 cm/year) lands – which occupy approximately 35 per cent of earth's land and about 20 per cent of the world's human population – eventually converting them to desert (Keller, 1999). The present rate of desertification is 6 million hectares per year (Barry and Chorley, 2003). It may be mentioned in this context that the drought areas in the world increased more than 50 per cent throughout 20th century and it is largely due to the drought conditions over the Sahel and southern Africa during the later part of the century (Hiscock, 2005). As an impact of global warming on the wetlands of the earth, now covering about 4–6 per cent of the land surface, will decrease during this century causing decrease of the methane flux to the atmosphere.

Alteration in the global oceanic circulation, caused by greenhouse effect, may initiate further changes in climate. For example, it has been already observed that, as a result of differences in north Atlantic circulation, winters in Europe have been warmer and wetter during the last decade, El Ninos have become more frequent and more intense, annual rainfall in USA has increased 5–10 per cent in comparison with the rainfall during the early 20th century, floods, erosion, landslides, droughts, thunderstorms, windstorms and hurricanes have been more frequent and more severe in recent years.

In light of the uncertainties surrounding global warming the predictions for change of the other climatic factors rendering some unproductive or poorly productive soils to productive agricultural crops are uncertain, although some scientists have suggested improvement of agricultural productions for some regions of the world. Due to global warming the people living in colder countries (*e.g.* Russia, Canada) will gain from a longer growing season and milder winter which would reduce expenditure on heating oil, natural gas and electricity.

Change in Global Hydrological Cycle

According to IPCC report of 2001 (Houghton *et al.*, 2001) climate change during the next 100 years is likely to cause intensification of the global hydrological cycle and major impacts on regional water resources. The expected impacts of global warming on hydrological regions are (Barry and Chorley, 2003): (i) more severe droughts and/or floods in some places, (ii) increase in precipitation intensities, (iii) greater impacts on drier areas, (iv) increase of evapotranspiration, (v) increase of the load of weathered products in the river along with that of rainfall, (vi) much great lowering of water levels in the lakes of dry regions due to high evaporation, (vii) increase in desiccation and soil erosion in semiarid regions, rangelands and Savannas, and (viii) a great change in world hydrological cycle. Figure 3.6 summarizes the likely impacts of climate change on natural hydrological systems (Hiscock, 2005). The general lowering of groundwater level with global warming is a common observation throughout the world and is a signal for future groundwater crisis in some regions. A large volume of run–off due to heavy rainfall during a short

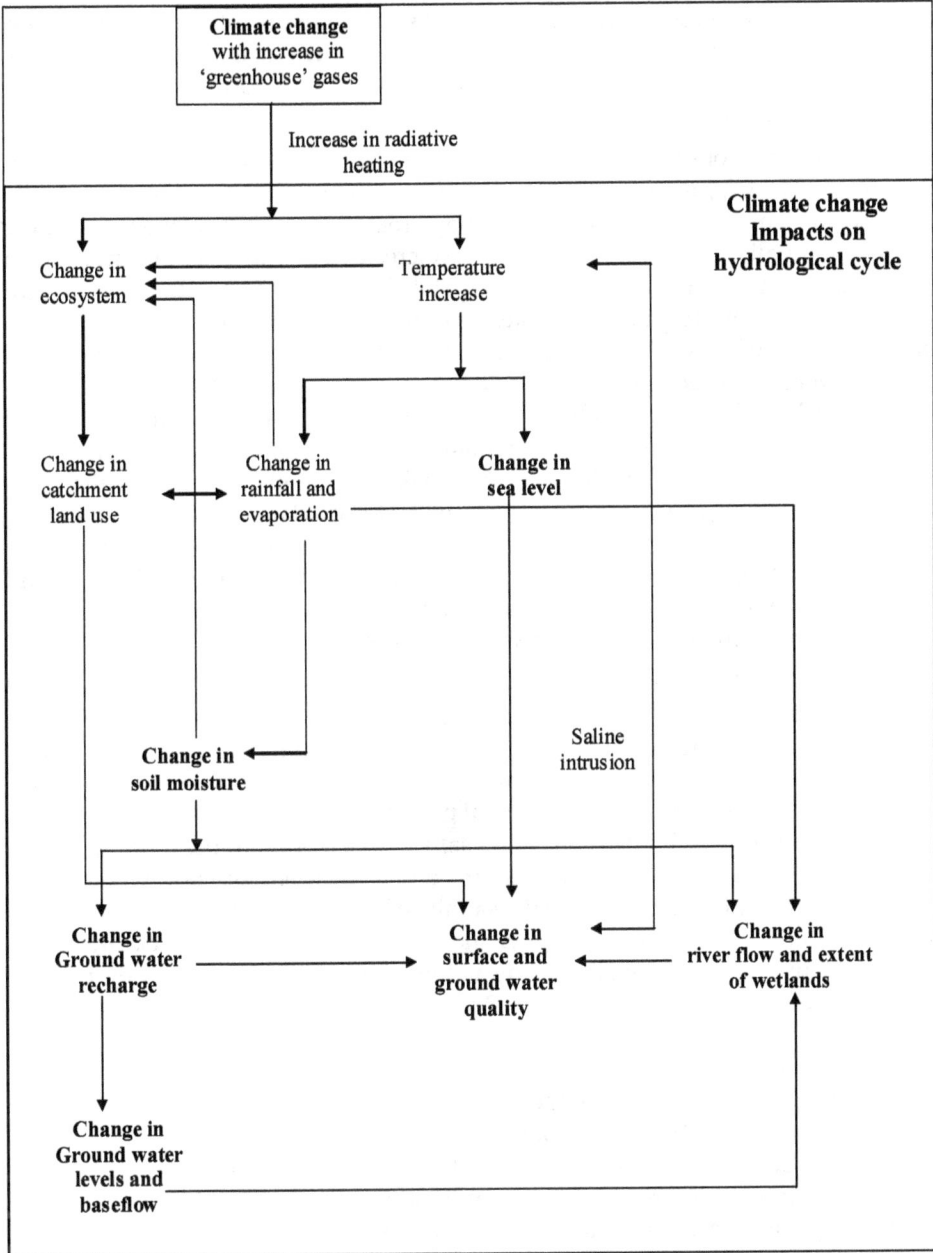

Figure 3.6: Impacts of increasing greenhouse gases on the natural hydrological cycle. The changes in the hydrological conditions are emphasized. Source: Amell, 1996 *in* Hiscock (2005).

precipitation season would cause flooding and may complicate the conjunctive use of surface water and groundwater. The changes in temperature, precipitation and sea–level rise (which affect volume of river flow and degree of saline intrusion) will affect the quality of water (Hiscock, 2005).

In regions of excessive rainfall the water table will rise resulting in easier access of groundwater from shallow tubewells. However, the areas of industry, municipality and agricultural fields (where excessive fertilizer is used) shallow–level groundwater will be subjected to more pollution.

Along shear zones/fault zones (along which one block of rocks moves passed the adjacent block forming fractures) the excessive rain water may percolate gradually downward, which, within a few years may create several slip planes at shallow depths eventually causing earthquakes. The increased load of water in reservoir of a dam causes more percolation which may lubricate and reactivate some ancient underlying fault resulting in earthquake. For example, in 1935 the excess accumulation of water in Lake Mead (Arizona) behind Hoover dam caused more than 600 local tremors over the subsequent 10 year period. Another such example is the Koyna dam earthquake in Maharashtra, India, where impounding of the reservoir was made in 1962 and the highest water level was retained from August 1967 to December 1967 which caused earthquake on 10 December, 1967. Similar examples of earthquakes come from the filling of a large artificial lake of the Kariba dam, Rhodesia and another artificial lake at Kremasta, Greece (Kovach, 1980). In this context, it may be mentioned that a very striking relationship was observed between the volume of disposal fluid injected to a depth of 3671m into Precambrian crystalline rocks (vicinity of Denver, Colorado, 1960s) and the number of earthquakes (Kovach, 1980).

It has been estimated that about 2220 km^3 groundwater discharges from inland catchments are added to the world's ocean per year. It has further been estimated that an additional direct groundwater discharges of approximately 222 km^3a^{-1} would take place to the oceans and seas assuming that greenhouse warming would lead to hypothetical 10 per cent increase in global precipitation of 2 mma^{-1} (Hiscock, 2005). As a result, a further load of $1.3x10^8$ t a^{-1} salts will be added to the dissolved solid content of 585 mg/l to the world's oceans and seas. It is uncertain whether this additional salt load would increase the salinity of the oceans, which depends on the balance of other factors such as inputs and outputs to and from the oceans under intensified hydrological cycle and any volume change in the oceans by thermal expansion and melting of ice (Hiscock, 2005).

Ozone Depletion and its Effects

Although ozone (O_3) is a minor constituent of the atmosphere (mainly present in the stratosphere and maximum concentration occurs around 25 km altitude) the discovery of ozone hole (a circular area in the atmosphere remarkably depleted in ozone) at Antarctica in 1985 by Farman and his associates (Midya and Jana, 2002), has been a great concern over the world because ozonosphere acts as a shield against the harmful UV–radiation coming to the earth's surface. Figure 3.7 shows the dramatic ozone depletion in Antarctica. Among the three types of UV–radiation (UV– A, UV– B and UV – C), UV– C (wavelength range 2000 – 2800 Å) and UV – B (wavelength 2800 – 3200 Å) are lethal to man and many living organisms and are totally absorbed by ozone layer whereas UV –A radiations (>3200 Å) is relatively harmless and is absorbed only slightly by atmospheric ozone. Midya and Jana (2002) have reviewed the atmospheric ozone depletion and its effects on environment. Ozone is depleted

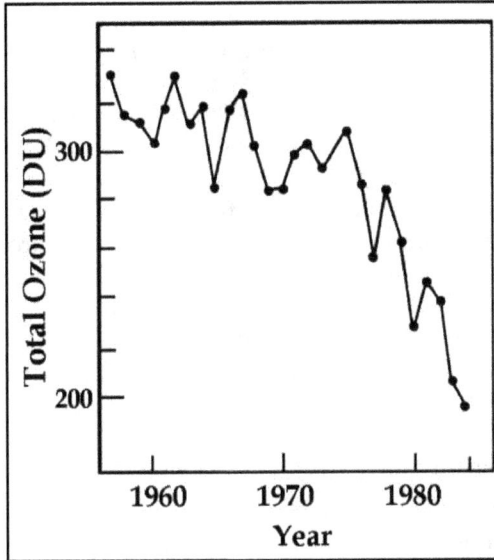

Figure 3.7: Ozone variation in spring time during 1957 – 1984 at Halley Bay (76°S, 27°W), a British Antarctic Survey Station. Source: Farman *et al. in* Midya and Jana (2002).

by UV– radiations, volcanic eruptions, HO_x, Cl – Br, CO and HO_x, NO_x, polar stratospheric clouds and greenhouse gases such as CO_2, CH_4, N_2O, Chloroflurocarbon – 11 ($CFCl_3$) and Chloroflurocarbon – 12. Direct exposure to UV–radiations causes acute sunburn, skin pigmentation and neomelanogenesis skin cancer, squamous cell carcinoma and basal cell carcinoma. The least frequent but more aggressive is cutaneous malignant melanoma. Squamous cell carcinoma has been reported in cattle, horses, cats, sheep, goats and dogs. The eye is the principal route of exposure to UV–radiation and its cornea is affected first, followed by the lens, the vitreous humor and the retina. The other ocular effects include photokeratitis, climate droplet keratopathy, pinguecule, pterygium and squamous cell carcinoma of the cornea and conjunctiva (Shaper *et al.*, 2002). Midya and Jana (2002) predicted in the 1992 Copenhagen Amendments that the calculated number of skin cancer caused by the ozone depletion will exceed 33,000 per year in the USA and 14,000 per year in northwest Europe around the year 2050.

The ozone depletion is observed everywhere which threatens the human species and other animals as well as plants (*e.g.*, bacterioplankton, picoplankton, cyanobacteria, phytoplankton and zooplankton). The function of cyanobacteria, which reduces atmospheric nitrogen into ammonium ions for aquatic eukaryotic phytoplankton, is hindered by UV–B radiation. The UV– radiation affects also the photosynthetic functioning of phytoplanktons, the biomass producer in aquatic ecosystem living at the top layer of the oceans and fresh water (euphotic zone) (Midya and Jana, 2002). UV–radiation changes also the tropospheric chemistry causing increase of HO_x, CH_4, CO and H_2O_2.

No record of life older than 3900 m.y. has yet being discovered. It means that possibly life did not appear before this time. One of the possible causes may be that the protective ozone layer was not formed and hence the harmful UV–radiations could hit the earth surface directly before 3900 m.y. The majority of research on early life comes from 3 localities: (i) the Isua Group rocks (3900–3700 m.y. old) in southwest Greenland including the island of Akilia; (ii) Barberton Greenstone Belt rocks (3550–3330 m.y. old), Kaapvaal craton of South Africa; (iii) Warrawoona Group of rocks (3515–3427 m.y. old). The earliest life probably consisted of anaerobic chemosynthetic microorganisms which are believed to have lived in a reducing environment (Taylor *et al.*, 2009) and hence free oxygen was not present in appreciable amount, not to speak of ozone. Alternatively, it has also been suggested that early life appeared a few centimeter below the surface of water where the UV–radiations could not penetrate. With younging of the geological time more and more free oxygen was formed in the atmosphere and when the protective ozone layer developed in the atmosphere life appeared in increasing amount.

Today, with increasing accumulation of greenhouse gases, particularly CFCs in the atmosphere the stratospheric ozone shield is being destroyed which has become a threat to man, a component of the geologic system. The Montreal Protocol of 1989, in which are now included 140 countries or so, decided to phase out CFC products by 1996 and set up a fund paid for by developed nations to help developing nations switch to ozone–safe chemicals (Press and Siever, 1999). The Montreal Protocol is a model for cooperative activity of scientists, industrial leaders and government officials, which can be applied to reduce the human–induced global warming and ozone depletion.

Remarks

Since the birth of the earth about 4600 m.y. ago from dusts or nebula, it took about 200 m.y. to form the primitive crust, by cooling and condensation of the 'boiling ball' of molten rock material. The first rock to form in the hot early earth was igneous and when the climate became cooler, suitable for weathering, sedimentary rocks were formed. An igneous rock, crystallized at depth (called plutonic rock), is coarse grained while the same magma when reaches the earth's surface gives rise to a fine grained rock (called volcanic rock). This is due to the change in environmental condition of crystallization of magma. The stepwise change of climatic conditions (temperature, precipitation etc.) from hot earth to relatively cool earth during the Archean facilitated more weathering of rocks eventually leading to increasing amount of different types of sedimentary rocks. The accumulation of some amount of free oxygen in the atmosphere and photosynthesis of algae in the ocean caused precipitation of Banded Iron Formation (BIF, chemical sediment) during the Archean. The appearance of adequate free oxygen in the atmosphere, *i.e.*, oxidizing condition, favoured formation of Red beds and evaporites in the Proterozoic time (2500–570 m.y.). The warm climate caused evaporation of inland sea water adjacent to coastal zone as a result of which direct precipitation of evaporite mineral deposits (*e.g.*, gypsum, halite) took place. The warm, humid and oxidizing conditions facilitated formation of arkose and warm–water carbonate rocks which are present in Proterozoic and Palaeozoic rock records. In the glacial climate, the movement of glacier produces a typical rock, till, which is composed of highly unsorted boulder to clay–sized particles. Varve typically formed in glacial lake due to seasonal fluctuation. Such glacial boulder beds or tills and varves have been occasionally recorded from Precambrian to Pleistocene time.

Palaeosoil was formed by weathering of rocks under suitable climatic conditions as we see today. The palaeosoils, residual mineral deposits and supergene enrichment mineral deposits have been recorded from different ages of earth's history, testifying the impact of palaeoclimate on the formation of such deposits (Bateman, 1956).

The stratigraphic records of sedimentary rocks indicate that palaeoclimatic conditions controlled not only the characters of the palaeolives (now preserved as fossils), but also their distribution on the earth. For example, the warm, humid climatic conditions (containing high CO_2 in the atmosphere) facilitated development of luxuriant *Glossopteris* flora and rich coal deposits (under swampy reducing environment) in Permian of the lower Gondwana regions. *Glossopteris* became almost extinct in the Triassic when the climate changed to arid condition giving rise to the appearance of *Dicroidium* flora which, in turn, was replaced by *Ptilophyllum* flora in the Jurassic Gondwana due to the reappearance of warm, humid climate. The warm climate favoured the appearance and development of some organisms such as hermatypic hexacorals, land–living reptiles, marine reptiles, plesiosaurus, robust labyrinthodont amphibians. The fossil *Lystrosaurus* present in Triassic and Jurassic rocks thrived in tropical to subtropical climates of the world. It was the impact of warm moist climate which facilitated the development of giant dinosaurs (reptilian fauna) of Late Jurassic.

The desert has been formed by the scanty rainfall but it has extended due to human–induced causes at the root of which lies the population explosion. The rocks in the desert climate underwent and are undergoing weathering mainly by physical process facilitated mainly by temperature variation between day and night over thousands of years. The landscape changes in the desert (sand dunes, ripple marks on the sand, barchans, etc.) take place mainly by wind action.

After the appearance of life in the Archean, several events of mass extinction took place in the geological history. The most severe mass extinction took place during Permo–Triassic boundary period (P/T) although much more attention has been drawn by Cretaceous–Tertiary boundary (K/T) period mass extinction. Although several theories have been proposed for mass extinctions (for example, bolide impact, volcanism, global warming/cooling, change in environment in the ocean) the effective cause was the climate change.

Glacial climates have repeatedly affected the earth since Late Archean (~ 3000 m.y. – 2500 m.y.) to Pleistocene (~ 1.8 m.y. ago). The Milankovitch Cycles are thought to be the main cause. The glacier movement changes the landforms of the Earth, for example, the V–shaped river valley is changed to U–shaped valley. Besides, the glaciers give rise to dome–shaped and channel–shaped landscapes as well as continental size ice–sheets (*e.g.,* Antarctic and Greenland) and ice–caps occupying flat–topped mountains – all depending upon the topography of the original landform.

In polar and semi–polar regions, glacial climate forms perennially frozen ground called permafrost (av. annual temperature $< -2°C$; for example, Alaska). The impact of glacial ice has been disastrous for human lives (for example, Titanic disaster of 1912 and Peru ice–avalanche of 1970). The subglacial volcanism may cause mudflows ('lahar') bringing about destruction of property and human lives.

The impacts of Pleistocene glaciation are (i) southward expansion of Arctic conditions in USA and northern Europe, (ii) fall of sea–level at least up to 75 m during maximum ice coverage, (iii) extensive development of glacial landforms, (iv) change of old drainage channels, (v) formation of great lakes of North America and (vi) transportation of fertile top soil from over the bed rock of many areas to elsewhere.

Global warming and cooling is a natural phenomenon throughout the geological history of the earth. The earth was much warmer in Late Cretaceous (65 m.y. ago) to Early Tertiary time than today. The clear trend of increasing global temperature during the last 150 years, its impacts on the geological system (*e.g.* sea–level rise, retreat of glaciers) observed over quite some years and predictions from different models regarding the probable disastrous effects in the near future have been of serious concern all over the world. The estimated lowest sea–level rise is likely to flood many coastal areas and the highest sea–level rise would cause disaster for most coastal cities where live half of the world's population. The increasing temperature will cause more rainfall, more expansion of desert, increase in frequency of droughts and thunderstorms, El Ninos and a great change in world hydrological cycle and major impacts on regional water resources. It is uncertain whether the agricultural crops will increase or decrease. If the increasing trend of global warming continues causing melting of all the glaciers of the Himalayas, it is likely that many glacial lakes of the Himalayas will explode causing floods in the icemelt–fed rivers of the Indo–Gangetic plains subsequently followed by drying of these river beds (because glacial melt water would no longer be available and there is uncertainty in the availability of sufficient rain water) and deforestation of the Himalayas leading to a great change in the ecological system and climatic change in all the surrounding Asian countries. The main cause of global warming in the last 100 years or so is the emission of increasing amount of greenhouse gases in the atmosphere to meet the demands for comforts of increasing human population. Our immediate task is to take appropriate steps to check emission of greenhouse gases and convince the world leaders to come to a consensus following the advice of IPCC.

Acknowledgement

I am very much grateful to Prof. Gautam Kr. Saha, Department of Zoology for inviting me to write this paper and for his encouragement and constant persuation to complete the paper. I am also grateful to Professor Subir Bera, Dr. Ruby Ghosh and Sandip More of the Department of Botany, Dr. Siddhartha Sengupta and Dr. Bapi Goswami, Associate Professors of Geology, Dr. Aniruddha Mukhopadhyay and Dr. Punarbasu Choudhury of the Department of Environmental Science, Calcutta University and Shreya Brahmma, Department of Zoology, Calcutta University for immense help during preparation of this paper. Thanks are also due to Library staff, especially Sri Basudeb Ghosh, Department of Geology, Calcutta University for cooperation during preparation of this paper.

References

Bandyopadhyay, S., Roy Chowdhury, T. and Sengupta, D.P. 2002. Taphonomy of some Gondwana vertebrate assemblages of India. *Sed. Geol.*, 147: 219–245.

Barry, R.G. and Chorley, R.J. 2003.Atmosphere, weather and climate. Routledge Taylor and Francis Group, 8th edn.: 353–390.

Bateman, A.M. 1956. Economic Mineral Deposits. John Wiley and Sons and Chapman and Hall, 2nd edn.

Blatt, H. 1997. Our Geological Environment. Prentice Hall.

Bowering, S.A., Williams, I.S. and Compston. 1989, 3.96 gneisses from the Slave Province, Northwest Territories, Canada. *Geology*, 17: 971–975.

Bowering, S.A. and Williams I.S. 1999. Priscoan (4.00–4.03 Ga) orthogneisses from northwestern Canada. *Contrib. Min. Petrol.*, 134: 3–16.

Caldeira, K., Rampino, M.R., Volk, T. 1900. Biogeochemical modelling at Mass Extinction boundaries: Atmospheric carbon dioxide and ocean alkalinity at K/T boundary. *In*: Kauffman, E.G. and Walliser, O.H. (eds.), Lecture notes in Earth sciences: Extinction events in Earth history. Springer–Verlag: 333–345.

Cloud, P.E. Jr. 1968. Atmospheric and hydrospheric evolution on the primitive earth. Science, 160: 729–736.

Cloud, P.E. Jr. 1973. Paleoecological significance of the banded iron formation. *Econ. Geol.*, 68: 1135–1143.

Cloud, P.E. Jr. 1976. Major features of crustal evolution.Trans. *Geol. Soc. S. Africa*, 79 (Annex): Pp.33.

Das, T.M. 2009. Global warming on Himalayan ecosystem with reference Glacial Lake Outburst Floods (GLOF) *News Letter, Centre for Applied Science and Technology* (CAST), 3 (1 and 2): 5–7.

Duff, P.M.D. (ed.) 1994. Holmes' Principles of Physical Geology. ELBS and Chapman and Hall, 4th edn.

Evans, A.M. 1993. Ore Geology and Industrial minerals: An introduction. Blackwell Science.

Farquhar, J. 2009. Atmospheric Evolution, Earth. *In* Gornitz, V. (ed.), *Encyclopedia of Paleoclimatology*, Springer: 61–65.

Frakes, L.A. 1979. Climates throughout geologic times. Elsevier.

Gilluly, J., Waters, A.C. and Woodford, A.O. 1959. Principles of Geology. Freeman and Company, 2nd edn.

Hiscock, K.M. 2005. Hydrogeology: Principles and Practice. Blackwell.

Hut, P., Alvarez, W., Elder, W.P. *et al.* 1987.Comet showers as a cause of Mass Extinctions. *Nature*, 329: 118–126.

Houghton, J. Y., Ding, Y., Griggs, J. *et al.* 2001. *Climate Change 2001. The Scientific Basis. The Third Assessment Report of the Inter Governmental Panel on Climate Change* (IPCC), Cambridge University Press, Cambridge.

Jansen, E., Overpeck, J., Briffa, K.R. *et al.* 2007. Palaeoclimate, *In* Climate Change, 2007. *The Physical Science basis, Contribution of Working Group to the Fourth Assessment Report of the Intergovernmental Panel on Climate Change.* Cambridge University Press, Cambridge, United Kingdom and New York, NY, USA: 433–497.

Johnsson, M.J. 2000. Chemical weathering and soils: Interface between the Geosphere, Hydrosphere, Atmosphere, and Biosphere. *In* Ernst, W.G. (ed.), *Earth Systems: Processes and Issues*: 119–132, Cambridge University Press.

Jones, R.W. 2006. Applied Palaentology. Cambridge University Press.

Keller, E.A. 1999. Introduction to Environmental Geology. Prentice Hall.

Keller,G., Stinnesbeck,W., Adatte, T. *et al.* 2003. Multiple impacts across the Cretaceous–Tertiary boundary. *Earth Sci. Rev.*, 62: 327–363.

Kerr, A.C. 1998. Oceanic plateau formation: a cause of Mass Extinction and black shale deposition around the Cenomanian–Turonian boundary. *Jour. Geol. Soc. London*, 155: 619–626.

Kovach, R.L. 1980. Earthquake Prediction and Modification. *In* Gass, I.G. *et al.* (eds.), Understanding the Earth, 2nd edn., ELBS, Open University: 327–331.

Lal, D. S. 2006. Climatology. Sharda pustak Bhawan, Allahabad.

Leary, P.N. and Rampino, M.R. 1990. A multi–causal model of Mass Extinctions: Increase in trace metals in the oceans. *In* Kauffman, *E.G.* and Walliser, O.H. (eds.), Lecture notes in Earth Sciences: Extinction events in Earth history, Springer–Verlag: 45–55.

Levin, H.L. 1988. The Earth through Time. Saunders College Publishing, 3rd edn: 489–531.

Mason, B. and Moore, C.B. 1985. Principles of Geochemistry.Wiley Eastern, 4th edn.

Merritts, D., Wet, A.D. and Manking, K. 1998. Environmental Geology: An Earth System Science approach. W.H. Freeman and Company.

Midya, S and Jana, P. K. 2002. Atmospheric ozone depletion and its effects on environment. *Ind. Jour. Physics*, 76 B (2): 107–138.

Mojzsis, S.J., Harrison, T.M. and Pidgeon, R.T. 2001. Oxygen–isotope evidence from ancient zircons for liquid water at the earth's surface 43,00 Myr ago. *Nature*, 409: 178–181.

Murck, B.W., Skinner, B. and Porter, S.C. 1996. Environmental Geology, John Wiley and Sons: 251–273.

Peck, W.H. and Valley, J.W. 2009. Archean environments. *In* Gornitz, V. (ed.), *Encyclopedia of Paleoclimatology and ancient environments*, Springer: 34–38.

Press, F. and Siever, R. 1999. Understanding Earth. Freeman and Company, 2nd edn: 611–682.

Rampino, M.R. 1987. Impact cratering and flood basalt volcanism. *Nature*, 327: 468.

Rampino, M.R. and Stothers, R.B. 1984.Terrestrial Mass Extinctions, comet impacts and the Sun's motion perpendicular to galactic plane. *Nature*, 308: 709–712.

Rampino, M.R. and Stothers, R.B. 1988. Flood basalt volcanism during the past 250 million years. *Science*, 241: 663–668.

Simonson, B. M. and Kaufman, A. J. 2009. Banded iron formations and the early atmosphere. *In* Gornitz, V. (ed.), *Encyclopedia of Paleoclimatology*, Springer: 85–89.

Taylor,T.N., Taylor,E.L. and Krings, M. 2009. Palaeobotany: The Biology and Evolution of Fossil Plants. Elsevier.

Wilde, P., Quinby–Hunt, M.S. and Berry, W.B.N. 1990. Vertical advection from oxic or anoxic water from the main pycnoline as a cause of rapid extinction or rapid radiations. *In* Kauffman, E.G. and Walliser, O.H. (eds.), *Lecture notes in Earth sciences: Extinction events in Earth history*. Springer–Verlag: 85–98.

Wilde,S.A., Valley, J.W., Peck,W.H. *et al.* 2001.Evidence from detrital zircons for the existence of continental crust and oceans on the Earth 4.4 gyr ago. *Nature*, 409: 175–178.

Willis, K.J. and McElwain, J.C. 2005. The Evolution of Plants. Oxford University Press.

www.bing.com/images/search=subglacial volcano

www.crystalinks.com/volcanosubglacial

www.en. wikipedia.org/wiki/subglacial volcano

www.free–extras.com/search/7/pictures of subglacial volcanoes.

Climate Change: Man and Environment (2012)   *Pages* 56–64
Editor: Goutam Kumar Saha
Published by: DAYA PUBLISHING HOUSE, NEW DELHI

4

Coal Mining and Climate Change: An Environmental Vista

☆ *Apurba Ratan Ghosh**

Introduction

Coal is an important fossil fuel contributing the world's major energy sources and coal mining is one of the core industries that contribute to the economic development of a country but deteriorates the environment consequently. Mining technology started development in its history through opencast method of mining for the minerals deposited. Mining is considered as a long–established human activity. In the process of development, coal mining became a major source of pollution of the environment. But at the same time it also assures the energy supply, which is indispensable for the development of a country. It is a major fuel used for generating electricity worldwide. Coal provides around 27 per cent of global primary energy needed and generates about 36 per cent of the world electricity. Coal is a major resource and most important primary source of energy in India, accounting for 0.8 per cent of the total reserves of the world, 5.7 per cent of the proven reserves and almost meets 67 per cent of the total energy reserves consumption of the country.

Impacts of Coal Mining and Climate Change

Mining operations of minerals is no longer an "ecofriendly" activity affecting at least of all of the components of the environment posing some serious impacts on the environment, which may be permanent or temporary, beneficial or harmful, repairable or irreparable and reversible or irreversible. Mining and its associated activities have

* Department of Environmental Science, The University of Burdwan, Burdwan.
  E-mail: apurbaghosh2010@gmail.com

several impacts on the different sectors of our environment, *viz.,* impacts on society, impacts on ecology, impacts on land, impacts on water regime and impacts on atmosphere including noise pollution. Industrial activities bring forth a change in the terrestrial ecology and subsequent effects on air and water quality of the surrounding areas. It poses a great threat on the floral and faunal composition of the medium. In the mining areas, during the production of coal, burning and cutting down of forest covers generate carbon dioxide, methane and nitrous oxide and due to changes in land use these greenhouse gases are released into the atmosphere.

Emission of methane from coal mining is one of the important sources of powerful greenhouse gases. Removal of methane from the atmosphere by chemical reactions is very difficult. The rapid rise in methane started more recently than the rise in CO_2, but methane contribution has been catching up fast. However, methane has an effective atmospheric lifetime of only 12 years, where as CO_2 survives much longer. In the mining environment this may be worsened by climate change. An increased temperature in mining zone influences the plant growth and human health and agricultural production may also be affected, particularly if combined with water depletion. In and around the mining industry higher concentration of CO_2 should stimulate photosynthesis, particularly in C3 plants, such as rice, barley, potato, etc. This is referred as "CO_2 fertilization" because it can increase yields of C3 crops by 15 per cent under optimal situation; where in case of sorghum and millet, the production may not be affected seriously, that is why in developing countries it plays an important role in food security. So, the negative effects of climate change can be limited by changes in crops and crop varieties. A change in hydrological cycle may result in wetter areas; a change in precipitation may also affect wetness, which then affect evapo–transpiration and cloud formation, which in turn affect precipitation. There is a very good link among climate change, water availability, food production, plant growth, and economic growth, but it is a very complex phenomenon.

Mining of minerals, especially by opencast method, is an activity, which degrades the land in many ways, and since mining is only an intermediate activity, it is the duty of the mining companies to bring the land back to the same status of utilization. The process of bringing the land back to the usable form comprises of reclamation, rehabilitation and mine closure and is theoretically a part of the mine plan. Invariably efforts are made to reclaim the land in such a manner that the end result is better aesthetic topography with land use commensurate with the post–mining activities of the area. In most of the situations it is desirable to reclaim the land to the uses, such as, afforestation, plantation, grassland, agriculture and pisciculture in the voids or locally called *khadan* in case of opencast method, *etc.*

Land degradation is the alteration and spoiling of land rendering it worthless for any useful purpose. The degree of damages the land suffers due to mining varies with the topographic setting and the nature and extent of mining operations. Other factors contributing to the land degradation in the mining areas are the wind velocity, wind direction and the nature and intensity of rainfall.

Land Degradation

According to Singh *et al.* (2007) the following types of land degradations can be expected to take place in the mining areas.

Loss of Vegetation

Loss of vegetation results in drastic changes in climatic conditions like rainfall, temperature and humidity, which results in disappearance of existing wild life, resident and migratory birds, *etc.*

Siltation

There are many instances of siltation of low–lying land around the dumps in the mining complexes, which have affected the fertility and usefulness of the land.

Pits/Excavations

Opencast mining causes extensive damages to the land physiography.

Waste Dumps

The waste/overburden dumps degrade the land for a time period of their existence.

Soil Quality

The standard practice of opencast mining involves separate removal, handling and storage of topsoil and subsoils so that these can be refilled during reclamation of the mined–out areas. Mining also disturbs soil sequences and due to loosening of the ground, the natural compactness of the soil reduces and as a result the soil becomes prone to erosion due to the rain and wind. The discharge of polluted effluents leaches the ground, alter the characteristics of the soil and thus affect their vegetation supporting capacity.

Damage to Agricultural Land and Creation of Wasteland

The Indian Bureau of Mines (IBM) opines that surface mining may result in constant decline of green and cultivable area by way of blockage of land for mining and allied activities. An estimate by the IBM indicates that this degradation is to the tune of 60 per cent by waste dumping, 23 per cent by pit excavation and 17 per cent by others (Singh *et al.*, 2007).

Soil Erosion, Landslides and Hydrological Imbalances

Locations of many mineral deposits in the country are such that the mining is started by chopping the hill top to the base. In such a mining invariably the overburden and waste material after removal is cast down the hill slope which destroys the vegetation over a large area. Deforestation and soil erosion due to such mining have caused land slides and hydrological imbalances in the Himalayan region. This has also caused drying of springs in some places.

Visual Intrusions

Formation of open pits, overburden dumps and other facilities for mining create scenic disturbances, which are important in the areas having large population and places of tourist interest.

Infrastructure

Construction of roads, colonies and other secondary and ancillary facilities in the mining complexes cause considerable damages to the land in the surrounding areas.

Tailing Ponds

Most of the metalliferous mines have tailing ponds, which require considerable expanse of land and also need land for the construction of dams for confining the tailings. The discharge from these ponds may damage the land lying on the down stream.

Ecological Impacts

Ecological impacts for mining industry can be studied in the following ways:

1. Ecological impacts of opencast mining:
   - Removal of all vegetation and subsequently the fauna.
   - Pollution of water affecting aquatic ecosystem.
   - Dusts and other gases cause damage in the atmosphere, depositing the particles on leaves of the plants and retard their growth.
   - The wildlife in the forests adjacent to the mining areas migrate. Faunal dispersal due to huge noise and vibrations caused by blasting and other operations of machines.
   - Scarcity of water regime affects growth of vegetation and agriculture in and around the areas.

2. Ecological impacts of underground mining :
   - Construction and infrastructural development cause changes in land use.
   - Changes in topography and drainage pattern due to subsidence.
   - Massive disturbances and damage to the surface and sub–surface and pollution in underground water bodies.
   - Infertility or damage to topsoil due to drainage of polluted water from the underground mines.
   - Generation of subsidence and heat.
   - Sudden collapse of land.

Atmospheric Pollution and Noise Pollution

A number of research programmes have already been undertaken by different research institutions, scientists in order to assess the air quality status in and around coal mining complexes in India. In most of the situation it was found that the concerned coal mine was surrounded by mining activities of other adjoining mines and there was hardly any demarcation between the residential and industrial areas. These studies revealed that mining activities contributed significantly to the deterioration of air quality in mining complexes, particularly in winter. The concentrations of suspended particulate matter (SPM), respirable particulate matter (PM_{10}), sulphur dioxide (SO_2), and oxides of nitrogen (NO_x) as measured for the air quality index (AQI) were found to be more or less to the status of moderately polluted areas during winter. It is imperative to indicate that the existence of mine fires severely worsened the air quality status in coal mining sector. It has been assumed that 3.95 per cent of the coal is lost in transit from the mines, of this 1.5 per cent is lost in the air and the balance 2.45 per cent is lost on the land (Ramaswamy and Erkman, 2001). Burning of coal also releases large amount of air pollutants, as for example, in United States every year burning of roughly one billion tons of coal produces 18 million metric tons of oxides of sulphur (SO_x), 5 million metric tons of nitrogen oxides (NO_x), 4 million metric tons of airborne particulate matter, 600,000 metric tons of HCs and carbon monoxide, 40 tons of mercury, and a trillion metric tons of CO_2 (Cunningham and Cunningham, 2007).

In townships around mines, the air quality status was observed equally bad, mainly due to frequent movement of petrol and diesel driven vehicles, congestion and traffic jam. The concentrations of pollutants like SPM, PM_{10}, SO_2 and NO_x were found higher, particularly during winter, than the prescribed limits of Indian standards. However, the overall quality status of the townships was found more or less to the status of heavily polluted areas to moderately polluted areas. Pollution of the surrounding atmosphere due to mining and associated activities is one of the major threats as recorded below:

☆ Production of dust increases the concentration of SPM level.

☆ Removal, handling, transportation and storage also increase in the SPM concentration.

☆ Use of heavy machineries operated by diesel cause increase in NO_x level.

☆ Drilling and blasting and explosive fumes also increase NO_x level.

☆ Minerals and rock mass having sulfur and its compounds may contribute SO_2.

☆ Some of the sedimentary rocks may have CH_4, this may contribute this gas to the surrounding air when mined operation continues.

☆ Fires in opencast mines contribute heat, SPM, CO_2 and CO.

☆ Use of petrol vehicles or equipments contributes HC and lead.

☆ Due to underground mining the exhaust air from the interior contains SPM, CO_2, CH_4, NO_x, SO_2 and other pollutants. Massive generators, boilers, etc., operated by diesel release SPM, NO_x and CO_2. Huge machines or equipment also produce noise in the areas.

Impacts of Other Activities

☆ Domestic burning of coal and other fuels, and open burning of coal produce CO_2, SPM.

☆ Petrol or diesel driven transport network generates SPM, NO_x, CO_2, CO and other atmospheric pollutants.

☆ Smoking produces SPM and CO_2

☆ Constructional activities contribute SPM to the atmosphere.

Effect on Water Regime

Mining activity either by opencast or by underground methods damages the water regime and thus causes a reduction in the overall availability of water in and around the mining areas. Mining and associated activities have quantitative as well as qualitative impacts on the water regime in and around the complexes. All the aquifers, including the water table aquifer, above the mineral deposit to be extracted are damaged because for exposing the mineral for extraction the over burden (OB) rocks are removed. Water in the nearby water bodies gets polluted due to leaching from overburden dumps, discharge of pumped out line water, and other activities in the vicinity of the water bodies. During rainy seasons the run–off water from the areas surrounding the mines may carry with it a large dose of suspended solids into nearby water bodies.

Underground mining activities also affect the water resources. The impacts depend on the magnitude of the subsidence movements. Due to underground mining also there is a disturbance to the underground water bodies resulting into reduction of the availability of water not only in the mining area but also in the neighbouring areas. During mining operations there is a development of huge cracks through which especially during rains the surface water finds way to the underground workings and carrying various pollutants from the surface. The polluted underground water when pumped out and discharged on the surface may pollute the surface water bodies. The effluents when discharged on the surface pollute the topsoil and subsoil and also the water table.

These activities enhance the depletion of surface water and groundwater resources in that area. The lowering of regional water table thus may require deepening of the wells in the surroundings. These wells may, hence, supply contaminated water as in the case of "salt–scalding" (Coates 1981), "seawater intrusion" or "upward groundwater flow from confined aquifer to unconfined aquifer". The regional lowering of water table if retains for number of years, the upper part of the aquifer will suffer from loss of pore water pressure. Groundwater resources are getting damaged due to lowering of water table due to pumping out of groundwater flowing at the excavation site and discharging it creating more of runoff and less of recharge. The polluted underground water causes pollution to the surface water when discharged.

Damage to Greenery

Loss of greenery due to open cast mining poses serious effect on the hydrological cycle as well as create imbalance in the cycle because the plants are climatic stabilizers and hence protect water resource; evapo–transpiration from plant leaves help running the hydrologic cycle properly and cause balanced rain.

Inadequate rainfall is caused due to decrease in surface water bodies and vegetation density leading to rise in atmospheric temperature. Deterioration of water quality may be resulted because of salt–scalding and seawater intrusion. This serious problem leads to push the total region towards desertification. Greeneries of a region influence the water resources in many ways. Roots of greenery anchor the soil particles, prevent erosion and hence siltation and drying up of surface water bodies.

Leaves of greenery protect land from scorching sun and lashing action of raindrops, thus prevents drying and erosion of land and hence siltation of surface water bodies. Tree litter makes a spongy cover on soil, and together with plant body makes a barrier on the way of surface runoff, hence decreases erosion and increases infiltration. Tree litter prevents evaporation. Plant roots make hair cracks on land that promotes infiltration. Trees, besides these abovementioned two facts together, help in raising water table in the region. Decayed vegetal matters add nutrients to topsoil, after getting rotten materials adds a fertile cover on land, hence promotes green cover over land and prevents erosion. It is estimated that one full–grown tree has cooling effect of 5 average air conditioners working 20 hours a day (Coates, 1981). Thus, plants are climatic stabilizers and hence protect water resource. Evapo–transpiration from plant leaves help running the hydrologic cycle properly and cause balanced rain. Damage to greenery will create loss of all these positive impacts and hence damage to surface water and groundwater potentialities.

The sequential effects are decreased infiltration, increased run–off, increased erosion, and evaporation of soil moisture, increased barren earth surface, its drying up, loss of groundwater through capillary rise and lowering of water table.

Impacts on Society

Impacts of mining were started on the ethnic society after the discovery of minerals and after establishment of its mining potentiality. People started buying the land and business resulting in to increase the value of the land. The impacts are as follows:

Displacement of the People

For opencast or underground mining it is required to clear the surface of large area adjacent to it, its building structures, vegetation, *etc.*

Loss of Livelihood

People loose their livelihood on the vest land, which is taken for the purpose.

Changes in Population Dynamics

Due to introduction of skilled, semiskilled manpower for the particular mining activities the ethnic population dynamics undergoes a dilution as well as a major change.

Cost of Living

The development of industrial and associated activities in the mining area greatly influence the level of the economic activities of the common people due to flow of more money and increased buying power of the people directly. This leads to an increase in the cost of living.

Water Scarcity

There is an overall reduction in the availability of water in and around mining areas.

Health Impacts

Due to abrupt changes in the environment with the introduction of pollutants in the air, water, noise and vibrations people living in and around the mining complexes get affected, causing serious problems like skin problem, lung diseases, respiratory problems, deafening, *etc.*

Infrastructure Facilities

With the development of mining industrial activities the surrounding areas become equipped with roads, schools, hospitals, commercial market, communication, *etc.* which tend to improve the quality of life of the complexes.

Employment Opportunities

It generates employment opportunities for the eligible, skilled people among the ethnic population. The Project Affected People (PAP) are provided with jobs, as well as training for self employment. Other people may be engaged in different mineral based activities.

Increase in Aspirations

When the ethnic people are exposed with the industrial activities, it increases their aspirations; in fact, it is also a necessary for the overall community development in the mining areas.

Addictions

Actually, the tribal people are becoming addicted due to increased economic activities.

Economic Disparity and Frustration

Industrial and economic activities in mining areas cause economic disparity and livelihood among the populations resulting into frustrations in the poorer clan of the people.

Remarks

It is well established that in the process of development coal mining becomes a major pollution source of our environment but it is indispensable because it contributes a large to the economic development of a country. Mining and its associated activities have several impacts on the different components of our environment. Coal India Limited has taken so many proper management strategies for designing and planning the mining activity in more safer ways like "Environmental and Social Mitigation Program", which has taken into account the environmental monitoring, reclamation, subsidence management, regional land and environmental management planning, self–development and training, *etc.* Production of mine wastes is relatively less in case of underground mining, although the amount of wastes will increase as the production increase. These wastes sometimes undergo spontaneous combustion, erosion from wind and water and cause acid drainage. Nowadays, the coal mining industries are facing a great challenge in reducing the amount of water use and to treat the water before discharging them into the environment. Both the opencast and underground mining have the ability to deteriorate the flow as well as quality of surface and subsurface water. It is told that proper reclamation could not only return a mining area to its pre–mining condition, but can also "improve" the original landscape. This obviously includes alteration of the surface topography in order to render drainage more effectively, diverting the rivers and streams, introducing new character into the landscape or making the area more suitable for agriculture. Loss of greenery along with cropland due to opencast mining is causing a serious threat to the hydrological cycle, adversely affecting the water table and surface water bodies and ultimately resulting in the change in topography. Lastly, the process of reclamation, rehabilitation and mining closure are made to reclaim the land in such a manner that the end result is to have a better aesthetic topography with land use that commensurate with the post–mining activities of the area. So reclaim through afforestation, plantation, grassland, agriculture, horticulture, pisciculture, etc., may render the better options to resist the land degradation and climate change.

References

Coates, D.R. 1981. *Environmental Geology*, John Wiley and Sons, New York. Pp. 731.

Cunningham, W.P. and Cunningham, M.A. 2007. *Principles of Environmental Science – Inquiry and Applications* (4th Edn.), Tata McGraw–Hill Publishing Company, New Delhi.

Ramaswamy R. and Erkman, S. 2001. *Industrial ecology in regional policy planning: Case study of the Damodar River Basin, India.*

Singh, R., Singh, P.K. and Singh, G. 2007. Evaluation of land degradation due to coal mining – A vibrant issue. In: *Proceedings of the First International Conference on MSECCMI*. New Delhi, India. Pp. 129 – 133.

Climate Change: Man and Environment (2012)   *Pages 65–86*
Editor: **Goutam Kumar Saha**
Published by: **DAYA PUBLISHING HOUSE, NEW DELHI**

5

Understanding Change in the Glaciers of the Indian Himalayas[#]

☆ *Divya Mohan and Shirish Sinha**

Introduction

The mankind has entered the 21st century in the midst of social, economic, developmental and environmental challenges that need to be addressed at a global scale. One of the most important issues confronting the planet is undoubtedly the threat of global warming and the most prominent indicators of climate change is the melting of ice mass and glaciers worldwide. Over the last few decades, changes in climate and local weather conditions have impacted the world's glaciers both in terms of structure and characteristics, reflected in the form of advancement or retreat of glacial snouts (UNEP, 2007). The changes in the length, width, area, volume and mass balance of the glaciers are among the most directly visible signals of global warming and these changes are the primary reasons why glacial observations have been used for climate system monitoring for many years (Haeberli, 1990; Wood, 1990), especially in areas where time series data on climate (mainly temperature and precipitation) is difficult to get and where climate change signals are not yet clear (Yadav *et al.,* 2004; Roy and Balling, 2005). Unnatural rates of glacial melting can have serious implications on the hydrology of the associated river systems and consequently on the livelihood of millions of people who are dependent on these rivers and their ecosystems. Warming has caused the snowline in the Himalayas to shift upwards, thus indicating an increase in the ablation area of the glaciers. This receding phenomenon has gathered pace in the past few years. Recent studies from

[#] This paper is drawn from the report "Witnessing Change: Glaciers in the Indian Himalayas", WWF India and BIT, 2009.

* Climate Change and Energy Programme, World Wide Fund for Nature – India, New Delhi.
  E-mail: shirish@wwfindia.net

466 glaciers of the Indian Himalayas indicate that there has been a 21 per cent reduction in the glacierised area–from 2,077 sq km in 1962 to 1,628 sq km in 2004 (Kulkarni *et al.*, 2007). Smaller glaciers of less than one sq km have reduced in area by as much as 38 per cent compared to a 12 per cent retreat of the larger glaciers. The melting process has two key adverse impacts on the environment. One is the loss in the reserve of freshwater and the other is a significant rise in sea levels (Kaser *et al.*, 2006). These ecosystems have an intricate web of interaction and therefore changes in any one of their components can have a chain of impacts on the other elements. The right approach, thus, in addressing these impacts is to have a better scientific understanding through long-term observations and analysis of the interactions of the different components of the mountain ecosystems with their climate, and then utilise this information to formulate effective adaptation and management strategies.

Himalayan Glaciers – An Overview

The Himalayas, the youngest and one of the most fragile mountain systems in the world, derive their name from a Sanskrit word, which means 'abode of snow'. The mountain ranges of the Himalayas stretch for a distance of about 2,400 km in an east–west direction in the shape of an arc along the northern border of India covering an area of about 500,000 km². The Himalayas have three parallel running series of mountains – the greater Himalayas or the Himadri range, which has some of the highest peaks of the world; the middle Himalayas or Himachal and the lower Himalayas or the Shivalik range. Plateaus and flat bottom valleys of thick gravel and alluvium are found in between the Himachal and Shivalik ranges (Jain *et al.*, 2007).

The Himalayas comprise approximately 33,000 sq km of glacierised area (Kaul, 1999; Dyurgerov and Meier, 2005) and its glaciers are a source of ten of the largest rivers in Asia. The rivers flow trans–boundary and meet the drinking water, irrigation, hydropower, fishery, inland navigation and other needs of more than 1.3 billion people living downstream. With about 9,575 small and large glaciers in the Himalayas (Singh *et al.*, 2009), they hold the largest reserves of water in the form of ice and snow outside the polar regions (Srivastava, 2001; Shukla and Siddiqui, 2001). The Himalayas are thus also referred to as the 'water towers' of Asia and a 'third pole' of the earth.

The glaciers of the Indian Himalayas are spread over different river basins including the Indus, Ganga and Brahmaputra. Inventories of the Himalayan glaciers by the Geological Survey of India (GSI) indicate that the Bhagirathi sub–basin has the largest glacierised area of about 755 sq km with as many as 238 glaciers including the Gangotri glacier (26–30 km). In comparison, the Brahmaputra basin has nearly 161 glaciers although it occupies a much smaller glacierised area of about 223 sq km. Some of the other important glaciers found in the Himalayas include Siachen (72 km), Zemu (26 km), Milam (19 km), Kedarnath (14.5 km) and Dokriani (5.5 km) (WWF, 2005).

The existence of the glaciers in the Himalaya is due to their orographic characteristics (high altitudes exceeding above 0°C isotherms) and local and regional climatic conditions. In addition to these, the Himalayas are influenced by both the Indian summer monsoon and the westerlies, though not homogeneously. Some of the Himalayan glaciers are nourished only by the Indian summer monsoon (in summer) when accumulation and ablation is concurrently taking place while some other glaciers are nurtured only by the westerlies (in winter) during the accumulation period. A few glaciers are nourished by both the monsoon and the westerlies. The dynamic monsoon system in turn is also influenced by the complex orographic characteristics of the Himalayas, coupled with snow and glacial environments. The radiation balance due to snow/ice cover provides feedback mechanisms for advection of water vapour from the surrounding oceans and maintains the seasonal cycles of monsoon. The Himalayas also play a critical role in the tropical summer monsoon climate in the Indian sub-continent by functioning as an effective meteorological barrier. They obstruct the advancement of the monsoon towards the north, thereby resulting in more rainfall on the southern slopes.

Himalayan Glaciers – Understanding Changes

The freshwater melt from the glacierised basins is a vital element in regulating the dry season flows of perennial Himalayan river systems. Being closer to the Tropic of Cancer, the Himalayan glaciers receive more heat than the arctic and temperate climate mountain glaciers, and hence they are very sensitive to the rising temperature or climate variability both at regional and global levels. The responses of various glaciers are different due to variations in mass balance and the climate change impacts they face. Both a short-term perturbation in inputs as well as a long-term change in precipitation are said to affect glacial retreat.

Some of the studies carried out in the Indian Himalayas clearly point out an increase in glacial melt (Kumar *et al.*, 2007). For instance, the Baspa basin of Himachal Pradesh has shown an increase in the winter stream flow by 75 per cent as compared to the rate in 1966. This is in tandem with the rise in average winter temperatures in the area, possibly illustrating the impacts of global warming in the form of increased snow ablation, which in turn has augmented the stream flow (Kulkarni and Bahuguna, 2002; Kulkarni *et al.*, 2002). Climate change impacts are also visible in the mass balance study of the Chhota Shigri glacier in the Chandra valley of Himachal Pradesh. The study shows that there has been a decrease in the Accumulation Area Ratio (AAR) of the glacier and it has had a negative mass balance in the years 2002–2005 (Kumar *et al.*, 2007, Berthier *et al.*, 2007).

Thus, climatic variability and growing impacts of climate change is posing pressure on our natural water supply. A holistic approach is required to manage the freshwater resources in the Himalayas and ensure environmental security in the region. For this, knowledge of glacial melt characteristics and their subsequent impacts on freshwater availability is essential. However, the current knowledge about the behaviour of glaciers in the Himalayan region is still limited.

Glacier Retreat in the Himalayas

In the Himalayan region, glaciers and snow cover have been thinning since the end of 19th century in line with the global trends. With significant snout fluctuations, most of the glaciers in the Himalayan mountain ranges have been retreating at accelerated rates in the last three decades (WWF, 2005) and their rate of retreat is much faster than that of glaciers in other parts of the world (Cruz *et al.*, 2007). These changes correspond to the rising surface temperature trends in the Himalayas, which have been reported to be higher than the global average warming (UNESCO, 2007; Jianchu *et al.*, 2007; Barnett *et al.*, 2005). A study of the temperature trends in the northwest Himalayan region shows that a significant warming of 1.6°C has occurred over the last century with warming in winter taking place at a faster rate, with the highest warming rates recorded in the period 1991–2002 (Bhutiyani *et al.*, 2007). This warming has been due to a rise in both maximum and minimum temperatures, though the maximum temperature have gone up more rapidly. Apart from warming, other factors like high human population density near these glaciers, deforestation and land use changes have also been responsible for the decline and shrinkage of glaciers (Cruz *et al.*, 2007). At current rates of retreat, the smaller valley type glaciers are more likely to decline at a faster rate in the future, with uncertain impacts on the downstream areas.

In India, although high mountain glaciers occur across all the Himalayan range states of Jammu and Kashmir, Himachal Pradesh, Uttarakhand, Sikkim and Arunachal Pradesh, only a few have been studied for long term monitoring or documented in terms of glacial retreat. The climatology and the topography of the Himalayan region have wide variations all across the arc (Dobhal *et al.*, 1995; 2007). Owing to these variations different parts of this region have been exhibiting diverse responses to the variability in climate. Singh and Bengtsson (2005) found that under a warmer climate, the melting from the seasonally snow covered part of the basin was reduced while, in contrast, it increased from the glacier–fed basin.

The dynamics of a glacier are influenced by climatic factors including temperature, precipitation, amount of snowfall and wind. In the Indian Himalayas, studies on snout recession/mass balance measurements have been carried out on a few glaciers and some studies have been able to generate long-term monitoring data (Kumar, *et al.*, 2007; Dobhal *et al.*, 2007; Wagnon *et al.*, 2007; Kulkarni, 2007). While studies by various institutions on the changing characteristics of glaciers in the Himalayas may be limited, this has generated an important debate on the role of climate change in the extent of glacier loss (WWF, 2005; GOI, 2009).

While the global view (IPCC, 2007) suggests that a large percentage of glaciers are declining worldwide, there are conflicting views also on the rate of retreat in the post industrial era. Some studies suggest that a few glaciers in the Himalayas may be showing reduced retreat patterns. For instance, the reduction in the retreat rate of Gangotri glacier for 2004–05 was obtained by Kumar *et al.* (2007). Similar results have been observed on the Gangotri glacier, which shows that the average rate of retreat for this glacier declined substantially (6.02 m/yr) during 1999 to 2006. On the basis of several observations, it can be said that the effects of climate change have variable impacts on glaciers depending on their size. Small glaciers are more likely to face the brunt of the changes in climate owing to their smaller accumulation zones. On the other hand large glaciers might sustain the impacts for a longer time due to their larger ice volume and bigger accumulation zone.

WWF–India has been raising awareness about climate change and the impacts on Himalayan ecosystem since 2005. In order to generate a deeper understanding of the potential impacts of glacial melt in the Himalayas, WWF initiated a project in 2006, which focused on studying two key glaciers from a climate change perspective and the subsequent impacts on freshwater availability. These glaciers were the 30 km long Gangotri glacier and the 4.2 km long Kafni glacier, both located in the state of Uttarakhand. This would help in developing regional climate projection models which are more accurate in providing information on future impacts.

Case Study I: Gangotri Glacier

Study Area

The Gangotri glacier, one of the largest ice bodies in the Garhwal Himalayas, is located in the Uttarkashi district of the state of Uttarakhand in India. It is one of the most sacred shrines in India, with immense religious significance. Being the main source of the river Ganga, it attracts thousands of pilgrims every year. The Gangotri glacier is a vital source of freshwater storage and water supply, especially during the summer season for a large human population living downstream. The discharge from the glacier flows as the river Bhagirathi initially before meeting the Alaknanda river at Devprayag to form the river Ganga. Snow and glaciers contribute about 29 per cent to the annual flows of the Ganga (up to Devprayag) and hence any impacts on these glaciers are likely to affect this large river system (Singh *et al.*, 2009).

The north–west facing Gangotri glacier is a valley type glacier originating in the Chaukhamba group of peaks. Numerous smaller glaciers join the main stem of the main glacier to form the Gangotri group of glaciers. The complete Gangotri glacier system along with its tributaries covers an area of 210.60 sq km (ETM+1999). The area and length of the main trunk of the glacier is 56.59 sq km and 29.13 km respectively. The average width of the glacier is 1.85 km. The glacier, lies between 79°4'46.13"E–79°16' 9.45"E and 30°43'47.00"N–30°55'51.05"N (ETM+1999). It has varying elevation of 4,015–6,145 m above sea level (SRTM data analysis). The snout of the glacier occurs at an altitude of about 3,949 m above sea level, and this is the place from where the Bhagirathi originates. Its snout position is at 73°4'47.26"E and 30°55'36.45"N (GPS observation, 2008).

Gangotri Glacier Retreat – A Historical Perspective

The Gangotri glacier has been receding since the last 'Little Ice Age', which ended in the 19th century. The tributary glaciers have also shrunk and some of them have even got separated from the main trunk of the glacier. This fact is evident by systematic studies going on since 1935 on the movement of the glacier snout, and by the presence of recessional features such as terminal and lateral moraines (Srivastava, 2001). In the past century, the retreat rate of the glacier has shown a rising trend. It has been observed by glaciologists that the snout of the Gangotri glacier has retreated by about 2 km in the last 100 years. Both the snout retreat and area vacated by the Gangotri glacier have been documented by several scientists.

Meteorological Conditions Around the Gangotri Glacier

Meteorological conditions play a pivotal role in governing the state of glaciers and their associated hydrological features such as water storage in downstream areas. Meteorological analysis of long-term data becomes a fundamental aspect in determining the structural changes taking place in glaciers.

Rainfall

Precipitation data collected from the region indicates that the area around Gangotri usually receives less than 15 mm of daily rainfall during the summer season (Singh *et al.*, 2005). There are only few unusual days when the rainfall patterns vary due to a storm or some unusual heavy rainfall event. The study also shows that August and September usually receive higher rainfall as compared to the other months. It has also been found that early morning and late evening are the most probable times for the occurrence of rainfall (Singh *et al.*, 2005).

Temperature

Temperature in the ablation season increases for a few months after which it starts decreasing. July has been found to be the warmest month on the basis of the mean–maximum and minimum temperatures. Diurnal variations in temperature show that the maximum temperature is observed around 1400 hours, while the minimum is observed in the early morning hours (Singh *et al.*, 2005). Changes in minimum temperature have been seen to be more significant than those in the maximum temperature. It has also been found that the 'maximum diurnal temperature range' occurs in May and October while August shows the least variation in temperature range. This is probably due to the presence of a cloud cover during the rainy season.

Hydrological Characteristics of the Gangotri Glacier

For the Gangotri glacier, the major sources of runoff are melting snow and ice. Since this area receives less rainfall, it does not contribute much to the runoff. Stream flow at the Gangotri glacier shows a wide variation depending on various factors. In the beginning and at the end of the ablation season, there is not much difference in the day and night time flow volume; however, as the peak melting time approaches there is a comparative reduction in the night time flow. Still, a significant flow is observed at night in spite of very little or no melting taking place at that time. This reflects the fact that the Gangotri glacier has strong melt water storage characteristics. Monthly variation of flow shows that the discharge starts rising from May and maintains a high flow during June to August with maximum average discharge in August. The discharge starts reducing from September onwards. Singh *et al.* (2006) have also reported similar results in the discharge flows from the glacier.

Methodology

The methodology for field studies at Gangotri involved a combination of primary field data through the use of Differential Global Positioning System (DGPS) and Remote Sensing Applications along with the existing secondary data. DGPS measurements have been extensively used for locating the snout position of the glacier during various field visits starting from 2006. Long-term observations are an important factor for accuracy in any glacial research as the snout position is quite dynamic and inter–seasonal changes are seen in the exact position of the snout. Snout position was recorded in the months of September and October of each year since 2006. For the collection of meteorological data, an automatic weather station (AWS) has been installed at Bhojwasa near the snout of the glacier. Apart from the glacial and meteorological monitoring, data has also been collected about the discharge patterns of the glacial melt water. The glacial area was calculated using satellite imageries of 1976, 1990, 1999 and 2006 by delineating the glacier boundary using ERDAS 9.0 and auto calculating the area using GIS methodology.

Results and Analysis

A comparative analysis of the glacier's snout position was carried out using data from secondary sources and interpretations from various satellite imageries over the past three decades. Satellite imageries available since 1976 formed the baseline for the analysis of the fluctuations in snout position together with DGPS observations since 2006. A study of data from all available sources illustrates that the main trunk of the Gangotri glacier has been in a continuous state of recession and fragmentation during the past century. The length of the glacier has been computed for different years based on available data. The trend shows that the length of the glacier has reduced by about 0.59 km in 33 years, from 1976–2009, with an average retreat rate of 17.59 m/year (Figure 5.1).

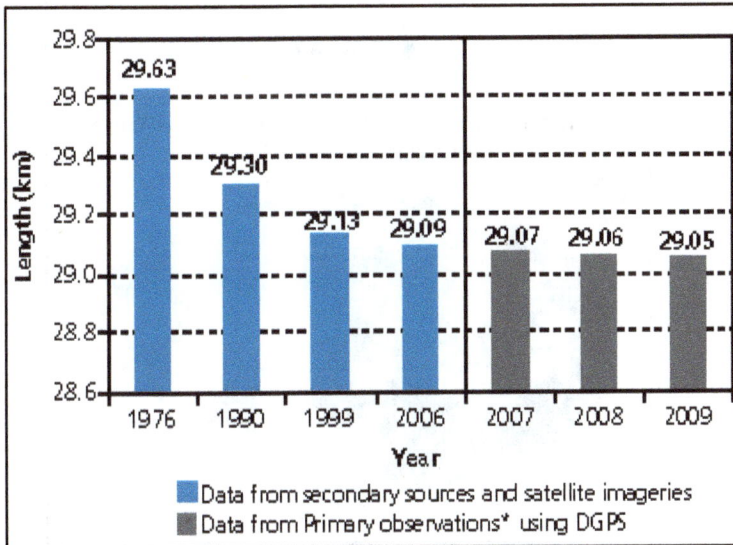

Figure 5.1: Length of the Gangotri glacier in different years.

Based on the comparison of satellite imageries of the Gangotri glacier for the years 1976, 1990, 1999 and 2006, our analysis shows that the glacier is not only receding in length but also in terms of glaciated area from all the sides. The possible reasons behind this retreat may be linked with two main factors: (a) reduction in snowfall and (b) an increase in the temperature of the region. Analysis shows that between 1976 and 2006, the glacier area has reduced by 15.5 km², with an average loss of 0.51 km² per year. This reduction in glacier area is 22.9 per cent over 1976 (Figure 5.2). The glacier area reduced by 7.2 Km² between 1976 and 1990 in fourteen years, with a 10.6 per cent reduction in the glacial area. In nine years, between 1990 and 1999, the glacial area reduced by 4.3 km² and resulting in 7 per cent reduction in area when compared to 1990. However, the rate has increased between 1999 and 2006, with glacial area reducing by 4.1 km² and a 7.25 per cent reduction in area over 1999.

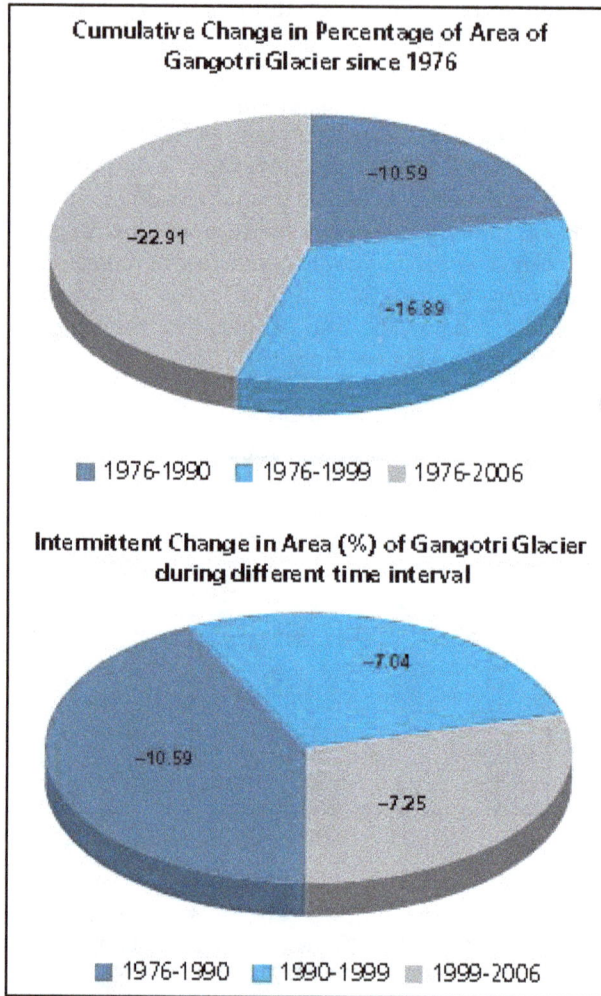

Figure 5.2: The cumulative and intermittent reduction in the area of Gangotri glacier since 1976.

With a reduction in the area and length of the Gangotri glacier, there has also been a retreat in the snout position. Data collected from various sources shows that this has been a continuous process; however, there have been fluctuations in the rate of retreat for different time intervals.

The calculation of the snout retreat by DGPS observation is made from point to point that gives the straight line distance of the two points that may include the shift of the water outlet point (left/right) in the curvilinear shape of the larger snout. The other way of measuring shift of the snout position in the glacier flow direction is to measure the perpendicular distance on the curvilinear shape of the snout. This measurement normally gives a lesser value if the snout point has moved either left or right rather than in the glacier flow direction as compared to the previous position. Based on this measurement we have observed that the snout retreat during 2006–07 was 17.85 m when compared to snout position of 2006 (Figure 5.3). In the subsequent years, the retreat rate has reduced–13.06 m in 2007–08 and 10.57 m in 2008–09. It can be said that retreat has taken place between 2007 and 2009, however, it is lower than the long-term average. Our observation from the glacier indicates that while there is a retreat, it is the fragmentation of the Gangotri glacier is of larger concern.

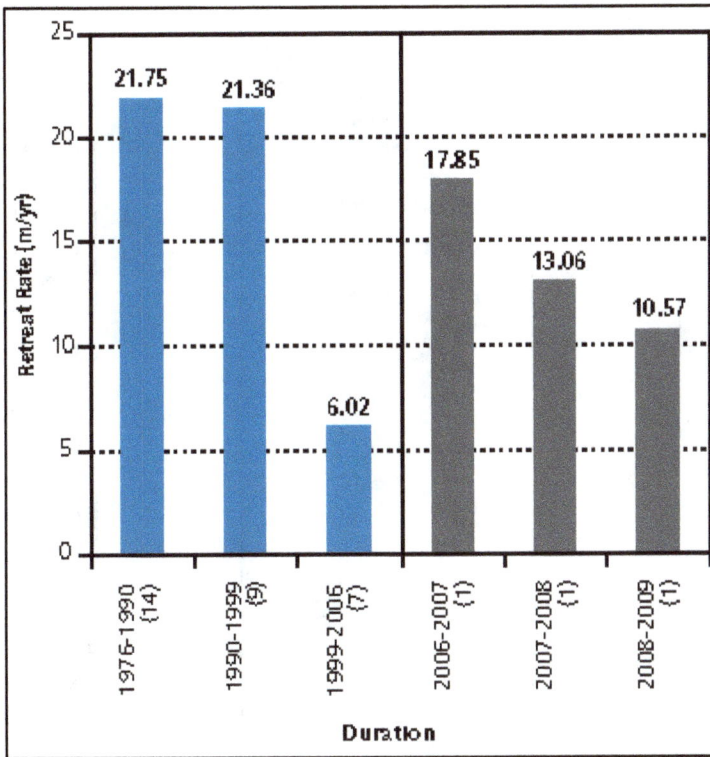

Figure 5.3: Retreat rate of Gangotri snout in different interval of time.

Hydro-meteorological Study

Trends emerging from the analysis of summer season discharge data for Gangotri show that there are variations on a diurnal and monthly basis. Flows recorded at different times of the day were found to be highest during the evening hours (1700 hours). The discharge is lowest during the morning hours (Figure 5.4). These results are also in agreement with the findings of an earlier study by Singh *et al.* (2006), in which they observed a high discharge rate during daytime and low rates during the night time for the Bhagirathi river at Bhojwasa.

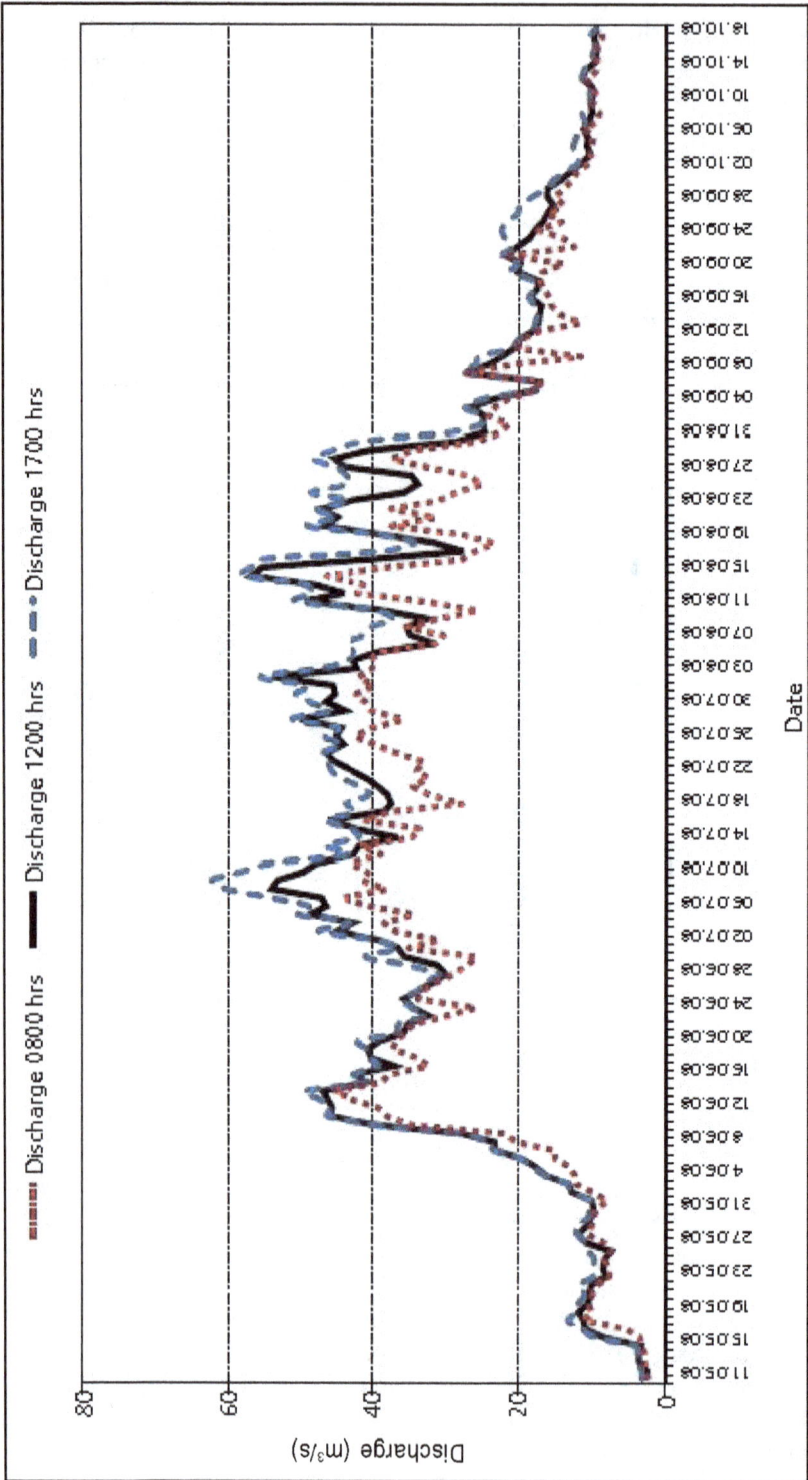

Figure 5.4: Diurnal variation in discharge at Bhojwasa, Gangotri 2008.

On a monthly scale, flows increase with the commencement of the summer season reaching a peak during July and August due to the combined impact of more melting of ice, which is driven by higher average temperatures and rainfall in the catchments. The discharge starts reducing in September and reaches a very low level throughout the accumulation season. During winter, the flow is reduced and keeps almost at a constant value. The combined impact of low winter temperature (sub zero) and higher albedo (90–95 per cent for fresh snow) reduces the winter discharge rate to quite a low level.

The daily average temperature and average discharge for the ablation and accumulation season at Bhojwasa are shown through Figures 5.5 and 5.6 respectively. A good correlation of 0.73 has been found to exist between the average temperature and average discharge in the ablation season. However, these parameters are not well correlated (0.46) in the accumulation season, which shows a lower influence of temperature on melting because most of the solar energy is reflected back to the atmosphere due to the snow–covered valley that has a high albedo. The preliminary trends observed during the study in 2008 seem to suggest that the volume of discharge and flow patterns is also being influenced by the temperature variations in the valley. This is in consonance with results obtained by other studies (Singh *et al.*, 2006) over the past decade. While these are initial results, comparisons with similar studies indicate a consistent pattern in the rate of discharge and temperature increase for the region.

Where the general trend of the impact of mid–day temperature on evening discharge in the glaciated basin is concerned, we have found a correlation of 0.63

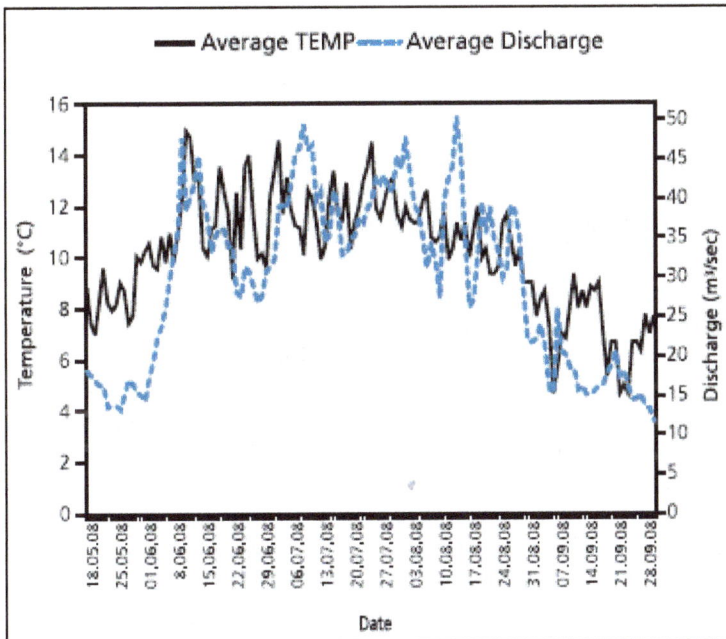

Figure 5.5: The average temperature and discharge during the ablation season at Bhojwasa.

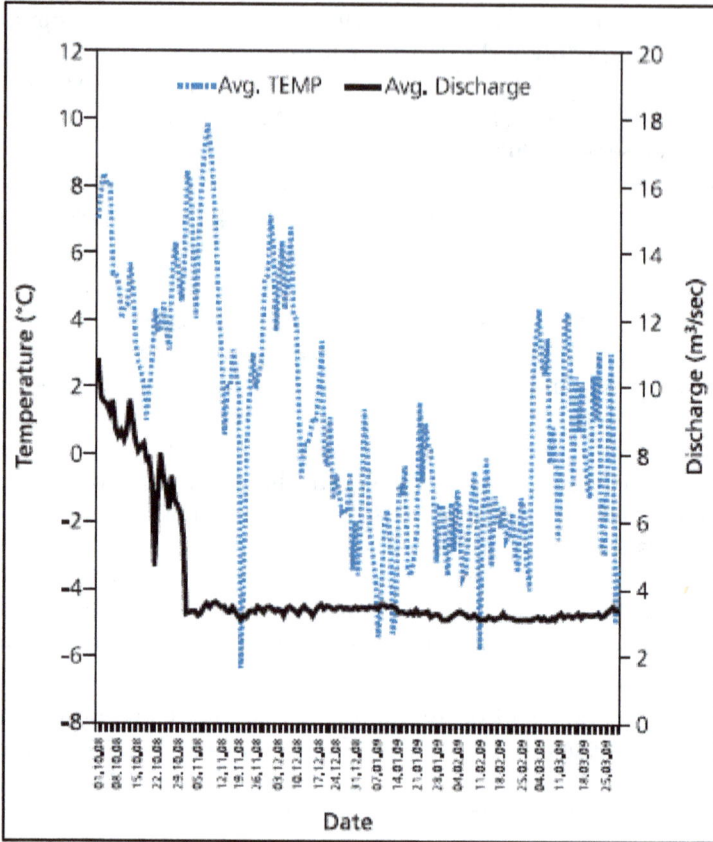

Figure 5.6: The average temperature and discharge during the accumulation season at Bhojwasa on Bhagirathi river.

between the 1200 hours temperature and the 1700 hours discharge at Bhojwasa on the Bhagirathi river (Figure 5.7). In the month of October the discharge rate falls as the average temperature starts reducing.

Trends in the Gangotri region (Figure 5.8) reflect that temperature start falling with the arrival of September. The maximum temperature reaches up to 18°C in June and July and remains mostly around 14°C during May to August. The morning temperature (at 0800 hours) remains at around 9°C till August and falls to zero in the mid of September. The winter temperature is quite low especially in December when the morning temperature is always sub–zero and can dip down to –15°C. The noon temperature remains at around 10°C. During January and February, the morning and evening temperature is always below zero while in the noon it just crosses the zero degree mark. During March, the morning temperature remains at below zero, while the noon and evening temperature varies around 5°C.

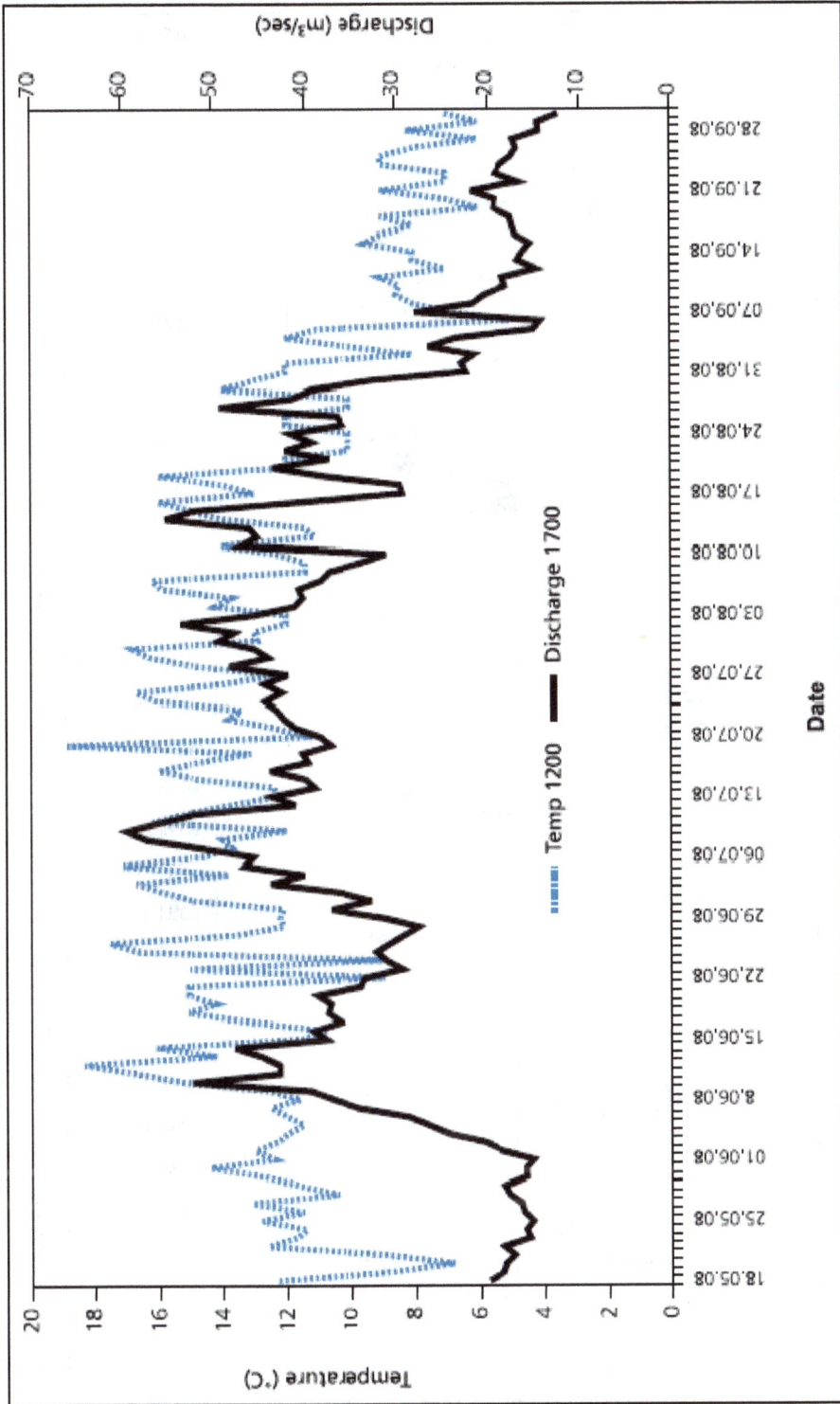

Figure 5.7: Influence of midday temperature (1200 hours) on evening discharge 1700 hours at Bhojwasa, Gangotri.

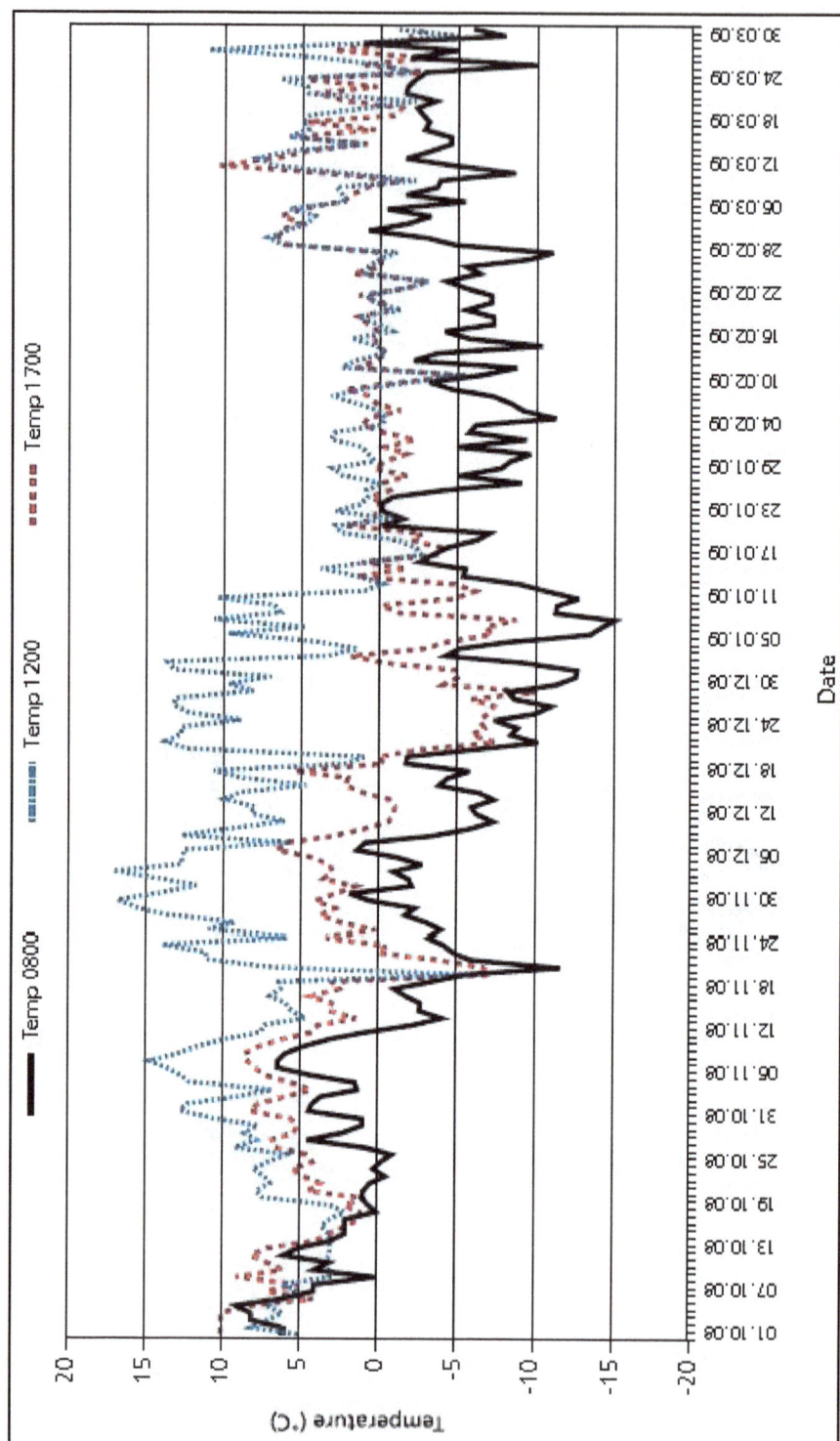

Figure 5.8: Temperature patterns at Gangotri glacier site during the summer season.

Case Study II: Kafni Glacier

Study Area

The Kafni glacier is located in the Pindar basin of the Kumaon Himalayas at the border of Bageshwar and Pithoragarh districts in Uttarakhand. The current snout position of the glacier lies at 30°13'12"N and 80°03'14"E. The north–south extending Kafni glacier originates from the southern slope of Nanda Kot–a major peak in the region. The glacier is the source of river Kafni, which originates from the ice cave formed at the snout. This river is a tributary of the Pindar river, which flows into the Alaknanda river system. The Kafni glacier consists of two tributary glaciers that are no longer connected with the main trunk. They now exist in the form of hanging glaciers and contribute to the river flow through their meltwater.

Methodology

The objective of selecting the Kafni glacier was to understand the impacts of the changing local climate on small glaciers. The glacier is being monitored for changes in length, area and volume using satellite imageries and DGPS. The recession pattern of the glacier has been studied by comparing the past satellite data (LANDSAT series of satellite imageries, 1976–1999) with the present data collected through DGPS during field visits in 2007 and 2008. The analysis has proved helpful in giving an overview of the changes in the snout position and glacierised area of the Kafni glacier in the last 30 years.

An automatic weather station has also been installed near the snout to understand the hydro–meteorological characteristics in the region and their linkages with the glacial melt. Along with weather monitoring, discharge patterns of the Kafni glacier have also been monitored to find the linkages between meteorological parameters and the glacial melt. Hydrological discharge data has been collected using float techniques.

Results and Analysis

A combination of data from primary and secondary sources indicates that the average retreat rate of the Kafni glacier has been 15.7m/year between 1976 and 2009. However, the snout position of the glacier in different years shows a variation in the retreat rate during different time periods, *i.e.* between 1976 and 2009. It is apparent from the analysis that the average retreat rate was higher at 21.36 m/yr from 1976–1990 but reduced to 10.61 m/yr during 1990–99. The average retreat increased to 12.99 m/yr between 1999 and 2006. Annual observations since 2006 shows that the Kafni snout retreated by 12.27 m/yr between 2006–07 and it continues to show a reducing trend – 10.6 m/yr in 2007–08 and 10.05 m/yr in 2008–09. While the retreat rate of the glacier shows a reducing trend, however, it is important to note that given that Kafni is a smaller glacier, even a small retreat has a significant impact on the mass balance of the glacier (Figure 5.9).

The characteristics and condition of the upper ablation area – rock exposed at several places and reduction in ice thickness – will show its impact in a few years on the snout and the total length of the glacier. The equilibrium in the retreat rate may be

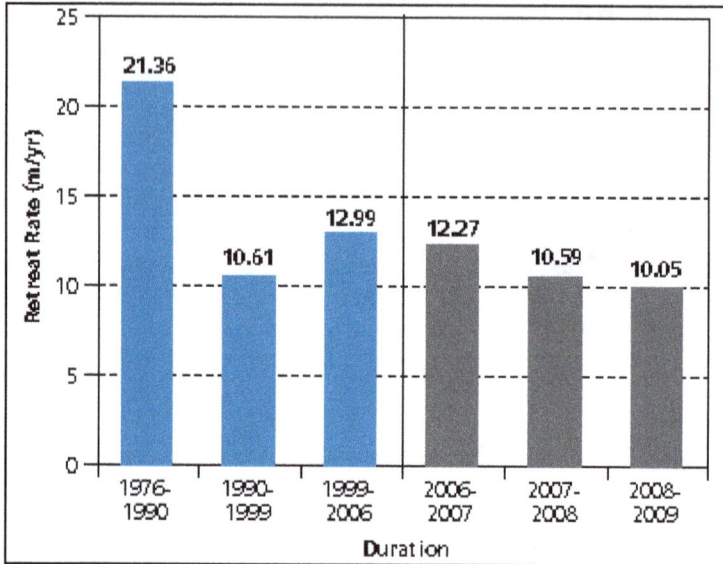

Figure 5.9: Retreat rate of Kafni snout in different intervals of time.

due to the thick debris on the lower to middle portion of the glacier, which reduces the solar energy transfer to the ice surface and hence a reduction in the melting. Kafni, being a smaller glacier has a lower accumulation zone and therefore is more sensitive to the variations in snowfall and any change in the equilibrium line. The tributary glacier of the Kafni (being quite small in size) is now hanging, meaning that it is not directly connected any more through ice mass to the main trunk of the Kafni. This indicates the loss of a huge ice volume in the glaciated catchment of Kafni. The glacier has vacated 14.76 per cent of its area during 1976–1990 and another 4.06 per cent during 1990–1999 as per analysis of this study taking the base area and year as 1976. The overall loss in the glaciated area of Kafni and its tributaries during 1976 to 1999 is 17.5 per cent (main trunk), 29 per cent (TG1) and 23 per cent (TG2) respectively. For quick referencing, the results of the Kafni glacier's snout retreat have been shown in Figure 5.10.

Hydro-meteorological Study

Initial trends about correlating meltwater discharge at Kafni with rainfall and temperature were established in 2008. Discharge data collected at the glacial snout in 2008 indicates a correlation of 0.51 between discharge and rainfall (Figure 5.10), which followed a uniform pattern throughout the entire summer season except for one extreme weather event during late September when high rainfall (about 165 mm) caused increase in discharge rate to ~30 m³/s. Towards October, trends indicated that the discharge rates dropped to 15 m³/s.

Similarly, a clear correlation was observed between increase in temperature and discharge rates although the relation does not appear to be as strong as in the case of rainfall and discharge. This shows the importance of rainfall in this valley. Usually

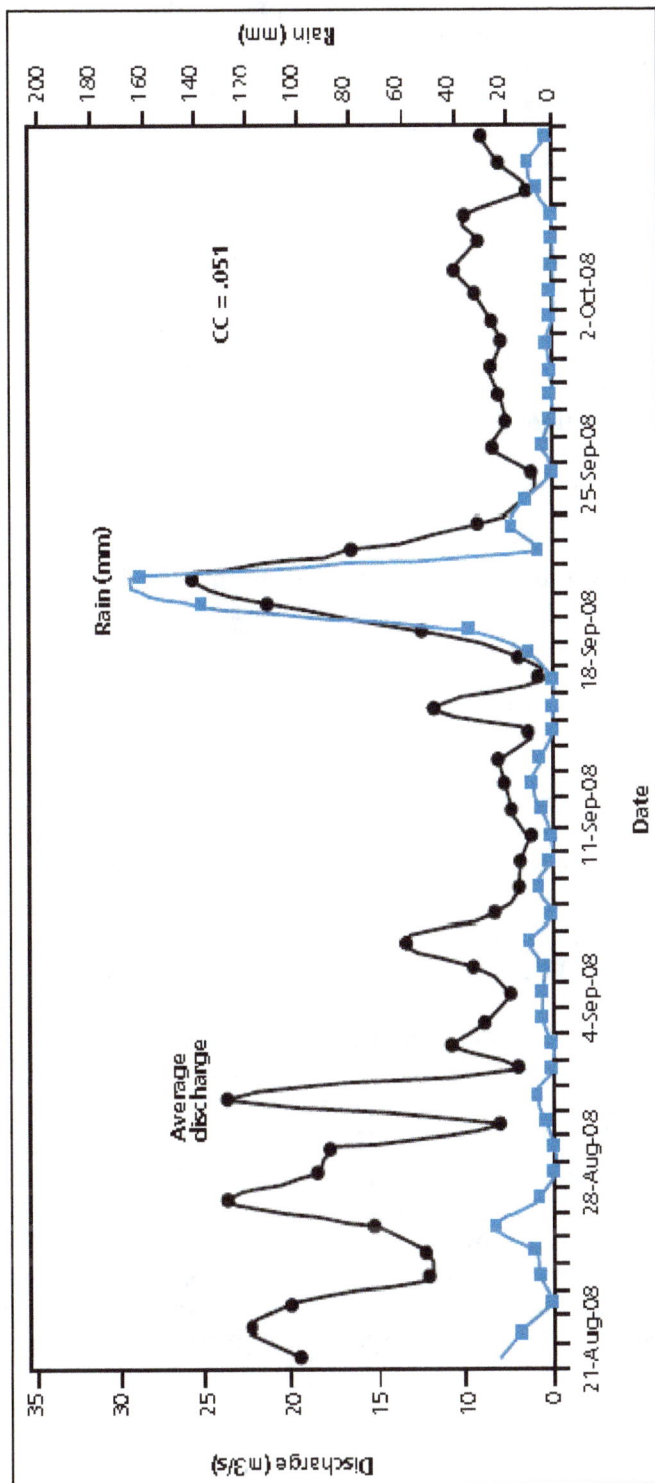

Figure 5.10: Rainfall and discharge correlations at Kafni in 2008.

the valleys in the Kumaon Himalayas get more rain during the monsoon season as compared to the Garhwal Himalayas. With higher average temperature in August during peak discharge season, actual discharge rates vary between 18–25 m³/s, corresponding with the temperature range of 25–30°C for the same month. In October, with the onset of the winter, discharge rates tapered off to around 10 m³/s in relation to average temperatures of about 4–8°C. Since this region was never equipped with a meteorological observatory, no analysis could be done from a hydro–meteorological point of view. This is for the first time that an attempt has been made in this part of the Himalayas to understand the hydro-meteorological characteristics of the Kafni glacier.

Remarks

The science of climate change is complex as it is not only about temperature variations but also the impact of local environmental factors; non- climate stressors play an equally critical role. The ecologically fragile Himalayan ecosystems harbour a diverse range of flora and fauna. Moreover, ecological services generated by these ecosystems support the wellbeing of communities in the mountains and lower plains.

Smaller Glaciers – More Vulnerable

The initial results from our field study indicate that the Himalayan glaciers are retreating, but at a reduced rate, and the larger glaciers like Gangotri are unlikely to disappear in near future, due to its large mass balance. The smaller glaciers like Kafni are retreating at a faster rate, and are not only losing more glaciated portion but also their tributary glaciers – a trend which has been observed across the Himalayas for many other smaller glaciers. Regional climate variations could threaten the fragile nature of these glaciers, which are likely to disappear at a much faster rate or be considerably reduced in length as compared to the larger ice bodies. These glaciers are perhaps more vulnerable to local climate variations and hence long-term and continuous assessment is required to monitor the hydro–meteorological parameters existing in their vicinity to develop predictive models for future water resource scenarios.

Larger glaciers like Gangotri show a continuous recessionary trend in recent years as evidenced from the present and other studies (Singh *et al.*, 2006; Nainwal *et al.*, 2008). However, at present this trend may have a limited influence on the water flows downstream primarily due to the large extent of the glacierised area. The movements of larger glaciers are dynamic and snout retreat is a delayed response, but the fragmentation of the larger glaciers needs to be continuously monitored.

Not only climatic variations, but there also are visible changes in the social and economic dimensions of the Himalayan region. Communities living closer to Gangotri have indicated changes in snowfall levels in the winter months resulting in less soil moisture, which in turn is changing cropping patterns and availability of water. Different glaciers in the same climatological set up respond differently to the changes in climate. Hence, it becomes difficult to predict the glacier retreat or advance scenarios with confidence.

Addressing the Data Challenge

The current observations at the Gangotri and Kafni glaciers strengthen the existing observations especially by providing the hydro–meteorological data. This database will encourage researchers in future to conduct studies in the region of Gangotri and Kafni, as well as other glaciers. This will further help in accurate assessments of the glacial melt and water flows based on long-term data. Better use of these technologies could provide an efficient and scientifically sound method to study different patterns and changes in the glacier systems on various spatial and temporal scales.

Since region specific climate models are limited, a long-term monitoring of meteorological parameters is crucial before arriving at definitive conclusion regarding the impact of global warming and climate change. It is also important to study the impact of local pollutants and climate variables together with non-climatic stressors.

As a way forward it is essential that more data is generated over a longer time period and that new mechanisms need to be created to ensure sharing of data to enable more in–depth research and analysis. This will enable development of more region–specific climate models to make projections for potential future impacts and preparation of appropriate response measures and mechanisms. It is also essential to carry out a long-term study on glacier mass balance and glacier dynamics to understand the impact of climate change in the Himalayas.

Non-climatic Stressors

Non-climatic stressors, such as rapid economic growth, including tourism and infrastructure development are equally increasing pressure on the Himalayan belt. Unplanned development has already resulted in severe pressures on both local ecology and communities. The impacts of climate change and changes in climate variability would further create additional stresses. The decline in glacial area and variations in annual runoff patterns in the future gains importance in the context of hydropower planning and development of the Himalayan states.

Adapting to the Change

Glacial retreat could pose the most far–reaching challenge in the Himalayan region. The dynamics of the monsoon are influenced by Himalayan systems which act as a reservoir to sustaining agriculture, providing freshwater and groundwater recharge and are home to a unique ecosystem with many endemic species (WWF, 2009). Adaptation to climate change, therefore, requires not just local action but also trans–boundary cooperative arrangements.

Future efforts in building the resilience of the local community and the ecosystems should take into account a concerted and integrated approach. There is an urgent need by communities, scientists and policymakers to take a closer look at the linkages between local impacts, scientific research, policy interventions and the larger understanding of using resource conservation technologies and practices for promoting societal benefits.

Way Forward

Science has provided evidence of changes happening in the glaciers, but probably not very accurately as there is inadequate recorded historical evidence. However, the growing body of anecdotal evidence and observations of the communities provide evidence of how communities are coping and managing with change. This needs to be supported with science and observed data. This study has provided data and information of the ground level parameters. In our view, this aligns with the growing emphasis on regional cooperation between Himalayan countries on the impacts of glacial retreat, as well as the national focus within India on the Himalayan ecosystems.

References

Barnett, T. P., Adam, J. C. and Lettenmaier, D. P. 2005. Potential impacts of a warming climate on water availability in snow–dominated regions. *Nature*, 438: 303–309.

Berthier, E., Arnaud, Y., Kumar, R. *et al.*, 2007. Remote sensing estimates of glacier mass balances in the Himachal Pradesh (Western Himalaya, India). *Jr. Remote Sensing of Environment*, 108: 327–338.

Bhutiyani, M.R., Vishwas S.K. and Pawar, N.J. 2007. Long-term trends in maximum, minimum and mean annual air temperatures across the Northwestern Himalaya during the twentieth century. *Climatic Change*, 85: 159–177.

Cruz, R.V., Harasawa, H., Lal, M. *et al.*, 2007. Asia. In: *Climate Change 2007: Impacts, Adaptation and Vulnerability*. (Parry, M.L., Canziani, O.F., Palutikof, J.P., *et al.* eds.) Cambridge University Press, Cambridge, UK. Pp. 469–506.

Dobhal, D.P., Kumar, S., and Mundepi, A.K. 1995. Morphology and glacier dynamics studies in monsoon–arid transition zone: An example from Chotta Shigri glacier, Himachal– Himalaya, India. *Current Science*, 68 (9): 936–944.

Dobhal, D. P., Gergan, J. T. and Thayyen, R J. 2004. Recession and morphogeometrical changes of Dokriani glacier (1962–1995) Garhwal Himalaya, India. *Current Science*, 86 (5): 101–107.

Dobhal, D.P., Gergan, J.T. and Thayyen, R.J. 2007. Mass balance and snout recession measurements (1991–2000) of Dokriani glacier, Garhwal Himalaya, India. Climatic and anthropogenic impacts on the variability of water resources. (Technical Document in Hydrology, UNESCO, Paris), *Hydro Sciences Montpellier*, 80: 53–64.

Dyurgerov, M., and Meier, M.F. 2005. *Glaciers and the Changing Earth System: A 2004 Snapshot (Occasional Paper 58)*. Institute of Arctic and Alpine Research, University of Colorado, Boulder.
Available at http://instaar.colorado.edu/other/download/OP58_dyurgerov_meier.pdf

GOI, 2009. *Himalayan Glaciers: A State of Art Review of Glacial Studies, Glacial Retreat and Climate Change, MoEF Discussion Paper*. Ministry of Environment and Forest, Govt. of India.

Haeberli, W. 1990. Glacier and permafrost signals of 20th century warming. *Annals of Glaciology*, 14: 99–101.

IPCC. 2007. Summary for Policymakers. In: *Climate Change 2007: The Physical Science Basis*. (Solomon, S., Qin, D., Manning, M. *et al.* eds.) Cambridge University Press, Cambridge, U.K.

Jain, S. K., Agarwal, P. K., and Singh, V. P. 2007. *Hydrology and Water resources of India.* Water Science and Technology Library, Springer, Dordrecht. Pp. 1258.

Jianchu, X., Shrestha, A., Vaidya, R. *et al.* 2007. *The Melting Himalayas. Regional Challenges and Local Impacts of Climate Change on Mountain Ecosystems and Livelihoods. ICIMOD Technical Paper,* ICIMOD, Kathmandu. Pp. 15.

Kaser, G., Cogley, J. G., Dyurgerov, M. B. *et al.* 2006. Mass balance of glaciers and ice caps: Consensus estimates for 1961–2004. *Geophysical Research Letters,* 33: L19501.

Kaul, M.K. (ed.).1999. *Inventory of the Himalayan Glaciers: A Contribution to the International Hydrological Programme,* GSI. Spl pub. 34. Pp. 165.

Kumar, R., Hasnain, S. I., Wagnon, P. *et al.* 2007. Climate change signals detected through mass balance measurements on benchmark glacier, Himachal Pradesh, India. Climatic and anthropogenic impacts on the variability of water resources (Technical Document in Hydrology, UNESCO, Paris), *Hydro Sciences Montpellier,* 80: 65–74.

Kulkarni, A.V. and Bahuguna, I.M. 2002. Glacial retreat in the Baspa Basin, Himalayas, monitored with satellite stereo data. *J. Glaciol.,* 48: 171–172.

Kulkarni, A.V., Bahuguna, I.M., Rathore, B.P. *et al.* 2007. Glacial retreat in Himalaya using Indian Remote Sensing satellite data. *Current science,* 92(1): 69–74.

Kulkarni, A.V., Mathur, P., Rathore, B.P. *et al.* 2002. Effect of Global warming on snow ablation pattern in the Himalayas. *Current Science,* 83: 120–123.

Nainwal, H.C., Negi, B.D.S., Chaudhary, M. *et al.* 2008. Temporal changes in rate of recession: Evidences from Satopanth and Bhagirath Kharak glaciers, Uttarakhand, using Total Station Survey. *Current Science,* 94(5): 653–660.

Roy, S.S. and Balling Jr., R.C. 2005. Analysis of trends in maximum and minimum temperature, diurnal temperature range, and cloud cover over India, *Geophys. Res. Lett.,* 32: L12702.

Shukla, S.P. and Siddiqui, M.A. 2001. Recession of the snout front of Milam Glacier, Gorigunga Valley, Pithorgarh district, Uttar Pradesh. In: *Proceedings of Symposium on Snow, Ice and Glaciers–A Himalayan Perspective* (Special Publication No. 53). Geological Survey of India, Lucknow.

Singh, P. and Bengtsson, L. 2005. Impact of warmer climate on melt and evaporation for the rainfed, snowfed and glacierfed basins in the Himalayan region. *Journal of Hydrology,* 300: 140–154.

Singh, P., Haritashya, U. K., Ramasastri, K. S. *et al.* 2005. Prevailing weather conditions during summer seasons around Gangotri Glacier. *Current science,* 88(5): 753–760.

Singh P., Haritashya, U. K., Kumar, N. *et al.* 2006. Hydrological characteristics of the Gangotri Glacier, central Himalayas, India. *Journal of Hydrology,* 327: 55–67.

Singh, P., Polglase, L. and Wilson, D. 2009. Role of Snow and Glacier melt runoff modeling in Hydropower projects in the Himalayan Region. *Hydrologic and Hydraulic Modelling,* 1: 366–371.

Srivastava, D. 2001. *Glaciology of Indian Himalaya* (*A Bilingual Contribution in 150 Year of Geological Survey of India, Special Publication No. 63*), Geological Survey of India. Pp. 215.

UNEP. 2007. *Global outlook for ice and snow.* UNEP/GRID–Arendal, Norway. Pp. 235.

UNESCO. 2007. *Case Studies on Climate Change and World Heritage.* UNESCO World Heritage Centre, Pp. 80.

Yadav, R.R., Park, W.K., Singh, J. *et al.* 2004. Do the western Himalayas defy global warming? *Geophys. Res. Lett.*, 31(17): L17201.

Wagnon, P., Linda, A., Arnaud, Y. *et al.* 2007. Four years of mass balance on Chhota Shigri Glacier, Himachal Pradesh, India, a new benchmark glacier in the western Himalaya. *Journal of Glaciology*, 53: 183.

Wood, F.B. 1990: Monitoring Global Climate Change: The Case of Greenhouse Warming. *Bull. Amer. Meteor. Soc.*, 71: 42–52.

WWF. 2005. *The Eastern Himalayas – Where World Collide.* Pp. 28.

WWF. 2009. *Witnessing Change: Glaciers in the Indian Himalayas.* Pp. 38.

Climate Change: Man and Environment (2012)   *Pages* **87–99**
Editor: **Goutam Kumar Saha**
Published by: **DAYA PUBLISHING HOUSE, NEW DELHI**

6

Predicted Global Forest Changes: Apprehensions and Realities

☆ *Tapan Kr. Mishra**

Introduction

The apprehension that climate change is one of the greatest challenges currently facing humankind is becoming true day by day. Increased severity and occurrence of natural disasters, changing weather patterns, retreating glaciers, polar ice melt, sea level rise and drought are just some of the consequences already being experienced by populations around the world. Climate change has dual, and sometimes conflicting, implications for forests. Forest ecosystems can act as a tool for mitigation and adaptation to climate change. Forests remove significant volumes of carbon dioxide from the atmosphere, acting as sinks, capturing carbon and storing it in the forest's biomass. Forests act as barriers in extreme weather events, preventing topsoil run–off in heavy rains and protecting people, animals and physical infrastructure from the effects of strong winds. However, just as forests have a positive role to play in efforts to combat climate change, forests are also highly vulnerable to changing climate conditions. The climate at a given location determines the type of forest – boreal, temperate, subtropical or tropical rain forest, that can become established (Figure 6.1). Likewise, when climate conditions change, forests must adapt. However, the time frame required for the adaptation process is usually far longer than the time scale allowed by changing climate conditions.

* Vidyasagar College, Kolkata. E-mail: mishratapan@hotmail.com

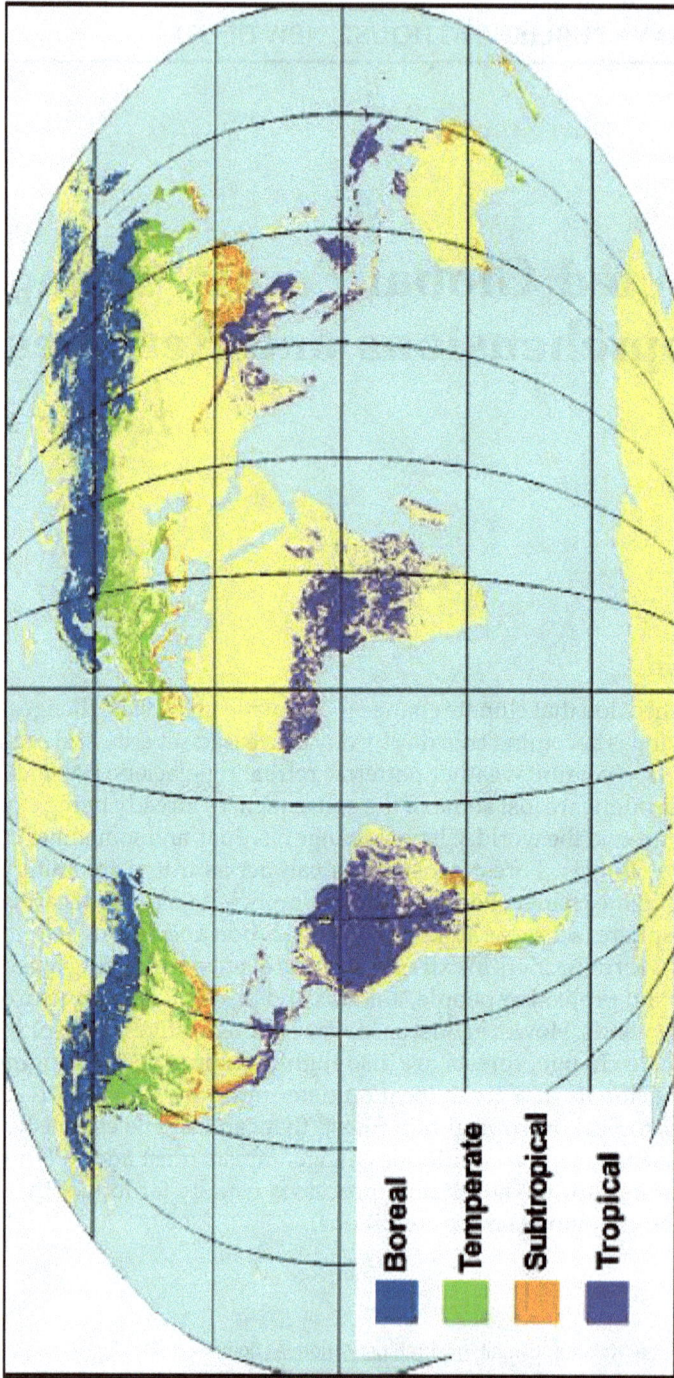

Figure 6.1: Distribution of the world's forests by major ecological zone.

Forests and woodlands are an important part of our landscape and provide many benefits to society. The tree species that are native to a particular part of the earth have adapted to the local climate, atmosphere and soils over many years. However, human activities have resulted in changes to the natural environment, especially over the last 200 years. As a result of changing climate conditions, forests' ability to adapt is compromised resulting in a loss of forest biodiversity and forests themselves, and along with them forests' ability to adapt to and mitigate the impacts of climate.

Changes in Forest Cover

All forest types will undergo some changes as a result of altered climate conditions; some of these changes are already occurring but widespread change is expected over the next 50–100 years (Alcamo *et al.*, 2007; Fischlin *et al.*, 2009). In many tropical forests, however, many rainforests may become dry tropical forests with reduced carbon storage capacity (Fischlin *et al.*, 2009). The diversity in these tropical regions suggests that some form of forest will continue to exist even with severe disturbance, but that many of the functions will change owing to the lack of resilience and new states, in general, will produce considerably less goods and services while supporting less biodiversity than at present.

Most evidence suggest that tropical forests may not be resilient to climate change over the long term, primarily owing to a predicted reduction in rainfall and increased drought (IPCC, 2007; Malhi *et al.*, 2009). In the short term, evidence suggests a positive effect of CO_2 fertilization on tropical forest production as a result of present climate change (Boisvenue and Running, 2006; Lewis *et al.*, 2009), although importantly this has involved some changes in species composition, indicating resilience to current change. Future capacity of these forests to maintain this service is highly uncertain (Cramer *et al.*, 2004) as a result of altered moisture regimes possibly leading to increased fire and drought (Malhi *et al.*, 2009). Loss of tropical forests will have consequences for global hydrology, among other consequences of global relevance (Fischlin *et al.*, 2009).

A study was carried out on devastating outbreak of mountain pine beetle in the province of British Columbia, Canada by Konkin and Hopkins (2009). The mountain pine beetle, *Dendroctonus ponderosae*, is a native bark beetle of the lodgepole pine (*Pinus contorta*) forests of western Canada that periodically outbreaks. However, since the late 1990s, populations have grown to an unprecedented scale, now attacking more than 13 million hectares of forest in the province of British Columbia. The epidemic may be attributed to multiple causes, including climate change and other factors such as forest management interventions. By 2015, the epidemic is expected to kill more than three–quarters of the pine volume in British Columbia representing over 900 million cubic meters of timber that was expected to contribute to the economic wealth of British Columbia's communities. In these communities climate change is no longer a theoretical matter – the impacts are real now. This epidemic creates challenges on multiple fronts, but it has also led to new economic opportunities for British Columbia. In addition, it has been a catalyst for increased collaboration among rural communities, natural resource industries and government agencies, and it has fostered new ways of thinking about forest management in the context of climate change and social objectives. Over the past decade, British Columbia has not experienced the extreme cold winter temperatures that curtailed previous outbreaks. Warming in British Columbia over the twentieth century (to 1995) was approximately equal to the global average of 0.6°C on the coast (British Columbia Ministry of Environment, 2002; IPCC, 2001), but two or three times higher in the interior. The higher winter temperature led to increased winter survival of the mountain pine beetle, which has culminated in the largest beetle epidemic in the province's recorded history.

From a broader perspective, climate change may lead to a sharp increase in rates of extinction. The study made by Thomas *et al.* (2004) focusing on five regions of the world suggests that if the climate continues to warm it could dramatically increase the number of species going extinct. Mid–range predictions suggest that 24 per cent of species in these regions will be on their way to extinction by 2050 due to climate change. This study also indicates that for many species, climate change poses a greater threat to their survival than the destruction of their natural habitat. Other observed impacts of climate change include changes in the timing of reproduction in certain species; in the length of the growing season in many regions; in the abundance of different species; and in the frequency of pest and disease outbreaks. For example, higher temperatures have led to an increase in the number of eggs laid by the spruce budworm, already one of the most devastating pests in North America's boreal forest. This could in turn contribute to more severe outbreaks of this pest (Gitay *et al.*, 2002). Climate change may also affect species at the level of cells and genes. Changes in the genetic makeup of species are expected as organisms adapt to new climatic conditions, and increases in temperature can also lead to increases in the rate at which cells use energy.

Impact on Indian Forests

India is a mega–biodiversity country where forests account for about 23.84 per cent (78.37 million ha) of the geographical area. Forests in India are extremely diverse and heterogeneous in nature, and it is difficult to classify them into a small number of categories. The sum total of the pan–Indian 'miscellaneous forest' category (with no dominant species) shows the highest (63 per cent) proportion. The miscellaneous forest area occurs under all the forest types. The other two most dominant forest types are *Shorea robusta* or *sal* (12 per cent) in the eastern part of central India and *Tectona grandis* or teak (9.5 per cent), spread across central India and the Western Ghats in southern India. The increase in precipitation can change the nature of the forest in terms of the floral species dominance, canopy cover, forest dynamics etc. It can rebuild the connections between fragmented ecosystems, support forest areas to encroach into grasslands, alter tree species dominance and thereby change forest class. On the other hand, reduction in precipitation can support shift towards deciduous category of forests, expansion of grass lands, and lead to forest fragmentation and raise frequencies of forest fire. All these can cause significant change in faunal species distribution, demography and composition.

The main factor that determines the *sal* or teak domination in the forests of central India is the time of occurrence of rainfall and period of rainfall besides other minor factors. Seeds of *sal* have a viability of 10 days or so and rains should coincide with this short period of viability. As the eastern part of the peninsula receives rains from the Bay of Bengal branch of the southwest monsoon, which arrives earlier than the Arabian Sea branch, the germination of *sal* seeds is favoured and the tree dominates the eastern half of the peninsula; consequently the western counterpart is marked by the dominance of teak. This example reveals how delicate the link could be between the climate and the geographic distribution of species within a bio–climatic framework as described in the forest classification done by Champion and Seth (1968). The classification states 16 major vegetation types starting from tropical wet evergreen forest to dry alpine forest. The country has largest extent of tropical moist deciduous forest followed by tropical dry deciduous forest. This classification shows that the rainfall pattern influences greatly the forest vegetation pattern though other factors also operate.

Rabindranath *et al.* (2006) suggested that the predicted increase in the precipitation in the forest areas in the Indian subcontinent is higher than that of the non-forest area. Their study with projected climate models available as 30–year long time series, predicts 2°C to 3.5°C increase in temperature and 250 mm to 500 mm increase in precipitation in the northeastern region. Increase in rainfall may not have significant impact on the forest areas of northeast which are already experiencing high rainfall but change in temperature regime may cause severe impact and significant changes (Ravindranath and Sukumar, 1998). Successive reports of IPCC have concluded that even moderate warming and climate change will impact forest ecosystems adversely.

Legris (1963) was one of the first workers to describe possible migration of forest types with a drop of 6°C temperature, as was the case during ice age or with an increase of rainfall during the pluvial phase. Ravindranath *et al.* (2006) pointed out that global assessments have shown future climate change is likely to significantly affect forest ecosystems. Their study made an assessment of the impact of projected climate change on forest ecosystems in India. The assessment was based on climate projections of Regional Climate Model of the Hadley Centre (HadRM3) using the A_2 (740 ppm CO_2) and B_2 (575 ppm CO_2) scenarios of Special Report on Emissions Scenarios and the $BIOME_4$ vegetation response model. Their main conclusion was that under the climate projection for the year 2085, 77 per cent and 68 per cent of the forested grids in India are likely to experience shift in forest types under A_2 and B_2 scenario, respectively. Indications are a shift towards wetter forest types in the northeastern region and drier forest types in the northwestern region in the absence of human influence. Increasing atmospheric CO_2 concentration and climate warming could also result in a doubling of net primary productivity under the A_2 scenario and nearly 70 per cent increase under the B_2 scenario. The trends of impacts could be considered as robust but the magnitudes should be viewed with caution, due to the uncertainty in climate projections. Given the projected trends of likely impacts of climate change on forest ecosystems, it is important to incorporate climate change consideration in forest sector long-term planning process.

Bose (2009) apprehends that the Sundarbans, south Asia's largest carbon sink which mops up carbon dioxide must survive to prevent global warming. Out of 60 varieties of mangrove species that are found in India, Sundarban accounts for 50, many of which are rare. It has a seemingly unlimited capacity to absorb pollutants from air and water. Though the economic and social benefit arising from this mangrove rehabilitation would be more meaningful to the local communities of this region but it acts as an eye opener to the whole world to understand the concept of forest conservation in mitigating climate change. All constituents of this complex ecosystem depend upon each other. Any damage to one part will damage and change the whole constitution of the ecology of this area. Sundarban reveals the extreme consequences of climate change and global warming, therefore mangrove plantation is the ultimate solution to this problem.

The state of Gujarat has the largest area of mangrove forests after West Bengal. While the mangroves of the Gulf of Kutch could possibly adapt to low or moderate sea level rise, a rise of more than one meter in the next century could cause serious losses (Singh, 2002). The mangroves are also threatened by the rise in temperature, which causes decreased tree height and leaf size. Besides sea level rise and temperature stress, the mangroves in the gulf towards Jamnagar and the Kutch coasts are also threatened by drought.

Species Shift and Extinction

Some important studies have been made on the ecological transformation or more appropriately the process of succession that one may encounter in future. Some studies also suggest that global warming is driving species ranges poleward and toward higher elevations at temperate latitudes, but evidence for range shifts is scarce for the tropics, where the shallow latitudinal temperature gradient makes upslope shifts more likely than poleward shifts. Based on new data for plants and insects on an elevation transect in Costa Rica, the potential for lowland biotic attrition, range–shift gaps, and mountain top extinctions under projected warming were assessed by Colwell *et al.* (2008). They concluded that tropical lowland biota may face an unparallel level of net lowland biotic attrition as compared to higher latitudes (where range shifts may be compensated by species from lower latitudes) and that a high proportion of tropical species soon faces gaps between current and projected elevation ranges.

Scientists made a study in Concord, Massachusetts on the phylogenetic patterns of species loss in Thoreau's woods. During the mid–19th century, the naturalist and conservationist Henry David Thoreau spent decades exploring the temperate fields, wetlands, and deciduous forests of Concord, Massachusetts, in the northeastern United States. He wrote extensively about the natural history of the area (Thoreau, 1897) and kept meticulous notes on plant species occurrences and flowering times (Miller–Rushing and Primack, 2008). Since then, several botanists have resurveyed the Concord area, thus providing a unique community–level perspective on changes in its floristic composition and flowering times during the past ≈150 years (Primack *et al.* 2009). Despite the fact that 60 per cent of all natural areas in Concord are underdeveloped or have remained well protected, a striking number of species have become locally extinct; 27 per cent of the species documented by Thoreau have been lost, and 36 per cent exist in such low population abundances that their extirpation may be imminent (Primack *et al*, 2009). Also, the species that have been lost are overly represented in particular plant families (Primack *et al.*, 2009), suggesting that extinction risk may be phylogenetically biased.

Although habitat loss due to succession and development (*e.g.*, loss of wetlands, abandonment of farms, reforestation, and construction of homes and roads) has contributed to the decrease in abundance for some species in Thoreau's Concord (Visser, 2008), climate change may also help to explain the seemingly nonrandom pattern of species loss among certain plant groups. It has been shown recently that the mean annual temperature in the Concord area has risen by 2.4°C over the past 100 years and that this temperature change is associated with shifts in flowering time. Species are now flowering in an average of 7 days earlier than in Thoreau's time. Along with changes in flowering phenology, species range is likely to be influenced by climate change (Visser, 2008). Thus, the Concord surveys provide a unique opportunity to examine the extent to which changes in abundance may be correlated with these climatologically sensitive traits. Also, by incorporating phylogenetic history into their analyses, they could test whether species that share similar traits were closely related (*i.e.* phylogenetic conservatism), and to what extent these traits correlated with decreases in abundance. Such findings could identify groups of closely related species that are at higher risk of extinction (McKinney, 1997; Purvis, 2004).

A study of extinction patterns of 25 large mammal species in India finds that improving existing protected areas, creating new areas, and interconnecting them will be necessary for many species to survive this century (Karanth *et al.*, 2010). They showed that forest cover and local human population densities are also key factors. Fostering greater human cultural tolerance for wildlife likewise will be critical. This study examined extinction probabilities for a range of species. It looked at species considered endangered or critically endangered in the IUCN Red List of Threatened Species–2008 (2009), including tigers, lions and elephants and it also looked at species of least concern, including jackals, wolves and other species. Their analysis revealed that protected areas were associated with lower extinction probabilities for 18 species. Higher proportion of forest cover was associated with lower extinction of seven species. On the other hand, time elapsed since the last historical citing was associated with higher extinction probabilities of 14 species, and human population density in a cell was associated with higher extinction probabilities of 13 species.

To identify factors critical to the species' survival and estimate their extinction probabilities, the team of researchers collected 30,000 records, including hunting, taxidermy and museum records dating back to 1850. They divided India's geographical area into a grid with 1,326 individual local "cells" and entered the historical data into each cell. They then used occupancy estimation models, based on observations of more than 100 local wildlife experts, to infer the current occurrence of species in each cell. One of the most important findings, of the study is that culturally tolerated species (animals that humans perceive as nonthreatening or beneficial) fared better overall. Using both historical data and current occupancy estimation models allowed the researchers to distinguish more clearly between when a species was truly absent or locally extinct in a cell, and when it likely still exists but hasn't been detected in recent surveys or field observations.

Another study by Cardelus *et al.* (2006) adds that for conservation to succeed, policymakers and land managers must also take into account rapid changes in land use, climate, population growth and spread, and economic development now occurring in India and southern Asia. To illustrate the potential for elevational range shifts in the tropics, Colwell *et al.* (2008) analyzed elevational range data for four large survey data sets of plants and insects (epiphytes, understorey Rubiaceae, geometrid moths, and ants). The data for all 1902 species were collected by the authors since 2001 from the Barva Transect, a continuously forested corridor ascending 2900 m up an elevational gradient from La Selva Biological Station, near sea level, to the top of Volcán Barva, in Costa Rica (Cardelus *et al.*, 2006). In many respects, the predictions illustrated in Figure 6.2 must be considered worst–case scenarios, even if warming occurs as assumed. Estimating elevational range limits from local inventory data is likely to underestimate regional elevation range, even accounting for local under sampling. The projections of Colwell *et al.* (2008) share with species distribution models (Williams *et al.*, 2003) the assumptions that the fundamental climatic niche of each species is fully expressed by current distributions; that the effects of climate outweigh any idiosyncratic effects of species interactions, dispersal limitation, demographic patterns, or historical contingency; that change will be too rapid for adaptation to warmer temperatures at lower range limits; and that habitats at the landscape scale are homogenous with regard to microclimate. In fact, species that currently occupy warmer microhabitats at their lower range limit, including lowland species, may shift to currently cooler (and wetter) refuges at the same elevation, in response to warming (Parmesan, 2006).

The biosphere is not likely to be a purely passive sink for carbon; the changes contributing to the terrestrial carbon sink are likely to be causing profound changes in the ecological balance of ecosystems, with consequences for ecosystem function and species diversity. Laboratory studies show that responsiveness to high CO_2 varies between species (Norby *et al.*, 1999); for example, at the most basic level, the CO_2 response is much higher in plants with a C3 photosynthetic mechanism (all trees, nearly all plants of cold climates, and most temperate crops including wheat and rice) than it is in those with a C4 mechanism (tropical and many temperate grasses, some desert shrubs and some important tropical crops including maize, sorghum and sugar cane). This has the potential to alter the competitive balance between trees and grasslands. Certain functional groups such as pioneers or lianas may also benefit disproportionately. There have been only a few systematic field studies that have looked for long-term trends in forest composition. For example, in the RAINFOR project (Malhi *et al.*, 2002), field researchers are re–censusing old–growth forest plots across the Amazon basin to look for evidence of shifts in forest biomass and composition.

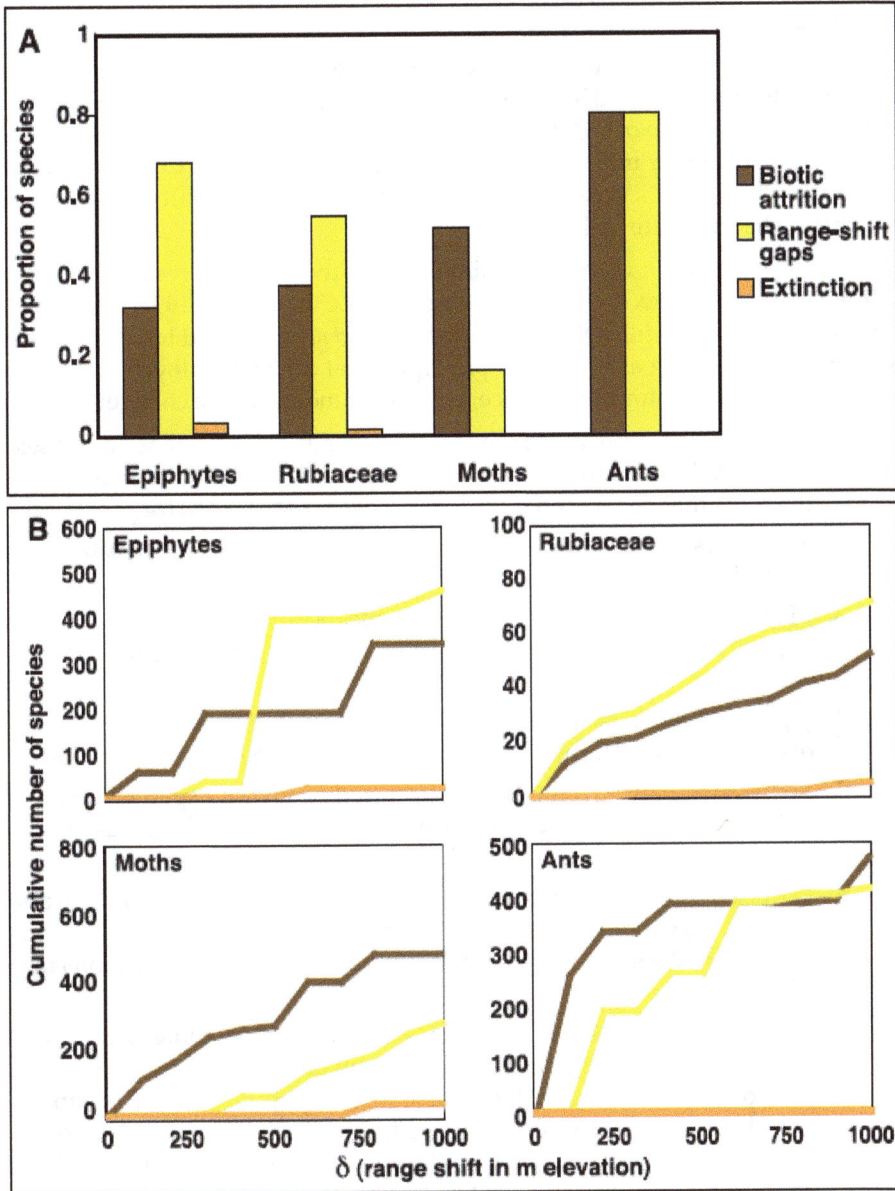

Figure 6.2: (A) Proportion of species in each of the four groups of Figure 6.2 subject to decline or disappearance in the lowlands (biotic attrition), faced by gaps between current and projected elevational range (range–shift gaps), and exposed to mountaintop extinction, given a 600–m upslope shift in all ranges. Proportions sum to greater than 1 because some species belong in two categories. (B) Cumulative number of species facing each of these three challenges as a function of warming–driven range shifts. The x axis represents model parameter d, measured in meters of elevation range shift, on a continuous scale of warming–driven isotherm shifts of up to 5°C (nearly 1000 m), the upper range of projections for central America for this century (32). The stair step patterns are a consequence of sampling at discrete sites on the gradient (5). (After Colwell *et al.*, 2008)

Remarks

It is a fact that absorbing carbon in trees clearly cannot 'solve' the global warming problem on its own as opined by Malhi *et al.* (2002). However, it is being a significant component in a package of CO_2 mitigation strategies, and providing an immediate carbon sink while other mitigation technologies are developed. Carbon absorbed early in the century has a greater effect on reducing end–of–century temperatures than carbon absorbed late in the century.

Caring for forests in ways that maintain their diversity and resilience is being made even more complex owing to climate change (Chapin *et al.*, 1997; Kellomaki *et al.*, 2008). Authors like Millar *et al.* (2007), Schaberg *et al.* (2008) and Innes *et al.* (2009) suggested the following as ecological principles that can be employed to maintain and enhance long-term forest resilience, especially under climate change:

1. Maintain genetic diversity in forests through practices that do not select only certain trees for harvesting based on site, growth rate, or form, or practices that depend only on certain genotypes (clones) for planting.

2. Maintain stand and landscape structural complexity using natural forests as models and benchmarks.

3. Maintain connectivity across forest landscapes by reducing fragmentation, recovering lost habitats (forest types), and expanding protected area networks.

4. Maintain functional diversity (and redundancy) and eliminate conversion of diverse natural forests to monotypic or reduced species plantations.

5. Reduce non-natural competition by controlling invasive species and reduce reliance on non-native tree crop species for plantation, afforestation, or reforestation projects.

6. Reduce the possibility of negative outcomes by apportioning some areas of assisted regeneration with trees from regional provenances and from climates of the same region that approximate expected conditions in the future.

7. Maintain biodiversity at all scales (stand, landscape, bioregional) and of all elements (genetic, species, community) and by taking specific actions including protecting isolated or disjunct populations of organisms, populations at margins of their distributions, source habitats and refugia networks. These populations are the most likely to represent pre–adapted gene pools for responding to climate change and could form core populations as conditions change.

8. Ensure that there are national and regional networks of scientifically designed, comprehensive, adequate, and representative protected areas. Build these networks into national and regional planning for large–scale landscape connectivity.

References

Alcamo, J., Moreno, J.M., Nováky, B. *et al.* 2007. Europe. In: *Climate Change 2007: Impacts, Adaptation and Vulnerability.* (Parry, M.L., Canziani, O.F., Palutikof, J.P., *et al.* (eds.)) Cambridge University Press, Cambridge, UK. Pp. 541–580.

Boisvenue, C. and Running, S.W. 2006. Impacts of climate change on natural forest productivity– evidence since the middle of the 20th century. *Global Change Biol.,* 12: 862–882.

Bose, S. 2009. Role of Indian Sundarban mangroves in mitigating climate impacts: an appraisal, *Climate Change: Global Risks, Challenges and Decisions. IOP Conf. Series: Earth and Environmental Science.,* 6:25. Available at: http://iopscience.iop.org/ 1755–1315/6/25/252017/pdf/1755–1315_6_25_252017.pdf

British Columbia Ministry of Environment. 2002. *Indicators of climate change for British Columbia.* Victoria, British Columbia, Canada. Available at: www.env.gov.bc.ca/ air/climate/indicat.

Cardelús, C.L., Colwell, R.K. and Watkins, J.E. 2006. Vascular epiphyte distribution patterns: explaining the mid–elevation richness peak. *J. Ecol.,* 94: 144–156.

Champion, H.G. and Seth, S. K. 1968. *Revised forest types of India,* Government of India Publication, Delhi. Pp. 404.

Chapin, F.S., Walker, B.H., Hobbs, R.J. *et al.* 1997. Biotic control over the functioning of ecosystems. *Science,* 277: 500–504.

Colwell, R.K., Brehm, G., Cardelus, C., *et al.* 2008. Global warming, elevational range shifts, and lowland biotic attrition in the wet tropics. *Science,* 322: 258–261.

Cramer, W., Bondeau, A., Schaphoff, S., *et al.* 2004. Tropical forests and the global carbon cycle: impacts of atmospheric carbon dioxide, climate change and rate of deforestation. *Phil. Trans. Roy. Soc. Lond. B.,* 359: 331–343.

Fischlin, A., Ayres, M., Karnosky, D., *et al.* 2009. Future environmental impacts and vulnerabilities. In: *Adaptation of forests and people to climate change: a global assessment report.* (Seppala, R., Buck, A. and Katila, P. eds.) IUFRO World Series, Helsinki. 22: 53–100.

Gitay, H., Suarez, A., Watson, R.T. and Dokken, D.J. (eds.) 2002. *Climate Change and Biodiversity (IPCC Technical Paper V).* IPCC, Geneva, Switzerland. Pp. 86.

Innes, J., Joyce, L.A., Kellomaki, S., *et al.* 2009. Management for adaptation. In: *Adaptation of forests and people to climate change: a global assessment report* (Seppala, R., Buck, A. and Katila, P. eds.), IUFRO World Series, Helsinki. 22: 135–169.

Intergovernmental Panel on Climate Change (IPCC). 2001. *Climate Change 2001: The Scientific Basis.* (Houghton, J.T., Ding, Y., Griggs, D.J. *et al.* eds.) Cambridge University Press, Cambridge, UK. Pp. 881.

Intergovernmental Panel on Climate Change (IPCC). 2007. Summary for policymakers. In: *Climate change 2007: Impacts, adaptation and vulnerability.* (Parry, M.L., Canziani, O.F., Palutikof, J.P., *et al.* eds.). Cambridge University Press, Cambridge, UK.

IUCN Red List of Threatened Species–2008. 2009 http://www.iucnredlist.org/ documents/redlist

Karanth, K., Norman, K., Christensen, L., *et al.* 2010. Online edition of the *Proceedings of the Royal Sciety* B 10.
http://rspb.royalsocietypublishing.org/content/early/2010/03/09/rspb.2009

Kellomaki, S., Peltola, H., Nuutinen, T., *et al.* 2008. Sensitivity of managed boreal forests in Finland to climate change, with implications for adaptive management. *Phil. Trans. Roy. Soc. B.,* 363: 2341–2351.

Konkin, D. and Hopkins, K. 2009. Learning to deal with climate change and catastrophic forest disturbances. *Unasylva.* 60: 231–232.

Legris, P. 1963. La Vegetation de Inde ecologie et Flora. *Inst. Fr. Pondicherry. Trav.Sec. Sci. Tech.,* 6: 1–589.

Lewis, S.L., Lopez–Gonzalez, G., Sonké, B. *et al.* 2009. Increasing carbon storage in intact African tropical forests. *Nature,* 457: 1003–1006.

Malhi, Y., Aragao, L.E.O.C., Galbraith, D. *et al.* 2009. Exploring the likelihood and mechanism of a climate–change–induced dieback of the Amazon rainforest. *Proc. Nat. Acad. Sci. USA.,* 106: 20610–20615.

Malhi, Y., Patrick, M. and Sandra, B. 2002. Forests, carbon and global climate. *Phil. Trans. Roy. Soc. Lond. A,* 360: 1567–1591.

Millar, C.I., Stephenson, N.L. and Stephens, S.L. 2007. Climate change and forests of the future: Managing in the face of uncertainty. *Ecol. Applic.,* 17: 2145–2151.

McKinney, M.L. 1997. Extinction vulnerability and selectivity: Combining ecological and paleontological views. *Annu. Rev. Ecol. Syst.,* 28: 495–516.

Miller–Rushing, A.J. and Primack, R.B. 2008. Global warming and flowering times in Thoreau's concord: A community perspective. *Ecology,* 89: 332–341.

Norby, R.J., Wullschleger, S.D., Gunderson, C.A. *et al.* 1999. Tree responses to rising CO_2 in field experiments: implications for the future forest. *Plant Cell Environ.,* 22: 683–714.

Parmesan, C. 2006. Ecological and evolutionary responses to recent climate change. *Annu. Rev. Ecol. Evol. Syst.,* 37: 637–669.

Primack, R.B., Miller–Rushing, A.J. and Dharaneeswaran, K. 2009. Changes in the flora of Thoreau's Concord. *Biol. Conserv.,* 142: 500–508.

Purvis, A., Agapow, P.M., Gittleman, J.L., *et al.* 2004. Nonrandom extinction and the loss of evolutionary history. *Science,* 288:328–330.

Ravindranath, N.H., Joshi, N.V., Sukumar, R. *et al.* 2006. Impact of climate change on forests in India. *Current Science,* 90(3): 10.

Ravindranath, N.H. and Sukumar, R. 1998. Climatic change and tropical forests in India. *Climatic Change,* 39: 563–581.

Schaberg, P.G., DeHayes, D.H., Hawley, G.J. *et al.* 2008. Anthropogenic alterations of genetic diversity within tree populations: Implications for forest ecosystem resilience. *For. Ecol. Manage.*, 256: 855–862.

Singh, H.S. 2002. Impact of climate change on mangroves. In. *Proc. South Asia Expert Workshop on Adaptation to Climate Change for Agricultural Productivity*, Ministry of Agriculture, Govt. of India, United Nations Environment Programme and Consultative Group on International Agricultural Research, New Delhi.

Thoreau, H.D. 1897. *Walden; or, Life in the Woods*, Houghton, Mifflin, and Company, Boston.

Thomas, C.D., Cameron, A.A., Green, R.E. *et al.* 2004. Extinction risk from climate change. *Nature*, 427: 145–148.

Visser, M.E. 2008. Keeping up with a warming world; assessing the rate of adaptation to climate change. *Proc. R. Soc. London Ser. B.*, 275: 649–659.

Williams, S.E., Bolitho, E.E. and Fox, S. 2003. Climate change in Australian tropical rainforests: an impending environmental catastrophe. *Proceedings of the Royal Society, Biological Sciences*, 270: 1887–1892.

Climate Change: Man and Environment (2012)   *Pages* 100–110
Editor: Goutam Kumar Saha
Published by: DAYA PUBLISHING HOUSE, NEW DELHI

7

Human-Altered Environment and Plant Community

☆ *Asok Kumar Mukherjee**

Introduction

There is unequivocal evidence that the earth's climate is warming at an unprecedented rate. Temperature increases are geographically inequitable. Some regions, particularly at high altitudes and latitudes, are warming more than other areas. Other climatic effects, including prolonged droughts in arid and semi–arid regions, increased flooding in mid to high latitudes and more extreme weather events are also increasing. Sea levels are rising. Climates are changing more rapidly than species can adapt and there is high risk of mass extinctions of biodiversity as the planet warms.

Ecosystems at Risk

An ecosystem is an array of living things (plants, animals and microbes) and the physical and chemical environment in which they interact. Healthy ecosystems provide the conditions that sustain human life through the provision of a diverse range of ecosystem services. Plant diversity underpins terrestrial ecosystems and they are often described according to the major vegetation type they consist of. Many ecosystems will be highly vulnerable to projected rates and magnitudes of climate change and the services lost through the appearance or fragmentation of ecosystems will be costly or impossible to replace. Forest ecosystems are particularly important, containing as much as two thirds of all known terrestrial species and storing about

* Department of Biotechnology, Bengal College of Engineering and Technology, Durgapur.
  E-mail: asok02@gmail.com

80 per cent of above–ground and 40 per cent of below–ground carbon. Deforestation is a major source of greenhouse gas emissions and contributes to loss of species as well as changes in regional and global climate. Reducing deforestation is therefore one of the most effective ways of reducing greenhouse gas emissions. Ecosystem responses to climate change will be complex and varied. Climatic changes will essentially affect all ecosystem processes but at different rates, magnitudes and directions. Responses will vary from the very short–term response of leaf–level photosynthesis to the long-term changes in storage and turnover of soil carbon and nitrogen stocks. All living organisms in terrestrial ecosystems ultimately depend directly or indirectly on photosynthesis for their energy requirements. Solar radiation, temperature, precipitation, air humidity and atmospheric CO_2 are the key ambient forces that drive ecosystem processes (Sorensen and Takle, 2005). Of these, changes in temperature, water availability and CO_2 levels are subject to change in the next 100 years.

Impact of Temperature Change on Plant Growth and Ecosystems

Plant growth and health may benefit from increased temperatures of global warming in that some regions will experience reduced incidence of damage from freezing and chilling. Plants in other regions may suffer from stress due to elevated temperatures. There is some evidence that extreme events (droughts, floods, high winds, etc.) may accompany global warming, in which circumstances plants may experience isolated highly damaging events. Net primary production (NPP) will generally be increased by moderate increases in temperature estimated to occur in the next 60 years, especially in boreal and mid–latitude regions. Estimates are that NPP will increase 1 per cent per degree centigrade in regions where the mean annual temperature is 30°C and 10 per cent in regions where the mean annual temperature is 0°C.

Global warming is dramatically revamping not only the ice but also tundra and forests at the top of the world, greening some parts and browning others. The alterations could exacerbate climate change. The satellite records revealed a startling result in the vast boreal forests south of and ringing the tundra (Sturm, 2010). Although studies confirmed that the tree line was continuing to move northward and to higher elevations, in many places the satellites indicated that behind this advancing front the forests were losing biomass and becoming less productive. The forests were browning – drying and dying – while the tundra was greening, a fact that seems to contradict the conventional wisdom concerning the forest response to climate warming (Sturm, 2010). Since about a decade gone by, Wilmking *et al.* (2004) started collecting a set of tree ring samples from near Fairbanks and south of the Brooks Range that have helped unravel the apparent contradiction. Instead of the customary positive correlation – higher temperatures in summer produced better growth and wide rings – they began to find stands in which higher temperatures had produced smaller rings and more slowly growing trees. In western Alaska, where it was wetter, they found the trees grew more vigorously as it warmed, but as they moved east into drier country, they discovered smaller rings, distressed trees and struggling, even dying, tree stands. The warmer summers were just too dry.

Dendrochronologists, Lloyd and Bunn (2007), using every boreal tree ring record they could uncover, confirmed that the browning of the boreal forests was a pan-arctic phenomenon and that although it predominated in spruce trees, it occurred in all boreal tree species. The exact causes of the declining tree growth are still being worked out, but drought and heat stress are two primary suspects, because browning has been observed more commonly in dry continental sites and in the southern part of each species' range. The trees have been getting hammered in two other ways as well, both thought to be linked to the warming climate – increased insect outbreaks and a rise in the frequency and size of forest fires. In Alaska, big forest–fire seasons seem to be coming about every five years rather than every 10, and infestations of insects such as the spruce bark beetle, which have ravaged more than 5,00,000 hectares of prime forest in Alaska so far, appear to be intensifying.

The changes taking place on the tundra and in the boreal forests present an ironical symmetry. The boreal forests have encroached on an estimated 11,600 sq km of the southern edge of the Alaskan tundra in 50 years, yet over the same period they have been drying out, burning up and suffering insect damage behind their advancing front. The outcome is going to be a conversion from forest to grassland (Sturm, 2010). At the same time, the tundra is becoming increasingly shrubby and jungle–like. Does the future have in store a switch, where the forest will begin to look a lot like tundra, and the tundra looks more and more like forest?

The problem with answering this question is our limited ability to understand the linked processes that are driving the vegetation changes, let alone predict their future course. Even though the arctic sea ice is a simple system of just water and ice that responds in principle to physical rules that can be coded into models, the ice has been declining at a rate that is twice as fast as that predicted by 13 of the scientific community's best large–scale models. Current predictions are for an ice–free arctic ocean in 40 years, but these predictions are more extrapolations of observed changes than model results. For the tundra, the boreal forests, with their great biological complexity and competing feedback mechanisms – some that dampen growth and some that accelerate it – the existing models are still too simplistic to produce accurate predictions.

Impact of Precipitation and Water Availability on Plant Growth and Ecosystems

Plant leaves have small openings called stomata that can be adjusted to regulate the exchange of water vapour and CO_2 with the atmosphere. Plants not under water stress keep their stomata open for optimum CO_2 exchange. Under stress, however, plants close their stomata to restrict water loss. They also may allow their leaves to droop to reduce light absorption or they may even shed leaves to reduce water loss. C_4 plants have higher water use efficiency (WUE) than C_3 plants. Higher atmospheric CO_2 levels will cause stomata to close slightly, increase WUE, and increase carbon gain for plants with limited water supply. Higher temperatures may lead to higher differences in water–vapour concentration inside and outside the stomata, however, and thereby lead to reduced WUE (Sorensen and Takle, 2005).

The Physiological Responses of Plants to Climate Change

The diversity and distribution of the world's terrestrial vegetation is the product of a complex suite of interactions between individual plants and a multitude of climatic and environmental variables. Plants are major regulators of the global climate, and their collective responses to increased atmospheric CO_2 concentrations have clearly played an important role in mitigating climate change up to this point. In looking to the future, it is increasingly critical to understand how plants respond on a basic level to the changes imposed upon them by continued increases in atmospheric CO_2, as well as the cascade of climatic and environmental changes triggered due to this increase.

Observing and Predicting Plant Responses

Understanding the effects of climate change on plant species and communities is a fairly recent conservation concern, but requires long-term data sets. Few such data sets exist, *viz.* long-term phonological records for a few plant species, but analysis can be hampered because data collection protocols and species selection generally were not set up to answer contemporary questions. Similarly, experimental approaches can be prohibitively expensive and lengthy, so research in this field relies heavily on modeling. Models can be used for predicting responses of single species, multi–species assemblages, global vegetation patterns and climate or hardiness zones. Models are only as good as the data and assumptions on which they are built and are continually improving as we refine and test them using data from past climate changes. While it remains important to scrutinize climate change predictions adequately, the scientific debate must not divert us from taking timely and appropriate action on both mitigation and adaptation. The extent of global change is still in our hands and scientific rigour should not replace action.

It is clear that different plant species will respond differently to climate change. Some species will stay in place and adapt to new climate conditions through selection or plasticity. Other species will move to higher latitudes or altitudes. Some species may become extinct. Because of this, plant community composition will be reorganized, new communities will emerge and others will be lost. One of the biggest concern of this community reshuffling is the disruption of food webs and coevolved mutualisms, such as the relationships between a plant and its pollinator or seed disperser. If species that rely on each other no longer co–occur in the same time or space, both may be driven to extinction. Diseases, pests and invasive species may spread into new ranges putting more pressure on fragile communities. Maintaining biodiverse communities will become an even greater conservation priority.

In an era of rapid climate change, species have three basic alternatives. They can migrate to appropriate environmental conditions, adapt to the new environmental conditions, or become extinct. In a changing environment, 'woody' species with fast generation times and wide ecological tolerances are more likely to adapt or migrate quickly and are more likely to flourish. Conservative species with specific habitat requirements or long generation times are more prone to the threat of extinction. At present an estimated one–quarter of vascular plant species are under threat in the wild.

With predicted temperature increases, changing hydrological cycles and other factors of climate change, as many as half of all plant species may be lost over the next century. This is a catastrophic scenario given the fundamental importance of plants to life on earth. As yet there is a lack of published information on plant extinctions directly due to climate change but with baseline information now being collected on the distribution, threat and ecology of various plant groups, monitoring schemes can be established. Plant species restricted to high–risk habitats, including montane, island or coastal habitats are likely to be the first casualties of climate change. Plant conservation action needs to be increased now to ensure that options are available for the future.

Direct Effects of CO_2

Photosynthetic rates in C_3 plants increase by 25–75 per cent for a doubling of CO_2. For C_4 plants the data are less conclusive and range from no response to an increase of 10–25 per cent. Results likely are temperature dependent. Increases in CO_2 with accompanying increment in photosynthetic rate and decreased water requirement, translate into increased growth and crop yield in C_3 plants, increased growth in C_4 plants, and increased tree seedling growth. The response to elevated CO_2 will be most pronounced in regions where water availability is a limiting factor. The actual growth enhancements expected in response to gradually increasing CO_2 concentrations are likely to have only a small and gradual impact on terrestrial ecosystems globally.

Climate Change, Plants and Livelihoods

The growth rates of terrestrial and aquatic plants are temperature dependent, with species (and genotypes) having optimal growth and competitive ability at particular temperatures, and thus in particular climates. This is likely the greatest cause of the geographic separation of species along continental climatic gradients, such as north–south gradients and elevation gradients. In addition, the geographic ranges and abundance of many terrestrial plants are limited by temperature extremes, especially by tissue damage associated with freezing or subfreezing temperatures. Moreover, within a region, differences in temperature–dependent growth could cause different plant species to be specialized on different portions of the growing season. In essence, plants may be limited by nutrients and other resources, by pathogens and herbivores, by disturbances, by dispersal abilities and by the physical environment including its climate. Plant abundance in both terrestrial and aquatic ecosystems is also limited by the densities and species identities of pathogens and herbivores, which in turn can be limited both by their predators and by dispersal. Thus, top–down forces can greatly constrain both terrestrial and aquatic ecosystems. Physical disturbances also limit terrestrial plant communities and sessile (benthic) freshwater and marine plant communities. For many terrestrial ecosystems, fire frequency has been a major constraint, as have been such physical disturbances as wind storms, landslides, mudslides, avalanches, clearings caused by gophers or other fossorial animals, disturbances caused by hooves, wallows, etc.

Anthropogenic Global Change and Plant Constraints

Many of the constraints are undergoing large, rapid changes because of human actions. Recent human activities have more than doubled the preindustrial rate of supply of nitrogen (N) to terrestrial ecosystems (Vitousek *et al.*, 1997). Nitrogen had a preindustrial terrestrial cycle that involved the annual fixation of about 90 to 140 Tg (teragrams) of N/yr (Vitousek, 1994; Vitousek *et al.*, 1997), with an additional 10 Tg of N/yr provided by atmospheric N fixation via lightening. Industrial N fixation for fertilizer currently totals about 88 Tg/yr. About 20 Tg/yr of N is fixed during the combustion of fossil fuels and about 40 Tg/yr of N is fixed by legume crops. In addition, land clearing, biomass burning and other human activities mobilize and release about an additional 70 Tg of N/yr. The projected expansion of global population to about 9×10^9 people by the year 2050 and shifts to diets higher in animal protein suggest that, by 2050, global food production will be double its current rate (Tilman, 1999). If so, anthropogenic terrestrial N inputs in 2050 would be about 3 to 4 times the pre–industrial rate. Much of this N would enter rivers and be carried to near–shore marine ecosystems. N would also be deposited atmospherically on non-agricultural terrestrial ecosystems.

Nitrate is readily leached from soil, carrying with it positively charged ions such as Calcium (Ca). Atmospheric N deposition may be depleting Ca and other cations in hardwood forests of the eastern United States (Likens *et al.*, 1998). This depletion of base cations could cause elements that had not been limiting in a region to become limiting. Plant species often have distributions constrained by soil pH and Ca. Phosphorus (P) is a commonly applied agricultural fertilizer and current P application is a doubling of the natural global rate for terrestrial ecosystems (Carpenter *et al.*, 1998). Projections to year 2050 are that agricultural P fertilization will be more than double. Much of this P may enter aquatic ecosystems, which can be P–limited. The accumulation of such greenhouse gases as CO_2 and methane may lead to global climate change, with the greatest changes, especially warmer winter temperatures.

Rather, we merely note that rainfall patterns, the frequency and severity of droughts and other aspects of climatic mean and variance, which all constrain plant communities, are also forecast to change. In addition, CO_2 is a plant nutrient, and elevated levels of CO_2 represent atmospheric eutrophication with a limiting plant resource. In total, human actions are modifying many environmental constraints that, in combination with intraspecific and interspecific trade–off, led to the evolution of extant plant species and thus influenced the composition, diversity and functioning of terrestrial and aquatic plant communities. If current trends continue, within 50 to 100 years the suites of factors constraining the structure of many plant communities may fall outside the envelope of values that existed both before the industrial revolution and when many of the plant species evolved.

Dynamics

Like all biological systems, plant communities are temporarily and spatially dynamic as they change at all possible scales. Dynamism in vegetation is defined primarily as changes in species composition and/or vegetation structure (Archibold, 1994).

Temporal Dynamics

Temporally, a large number of processes or events can cause change, but for sake of simplicity they can be categorized roughly as either abrupt or gradual. Abrupt changes are generally referred to as disturbances; these include things like wildfires, high winds, landslides, floods, avalanches etc. Their causes are usually external (exogenous) to the community – they are natural processes, occurring (mostly) independently of the natural processes of the community (such as germination, growth, death, etc). Such events can change vegetation structure and species composition very quickly and for long time periods, and they can do so over large areas. Very few ecosystems are without some type of disturbance as a regular and recurring part of the long-term system dynamic. Fire and wind disturbances are particularly common throughout many vegetation types worldwide. Fire is particularly potent because of its ability to destroy not only living plants, but also the seeds, spores, and living meristems representing the potential next generation, and because of fire's impact on faunal populations, soil characteristics and other ecosystem elements and processes.

Temporal change at a slower pace is ubiquitous; it comprises the field of ecological succession. Succession is the relatively gradual change in structure and taxonomic composition that arises as the vegetation itself modifies various environmental variables over time, including light, water and nutrient levels. These modifications change the suite of species most adapted to grow, survive and reproduce in an area, causing floristic changes. These floristic changes contribute to structural changes that are inherent in plant growth even in the absence of species changes (especially where plants have a large maximum size, *i.e.* trees), causing slow and broadly predictable changes in the vegetation. Succession can be interrupted at any time by disturbance, setting the system either back to a previous state, or off on another trajectory altogether. Because of this, successional processes may or may not lead to some static, final state (Lehman and Tilman, 1997; Tilman and Lehman, 2001). Moreover, accurately predicting the characteristics of such a state, even if it does arise, is not always possible. In short, vegetative communities are subject to many variables that together set limits on the predictability of future conditions.

Spatial Dynamics

As a general rule, the larger an area under consideration, the more likely the vegetation will be heterogeneous across it. Two main factors are at work. First, the temporal dynamics of disturbance and succession are increasingly unlikely to be in synchrony across any area as the size of that area increases. That is, different areas will be at different developmental stages due to different local histories, particularly their times since last major disturbance. This fact interacts with inherent environmental variability (*e.g.* in soils, climate, topography, etc.), which is also a function of area. Environmental variability constrains the suite of species that can occupy a given area, and the two factors together interact to create a mosaic of vegetation conditions across the landscape. Only in agricultural or horticultural systems does vegetation ever approach perfect uniformity. In natural systems, there is always heterogeneity, although its scale and intensity will vary widely. Natural grassland may be homogeneous when compared to the same area of partially burned forest, but highly diverse and heterogeneous when compared to the wheat field next to it.

At regional and global scales there is predictability of certain vegetation characteristics, especially physiognomic ones, which are related to the predictability in certain environmental characteristics. Much of the variation in these global patterns is directly explainable by corresponding patterns of temperature and precipitation (sometimes referred to as the energy and moisture balances). These two factors are highly interactive in their effect on plant growth, and their relationship to each other throughout the year is critical.

Soil Processes and Carbon Dynamics

Temperature changes will have only minimal effects on reaction rates for inorganic processes in soils, but changes in soil moisture could have significant effects on rates of diffusion and supply of nutrients to plants. Both NPP and organic matter decomposition will increase likely under increasing temperature. If moisture is readily available, decomposition of organic matter is likely to be enhanced more than NPP under global warming, thereby adding more CO_2 to the atmosphere. However, if moisture becomes more limiting then decomposition will be reduced. Models that take both temperature and moisture into account suggest that increased NPP would lead to increases in soil carbon under increasing atmospheric CO_2 (Sorensen and Takle, 2005). Land use is a much more important factor than changes in NPP for determining soil carbon. Typically about half of the native carbon is lost from soils when they are put under cultivation over a period of 50–100 years. Minimum tillage practices reduce carbon loss from soils.

Climate change, specifically changes in temperature and water availability, could change soil microbial and faunal populations, but changes in land–use practices are likely to have much greater impact. However, another element of global change, namely increased deposition of nitrogen from industrial NO_X emissions, is being more widely associated with major losses of fungi in the root zone in some (particularly forest) biomes.

Ecological Processes and Community Dynamics

Organisms interact with their physical environment and with other organisms to form a complex set of dependencies and interrelationships sometimes called the "web of life". This interconnectedness makes the study of impacts of changes in external factors on ecoystems very difficult. The combinations of all environmental facts and interactions with other organisms determine the preferred places for each organism to live, *i.e.*, its "niche". Some niches are more vulnerable to climate change than others. Interactions within ecosystems include competition, herbivory and actions of parasites, disease and mutualists (ecosystem components that provide mutual benefit such as pollinating bees and flowering plants).

The collection of different species that interact in a variety of ways in a defined patch of land is called a community. Communities are always changing and are subject to "succession", which may be a complete changeover to another collection of plants or a more incremental series of species losses and gains. Loss of one species may provide opportunities for changes in populations of existing species or gain of new species. Communities may migrate and disperse as their environmental conditions change. The rate of change compared to the ability of the community to move determines whether the community will survive under such changing conditions.

Losses of diversity and shifts in species composition have, at their core, a conceptually simple basis (Tilman, 1990; 1994; 1997). The plant species that coexist in the unfertilized control plots do so for a variety of reasons, including interspecific trade–off in their ability to compete for limiting resources (Tilman, 1982), or trade–off between competitive ability versus local dispersal ability (Hastings, 1980; Tilman, 1994) or a trade–off between competitive ability versus resistance to herbivory or disease (Tilman, 1982; Tilman *et al.*, 1994).

The High Dimensionality of Environmental Change

The greater the dimensionality of a habitat is (*i.e.*, the greater its number of constraints), the more its diversity and composition would be impacted by a given amount of environmental change in each variable. Human actions are changing many environmental constraints simultaneously, including N, P, Ca, CO_2, pH, fire frequency, trophic structure and climate. The high dimensionality of these changes may lead to much greater impacts on plant communities than anticipated from a consideration of only one or a few of these factors.

A simple example illustrates this. Consider a habitat in which there are three constraints, factors 1, 2 and 3. The low and high values of these factors might map into a cubic trait space for competitive coexistence. If the values of factor 1 were shifted up by 50 per cent, but nothing else changed, the old trait space and the new trait space would share 50 per cent of their volume, indicating that this change would eliminate about half of the original species and create vacant niches that could be colonized by a comparable number of species, should they exist regionally. If both factor 1 and 2 were increased 50 per cent, the new trait space would overlap with only 25 per cent of the old (*i.e.*, $1/2 \times 1/2 = 1/4$). If each of the three factors were shifted by $1/2$, new trait space would overlap with only $1/8$ of the original. In this case, $7/8$ of the original species would be driven locally extinct. Comparably, if each of three variables were to be shifted by $2/3$, the resultant trait space would overlap only $1/27$ of its original volume, and $26/27$ of the original species would be lost, on average.

In the short–term, such shifts in environmental constraints would eliminate many species and favour once–rare species. The longer–term dynamics of these terrestrial plant communities would depend on the dispersal rates of species both within a region and from other regions, if any, that formerly had characteristics similar to those that occur in the human–impacted region. They also would depend on the evolutionary responses of the species that remain in these habitats.

Remarks

Anthropogenic changes in environmental limiting factors are likely to cause significant loss of plant diversity, leaving many niches empty and creating plant communities dominated by weedier species (poor competitors but good dispersers). The extent of this effect will depend both on the number of constraints that are changed (*i.e.*, dimensionally) and on the magnitude of such changes (Tilman and Lehman, 2001). Because the impacts of multidimensional environmental changes are expected to be multiplicative, a series of relatively small changes may be as important as a single change.

The vacant niches of a region experiencing a major change in an environmental constraint, such as a high rate of N deposition, indicate several things about such habitats. First, species that have traits and which fall within the newly created vacant niches should be able to invade into, spread through and persist if propagules are regionally available. Secondly, any heritable variation within existing species that allowed individuals to fill the vacant niches would be favoured.

References

Archibold, O.W. 1994. *Ecology of World Vegetation*. Springer Publishing, New York.

Carpenter, S.R., Caraco, N.F., Correll, D.L. *et al.* 1998. Non point pollution of surface waters with phosphorus and nitrogen. *Ecol. Appl.*, 8: 559–568.

Hastings, A. 1980. Disturbance, coexistence, history and competition for space. *Theor. Popul. Biol.*, 18: 363–373.

Lehman, C. and Tilman, D. 1997. Competition in spatial habitats. In. *Spatial Ecology: The Role of Space in Population Dynamics and Interspecific Interactions* (Tilman, D. and Kareiva, P. eds.). Princeton Univ. Press, Princeton. Pp. 185–203.

Likens, G.E., Driscoll, C.T., Buso, D.C. *et al.* 1998. The biogeochemistry of calcium at Hubbard Brook. *Biogeochemistry*, 41: 89–173.

Lloyd, A.H. and Bunn, A.G. 2007. Response of the circumpolar boreal forest to 20th century climate variability. *Environmental Research Letters*, 2(4): 1–13.

Sturm, M. 2010. Arctic Plants Feel the Heat. *Scientific American*, 46–53.

Sorensen, E.K. and Takle, E.S. 2005. Investigating knowledge building dialogues in networked communities of practice. A Collaborative learning endeavor across cultures. *Interactive Educational Multimedia*, 10: 50–60.

Tilman, D. 1982. *Resource Competition and Community Structure. Monographs in Population Biology*. Princeton Univ. Press., Princeton. Pp. 296–310.

Tilman, D. 1990. Constraints and tradeoffs: toward a predictive theory of competition and succession. *Oikos*, 58: 3–15.

Tilman, D. 1994. Competition and biodiversity in spatially structured habitats. *Ecology*, 75:2–16.

Tilman, D. 1997. Community invasibility, recruitment limitation and grassland biodiversity. *Ecology*, 78: 81–92.

Tilman, D., Dodd, M.E., Silvertown, J. *et al.* 1994. The park grass experiment: insights from the most long-term ecological study. In: *Long-term Experiments in Agricultural and Ecological Sciences*, (Leigh, R. A. and Johnston, A. E. eds.). CAB International, Wallingford, Oxon, UK. Pp. 287–303.

Tilman, D. 1999. Global environmental impacts of agricultural expansion: the need for sustainable and efficient practices. *Proc. Natl. Acad. Sci.*, 96: 5995–6000.

Tilman, D. and Lehman, C. 2001. Human–caused environmental change: Impacts on plant diversity and evolution. *Proc. Natl. Acad. Sci.*, 98(10): 5433–5440.

Vitousek, P.M. 1994. Beyond global warming: ecology and global change. *Ecology*, 75: 1861–1876.

Vitousek, P.M., Aber, J.D., Howarth, R.W. *et al.* 1997. Human alteration of the global nitrogen cycle: sources and consequences. *Ecol. Appl.*, 7:737–750.

Wilmking, M., Juday, G. P., Barber, V. A. *et al.* 2004. Recent climate warming forces opposite growth responses of white spruce at treeline in Alaska through temperature thresholds. *Global Change Biology*, 10(10): 1724–1736.

Climate Change: Man and Environment (2012) Pages 111–120
Editor: Goutam Kumar Saha
Published by: DAYA PUBLISHING HOUSE, NEW DELHI

8

Impact of Climate Change on Diversity and Distribution of Fungal Community

☆ *Prakash Pradhan and Krishnendu Acharya**

Introduction

Climate may be thought of as a description of the regularities and extremes in weather for a particular location, which is governed naturally by a set of known variables mainly related to changes in ocean currents or sea–surface temperatures, volcanic eruptions, alterations in the sun's output of energy and many more complex phenomenon (Taylor and Taylor, 1997). However, the industrial revolution and subsequent anthropogenic burden of greenhouse gases has played key role in driving the earth's climate change. Global warming is the most substantial manifestation of climate change, which can be attributed to increased temperature and elevated CO_2 (Parmesan and Yohe, 2003). There has been 0.3°C to 0.6°C rise in global mean temperature since the late 19th century and global sea levels have also risen between 10 and 25 cm (Taylor and Taylor, 1997). The Intergovernmental Panel on Climate Change (IPCC) in 2001 has estimated that the globally averaged surface temperature will increase by between 1.4 and 5.8°C over the period 1990 to 2100.

Recent climate change has resulted in changes in the timing of phenological events in many organisms (Fitter and Fitter, 2002; Root *et al.*, 2003; Menzel *et al.*, 2006), global meta–analysis involving 1700 species have documented significant range shifts averaging 6.1 km per decade towards poles and significant mean advancement

* Department of Botany, University of Calcutta, Kolkata. E-mail: krish_paper@yahoo.com

of spring events by 2.3 days per decade (Parmesan and Yohe, 2003). Although, fungi are among the most diverse, important and omnipresent groups of organisms on earth, relatively little attention has been paid to how fungi and other micro–organisms respond to climate change, therefore related studies are mostly inadequate worldwide (Hawksworth, 1991; Cannon, 1997; Guzman, 1998; Piepenbring, 2007). Along with other microorganisms, fungi not only make conditions suitable for the evolution and existence of macroscopic life forms, but also continue to drive many of the ecological processes like bioremediation, biogeochemical cycle, nutrient recycling, litter decomposition, soil formation, indicating perturbation within the environment and ecosystem maintenance related to anthropogenic activities (Hawksworth, 1995; Guzman, 1998; Hunt, 1999; Pilz and Molina, 2001; Gates *et al.*, 2005).

Mycorrhizal fungi are the key functional components of a forest ecosystem where they form symbiotic associations with the roots of 75–80 per cent of vascular plants, enabling plants towards better nutrient uptake, which is especially crucial under adverse edaphic conditions (Watling, 1997; Brown *et al.*, 2006). Furthermore, fungi (in particular the mushroom–forming species) serve as valuable food sources for numerous invertebrate and vertebrate forest inhabitants (Hawksworth, 1991). Besides that, fungi find utilization in industry, agriculture, medicine (Cowan, 2001), food and textile industries (Keizer, 1998; Pilz and Molina, 2001).

Climate is a complex and crucial factor in fungal growth, fruiting and distribution. Seasonal and spatial distribution of macrofungi, the number of sporocarps (O'Dell *et al.*, 1999) and fungal pathogenesis (Burdon, 1987) are influenced by the temperature and moisture, where high levels of precipitation resulting in high humidity and soil moisture, along with warm temperatures, favour both fungal productivity as well as disease incidents from pathogenic species. The major concern arising from the climate change issue is the impact it may have on fungi and its reciprocal interactions with the environment, which ultimately would pose the questions on the planet's ecosystem and human sustainability.

Fungal Fruiting in Response to Precipitation and Temperature Change

The influences of precipitation and temperature on fungal fruiting have been studied by authors in different parts of the world (Wilkins and Harris, 1946; Mehus, 1986; Dahlberg, 1991; Ohenoja, 1993; O'Dell *et al.*, 1999; Straatsma *et al.*, 2001). Due to difficulties in collecting phenological data pertaining to the ephemeral nature of fungal fructification, most of the studies regarding the effect of climate change on fungi are based on either herbarium data or climate change simulations. Herbarium data gives potential phenological information about the fungi, whereas it also enables us to understand and predict climate–induced ecological changes in the future (Lavoie and Lachance, 2006; Primack *et al.*, 2004).

One of the important effects of climate change on fungi is the change in phenology. Some studies from Norway and the UK have reported shifts in the phenology of fungal fruiting over the last 50 years, as depicted in Figure 8.1 (Kauserud *et al.*, 2008). In UK, an extension of autumnal fruiting season in both directions has been observed

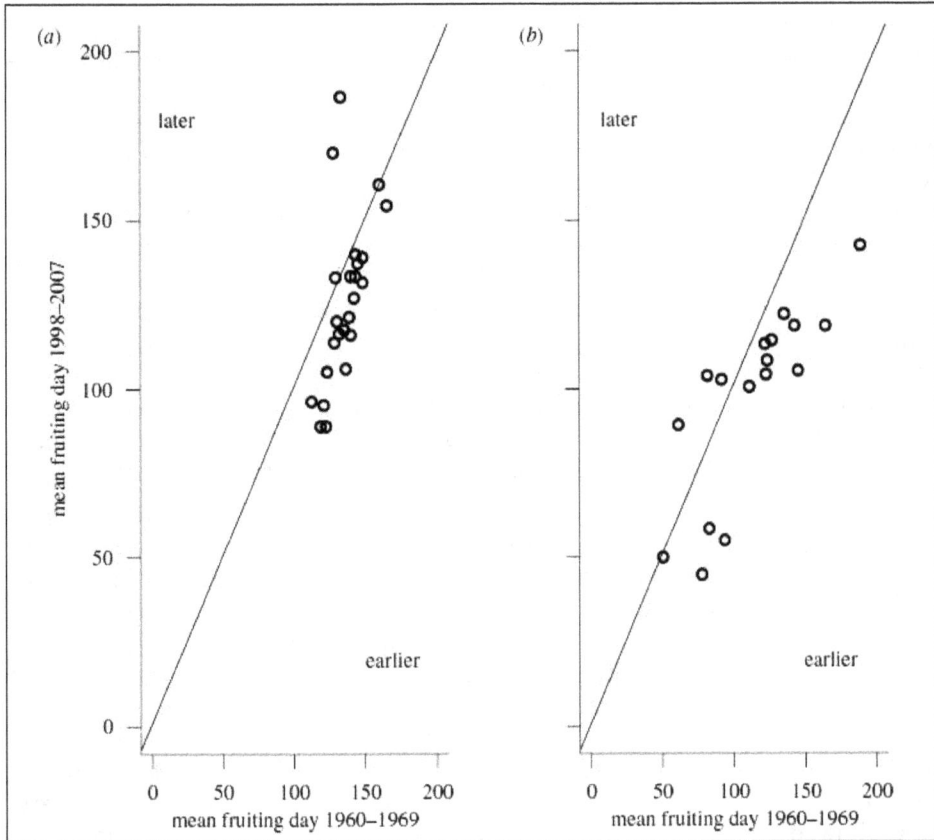

Figure 8.1: The change in mean fruiting date for individual species during period 1960–2007 in (a) Norway and (b) the UK. Points below the diagonal line indicate that fruiting took place earlier during 1988–2007 than during 1960–1969. The vertical distance from the diagonal to each point is the change in mean fruiting day between the two time periods (after Kauserud *et al.*, 2009).

(*i.e.*, earlier onset and later finish) both for mycorrhizal and saprotrophic species (Gange *et al.*, 2007). The extended activity of fungi to winter and spring is viewed by Moore *et al.* (2008) as a result of gain of more thermal energy before fruiting which would help in more rapid nutrient acquisition and therefore helping fungi to fruit earlier in the year. However, Kauserud *et al.* (2009) argue that a warm October may be unfavorable for mycelial growth, leading to a delay in resource acquisition and fruiting.

The time and resources required for mycelial growth and subsequent fruit body production are unknown for the vast majority of fungal species. In particular, the July temperature and precipitation during August is found to promote an earlier average fruiting (Moore *et al.*, 2008). Studying the effects of climatic predictors on the fruiting time during the study period 1960–2007, Kauserud *et al.* (2009) suggested that the higher July and winter temperature were responsible for driving earlier spring fruiting both in Norway and UK. 1°C increase in July temperature caused nearly three days earlier fruiting and the annual linear trend was estimated to be 20.37±0.07, suggesting on average 18 days earlier fruiting over the entire study period and 3.8 days per decade. In Norway, 1°C warming of January climate has been linked to approximately one–day earlier fruiting (average fruiting time normalized on species and geography) whereas in UK the warming of January and February by 1°C amount to approximately three days earlier fruiting (Kauserud *et al.*, 2009). Many species of fleshy fungi that were formerly reported to fruit only in autumn are currently producing sporocarps in spring (Gange *et al.*, 2007; Moore *et al.*, 2008). This indicates that the state of below–ground mycelia of fungi (including the fruiting potential) is determined by climatic conditions over more than a single year. A delayed effect of August precipitation the year before was also found, indicating that a wet summer is associated with earlier fruiting the next season. Thus, in general, the results suggest that a warm and wet summer produce earlier fruiting next year, and in combination with a cooler autumn but warm winter will give early average fruiting dates (Kauserud *et al.*, 2009).

Temperature is found to play an important role in influencing evapo–transpiration (ETP) processes, with rise in temperatures causing higher ETP rates. De Aragón *et al.* (2007), argue that it is the difference between rainfall and ETP which gives better estimation of the water available to plant and fungi than the rain gauge reading. ETP rates higher than the rainfall may pose threat of physiological drought to plant and fungal communities (De Aragón *et al.*, 2007). Temperature not only has direct effects on microbial activity, but eventually it can also affect the temperature dependency of the community. This would result in communities performing better over time in response to increased temperatures. Bárcenas–Moreno *et al.* (2009) have shown that for soil fungi growing in maritime climate with mean soil temperature of 10°C showed optimal growth rates around 30°C, which decreased rapidly with increasing temperature, with no significant fungal growth at 45°C and above, concluding that the environmental temperatures above optimum have greatest effect on temperature response.

The studies of individual climate variables have shown a clear relationship with mushroom production and rainfall. De Aragón *et al.* (2007) through their five year inventory (1997–2001) in the forests of pre–Pyrenees mountains, Spain, have shown that there is positive correlation between total sporocarp production and mean annual rainfall (R^2=0.27; $P<0.001$) and the best climate equation (R^2=0.66; $P<0.001$) for annual sporocarp estimation was based on the difference between monthly mean precipitation and accumulated monthly mean evapo–transpiration for the months of September and October and the monthly mean minimum soil temperature in August. They also observed that timing of rainfall, soil water availability and temperatures in the month of August were influencing overall sporocarp production in Spanish set of conditions. Species richness, like total production is influenced by climate and forest stand characteristics. O'Dell *et al.* (1999) observed that with an increase in mean annual precipitation there was an increase in number of species. Keizer and Arnolds (1994) observed an increase in fungal diversity with an increase in forest stand age. In many mushrooms, fruiting can be induced experimentally after vegetative growth by reducing the temperature by at least 5°C and this might be an important environmental cue that has been delayed because of global warming (Kües and Liu, 2000).

Elevated CO_2 and Ectomycorrhizal Fungi (ECM)

Anthropogenic global carbon dioxide (CO_2) emissions have been accelerating, with their growth rate increasing from 1.1 per cent per year between 1990 and 1999, to more than 3 per cent per year between 2000 and 2004 (Tin, 2008). Increased levels of CO_2 in the atmosphere are known to affect both host plants (Ceulemans *et al.*, 1999; Norby *et al.*, 1999) and mycorrhizal fungi (Hodge, 1996; Fitter *et al.*, 2000). Increased production of mycelium due to elevated CO_2 has been reported for several ECM species (Ineichen *et al.*, 1995; Rouhier and Read, 1998a) as well as for arbuscular mycorrhizal (AM) fungi (Rouhier and Read, 1998b). The increase in fungal biomass is sustained by the extra carbon fixed by and supplied from the host plant under enriched CO_2 regimes (Godbold *et al.*, 1997).

In a study, the changes in the ectomycorrhizal fungal community structure in a 37 year old Norway spruce forest in Sweden after factorial combinations of elevated CO_2 (700 ppm) and balanced nutrient addition during 1998–2000, Fransson *et al.* (2001) noted that there was an increase in the mean value of morphotypes per tree, from 6.4 ± 0.8 in 1997 to 7.1 ± 0.3 in 2000. The level of colonization was also found to increase in 2000 when the mean colonization level was 97.9 per cent ± 1.1 compared to 85.2 per cent ± 4.3 in 1997. Significantly more unidentified ECM morphotypes were found on the elevated CO_2 trees than control, thus increasing the ECM fungal diversity. The shift in community structure was also linked to the change in abundance of a few common morphotypes (Godbold *et al.*, 1997). Rey and Jarvis (1997) indicated the shift in mycorrhizal species composition in birch towards later successional stages after CO_2 treatment. It was interpreted as an acceleration of tree ontogeny leading to the trees supporting ECM fungal species with a higher carbon demand, which was in accordance with the conclusions of Godbold *et al.* (1997).

Under elevated CO_2, trees were found to invest more carbon into fine roots and there was increased production of mycelia contributing to more efficient nutrient acquisition by plants (Rey and Jarvis, 1997). About three times as much carbon is stored in terrestrial vegetation and soil than is stored in the atmosphere (Houghton *et al.*, 1996). Fungal mycelia network, alive or dead, may account for a very large proportion of the soil microbial biomass. There are several ways in which mycorrhizal carbon may enter the soil: (a) hyphal turnover; (b) grazing of the extra–radical hyphae by soil fauna (and subsequent defecation); and (c) passive carbon losses from the mycelial network (similar to leaching and exudations from roots) as suggested by Olsson *et al.* (1999). However, the turnover of mycelium is likely to be faster compared to roots and an increase in the amount of mycelium could lead to a faster turnover of carbon within the plant–fungus system (Fitter *et al.*, 2000).

Climate Change and Distribution of Macrofungi

Kauserud *et al.* (2008) showed that geographic location was an important factor determining the fungal fruiting time. They found that appearance of fruit bodies was considerably earlier (in the range of 10–20 days) in northern, continental and alpine regions of Norway compared with more southern and oceanic regions. This latitudinal pattern correlates well with the general trends known from plant phenology (Ovaska *et al.*, 2005). In Canadian context, Taylor and Taylor (1997) suggested that global warming could set southern fungal species north bound. The retreating glaciers would create new habitats for some alpine species although some may lose habitat at lower elevations from encroaching tree lines. However, loss of species from pioneering communities is predicted if the glaciers melt completely. In forests, where tree and shrub species are affected (such as by replacement of climax forests with young forests through short rotation timber harvest), their associated mycological communities might well also be affected (Taylor and Taylor, 1997).

Remarks

Our knowledge of fungi and their requirements is so limited that it is very difficult to say which, if any, would be negatively affected by climate change. Major oscillations in climates will certainly increase the likelihood of community disruption for species that can only be found under particular climatic conditions, whereas, rise in global temperature would cause the replacement of cold–adapted species by warm–adapted species (Tin, 2008). Increased climatic stress will also increase the susceptibility of trees to facultative or 'opportunist' pathogens like *Armillaria gallica*, which appear to be almost entirely dependent on host stress to impair host resistance before they can infect. High summer temperatures (the preceding year) and winter temperatures were associated with early fruiting in countries like Norway and in the UK, which point towards global warming as an underlying cause for the observed displacements in fruiting time (Kauserud *et al.*, 2009). The magnitude of the response amounts to nearly 1–3 day shift towards earlier fruiting in response to an increase of 1°C during the winter/spring period. Alleviated CO_2 would promote mycelial ramifications in soil and their subsequent carbon sequestration activity would have negative feedback effect on the rising atmospheric CO_2 level.

Most of the disturbances in fungal population and diversity are the result of problems deep–rooted in the fabric of modern society. We now are aware that the lack of stakeholder's attention, simple fragmentation, as well as outright destruction of our natural areas is leading to an ever increasing decline in fungal diversity worldwide (Bunyard *et al.*, 1996). Since range shift is an international issue, trans–boundary multinational management system is required to be set up which will require new administrative structures, new political agreements, jointly implemented research agendas, technology transfer and training. However, the recent exclusion of fungi as a priority for conservation in New Zealand and Australia, despite their importance to biodiversity, ecosystem functioning and humanity, is a serious blow to global fungal conservation initiatives (Buchanan and May, 2003). So, there is a need to amend the approach of policy makers about fungi. Loss of biodiversity is a silent crisis, therefore, urgent issues like inventorisation of risk areas and studies related to synergy of changing climate and other factors upon mycological realm needs to be conducted for arriving at a definite conclusion.

References

Bárcenas–Moreno, G., Gómez–Brandónw, M., Rouskz, J. *et al.* 2009. Adaptation of soil microbial communities to temperature: comparison of fungi and bacteria in a laboratory experiment. *Glob. Change Biol.*, 15: 2950–2957.

Brown, N., Bhagwat, S. and Watkinson, S. 2006. Macrofungal diversity in fragmented and disturbed forests of the Western Ghats of India. *J. Appl. Ecol.*, 43(1): 11–17.

Buchanan, P.K. and May, T.W. 2003. Conservation of New Zealand and Australian fungi. *New Zealand J. Bot.*, 41: 407–421.

Bunyard, B.A., Nicholson, M.S. and Royse, D.J. 1996. Phylogeny of the genus *Agaricus* inferred from restriction analysis of enzymatically amplified ribosomal DNA. *Fungal Genet. Biol.*, 20: 243–253.

Burdon, J. 1987. *Diseases and Plant Population Biology*. Cambridge University Press, Cambridge.

Cannon, P.F. 1997. Strategies for rapid assessment of fungal diversity. *Biodiv. Conserv.*, 6: 669–680.

Ceulemans, R., Janssens, I.A. and Jach, M.E. 1999. Effects of CO_2 enrichment on trees and forests: Lessons to be learned in view of future ecosystem studies. *Ann. Bot.*, 84: 577–590.

Cowan, A. 2001. Fungi – Life Support for Ecosystems. *Essential ARB*, 4: 1–5.

Dahlberg, A. 1991. *Dynamics of ectomycorrhizal fungi in a Swedish coniferous forest: a five year survey of epigeous sporocarps*. Swedish Univ. Agricultural Sciences, Dept. Forest Mycology and Pathology: Uppsala. Pp. 23.

De Aragón, J.M., Bonet, J.A., Fischer, C.R. *et al.* 2007. Productivity of ectomycorrhizal and selected edible saprotrophic fungi in pine forests of the pre–Pyrenees Mountains, Spain: Predictive equations for forest management of mycological resources. *Forest Ecol. Manag.*, 252: 239–256.

Fitter, A.H. and Fitter, R.S.R. 2002. Rapid changes in flowering time in British plants. *Science*, 296: 1689–1691.

Fitter, A.H., Heinemeyer, A. and Staddon, P.L. 2000. The impact of elevated CO_2 and global climate change on arbuscular mycorrhizas: A mycocentric approach. *New Phytol.*, 147: 179–187.

Fransson, P.M.A., Taylor, A.F.S. and Finlay, R.D. 2001. Elevated atmospheric CO_2 alters root symbiont community structure in forest trees. *New Phytol.*, 152(3): 431–442.

Gange, A.C., Gange, E.G., Sparks, T.H. *et al.*, 2007. Rapid and recent changes in fungal fruiting patterns. *Science*, 316: 71.

Gates, G.M., Ratkowsky, D.A. and Grove, S.J. 2005. A comparison of macrofungi in young silvicultural regeneration and mature forest at the Warra LTER siet in the southern forests of Tasmania. *Tasfor.*, 16: 127.

Godbold, D.L., Berntson, G.M. and Bazzaz, F.A. 1997. Growth and mycorrhizal colonization of three North American tree species under elevated atmospheric CO_2. *New Phytol.*, 137: 433–440.

Guzman, G. 1998. Inventoring the fungi of Mexico. *Biodiv. Conserv.*, 7: 369–384.

Hawksworth, D.L. 1991. The fungal dimension of biodiversity: Magnitude, significance, and conservation. *Mycol. Res.*, 6: 641–655.

Hawksworth, D.L. 1995. Challenges in mycology. *Mycol. Res.*, 99: 127–128.

Hodge, A. 1996. Impact of elevated CO_2 on mycorrhizal associations and implications for plant growth. *Biol. Fertility Soils.*, 23: 388–398.

Houghton, J.T., Meira Filho, L.G., Callander, B.A. *et al.* (eds.). 1996. *Climate change 1995: The science of climate change.* Cambridge University Press, Cambridge, UK.

Hunt, G.A. 1999. Assessing macrofungi of special concern for conservation in forested ecosystems. *Proc. Biology and Management of Species and Habitats at Risk.* University College of the Cariboo, Kamloops. 2: 779.

Ineichen, K., Wiemkenm, V. and Wiemken, A. 1995. Shoots, roots and ectomycorrhiza formation of Pine–seedlings at elevated atmospheric carbon dioxide. *Plant Cell Environ.*, 18: 703–707.

Intergovernmental Panel on Climate Change (IPCC). 2001. *Climate Change 2001: The Scientific Basis.* (Houghton, J.T., Ding, Y., Griggs, D.J. *et al.* eds.) Cambridge University Press, Cambridge, UK. Pp. 881.

Kauserud, H., Heegaard, E., Semenov, M.A., *et al.* 2009. Climate change and spring–fruiting fungi. *Proc. Royal Soc. B.*, 1–9.

Kauserud, H., Stige, L.C., Vik, J.O. *et al.* 2008. Mushroom fruiting and climate change. *Proc. Natl. Acad. Sci.*, USA, 105: 3811–3814.

Keizer, G.J. 1998. *The complete encyclopedia of mushrooms.* Rebo publishers, Netherland. Pp. 268.

Keizer, P.J. and Arnolds, E. 1994. Succession of ectomycorrhizal fungi in roadside verges planted with common oak (*Quercus robur* L.) in Drenthe, The Netherlands. *Mycorrhiza*, 4: 147–159.

Kües, U. and Liu, Y. 2000. Fruiting body production in basidiomycetes. *Appl. Microbiol. Biotechnol.*, 54: 141–152.

Lavoie, C. and Lachance, D. 2006. A new herbarium–based method for reconstructing the phenology of plant species across large areas. *Am. J. Bot.*, 93: 512–516.

Mehus, H. 1986. Fruit body production of macrofungi in some North Norwegian forest types. *Nord. J. Bot.*, 6(5): 679–702.

Menzel, A., Spark, T.H., Estrella, N. *et al.* 2006. European phenological response to climate change matches the warming pattern. *Glob. Change Biol.*, 12: 1969–1976.

Moore, D., Gange, A.C., Gange, E.G. *et al.* 2008. Fruit bodies: their production and development in relation to environment. In. *Ecology of saprotrophic basidiomycetes*

(Boddy, L., Frankland, J.C. and Van West, P. eds.). Elsevier Science, Amsterdam, The Netherlands. Pp. 79–102.

Norby, R.J., Wullschleger, S.D., Gunderson, C.A. *et al.* 1999. Tree responses to rising CO_2 in field experiments: Implications for the future forest. *Plant, Cell and Environment*, 22: 683–714.

O'Dell, T.E., Ammirati, J.F. and Schreiner, E.G. 1999. Species richness and abundance of ectomycorrhizal basidiomycete sporocarps on a moisture gradient in the *Tsuga heterophylla* zone. *Can. J. Bot.*, 77: 1699–1711.

Ohenoja, E. 1993. Effect of weather conditions on the larger fungi at different forest sites in Northern Finland in 1976–1988. *Scientiae Rerum Naturalium*, 243: 11–69.

Olsson, P.A., Thingstrup, I., Jakobsen, I. *et al.* 1999. Estimation of the biomass of arbuscular mycorrhizal fungi in a linseed field. *Soil Biol. Biochem.*, 31: 1879–1887.

Ovaska, J.A., Nilsen, J., Wielgolaski, F.E. *et al.* 2005. Phenology and performance of mountain birch provenances in transplant gardens, latitudinal, altitudinal and oceanity–continentality gradients. In: *Plant Ecology, Herbivory and Human Impact in Nordic Mountain Birch Forests, Ecol. Stud.* (Wielgolaski, F.E. ed.) 180: 99–115.

Parmesan, C. and Yohe, G. 2003. A globally coherent fingerprint of climate change across natural systems. *Nature*, 421: 37–42.

Piepenbring, M. 2007. Inventoring the fungi of Panama. *Biodiv. Conserv.*,16(1): 73–84.

Pilz, D. and Molina, R. 2001. Commercial harvests of Edible Mushrooms from the forests of the Pacific Northwest United States: Issues, Management and Monitoring for sustainability. *Forest Ecology and Management*, 1–14.

Primack, D., Imbres, C., Primack, R.B. *et al.* 2004. Herbarium specimens demonstrate earlier flowering in response to warming in Boston. *Am. J. Bot.*, 91: 1260–1264.

Rey, A. and Jarvis, P.G. 1997. Growth response of young birch trees (*Betula pendula* Roth.) after four and a half years of CO_2 exposure. *Ann. Bot.*, 80: 809–816.

Root, T.L., Price, J.T., Hall, K.R. *et al.* 2003. Fingerprints of global warming on wild animals and plants. *Nature*, 421: 57–60.

Rouhier, H. and Read, D.J. 1998a. Plant and fungal responses to elevated atmospheric carbon dioxide in mycorrhizal seedlings of *Pinus sylvestris*. *Environ. Ex. Bot.*, 40: 237–246.

Rouhier, H. and Read, D.J. 1998b. The role of mycorrhiza in determining the response of *Plantago lanceolata* to CO_2 enrichment. *New Phytol.*, 139: 367–373.

Straatsma, G., Ayer, F. and Egli, S. 2001. Species richness, abundance and phenology of fungal fruit bodies over 21 years in a Swiss forest plot. *Mycol. Res.*, 105(5): 515–523.

Taylor, E. and Taylor, B. (eds.) 1997. Responding to global climate change in British Columbia and Yukon. In. *Contribution to the Canada country study: climate impacts and adaptation.* Simon Fraser University, Canada. Pp. 363.

Tin, T. 2008. Climate change: faster, stronger, sooner. *A European update of climate science*. WWF. Pp. 8.

Watling, R. 1997. Pulling the threads together: habitat diversity. *Biodiv. Conserv.*, 6: 753–763.

Wilkins, W.H. and Harris, G.C. 1946. The ecology of the larger fungi. V. An investigation into the influence of rainfall and temperature on the seasonal production of fungi in a beechwood and a pinewood. *Ann. Appl. Biol.*, 33: 179–188.

Climate Change: Man and Environment (2012)   *Pages* 121–137
Editor: **Goutam Kumar Saha**
Published by: DAYA PUBLISHING HOUSE, NEW DELHI

9

Climate Changes and Crop Protection: An Overview

☆ *Ashim Chowdhury[1], Wasim Akhtar[1],*
Sujay Saha[2] and Ruchi Garg[2]*

Introduction

With the onward march of the dial hour, as human civilization makes its existence felt in the twenty first century, there has been a paradigm shift in the arena of population, demand for food and food production technologies. Human activities, driven by demographic, economic, technological and social changes do have a major effect on the interactive global systems constituting of land, water, vegetation and climate. Global population on the rise can be visualized by the changes in the land use pattern and rapid urbanization. Shrinkage in the land resources meant for agriculture is of a major concern. Globally, some 25 crops stand between the rapidly expanding world population and starvation (Wittwer, 1995). As it is evident that atmospheric concentration of greenhouse gases is on the rise due to anthropogenic intervention, they will have a profound effect on climate change. Uncertainly of climatological factors will further accentuate the challenges of increasing sustainable agricultural production. Climate changes will have many facets (Mahlman, 1997) – except accumulation of greenhouse gases, changes in other radioactively active substances and increased cloudiness caused by greater evaporation in a warmer climate may offset some of the greenhouse effect. The expected rate of increase of the temperature

1 Institute of Agricultural Sciences, University of Calcutta, Kolkata.

2 Institute of Vegetable Research, Varanasi.

* E-mail: ashimchowdhury@gmail.com; sujoyta@gmail.com

of the earth's surface is $0°$ C per decade (IPCC, 2001) while the global mean concentration of water vapour in the lower troposphere would increase at the rate of 6 per cent per $1°C$ warming, which in turn would lead to an increase in global precipitation of 0.5 per cent/$°C$ warming (Mahlman, 1997).

Farmers differ in their opinion regarding climate changes, pest and crop protection, but they unanimously agree to the fact that there will be an increase in insect population, fungal disease and weeds. Hence the risk of loss due to these biotic factors is likely to increase. The range of many insects will change or expand and new combination of disease and pests may emerge as natural ecosystems respond to shift in temperature and precipitation profiles. The effect of climate on pest may add to the effect of other factors such as the over use of pesticides and the loss of biodiversity, which contribute to plant pest and disease outbreak. Constraints from diseases, insects or weeds have not been considered to any significant extent in any impact assessment of natural or managed ecosystems despite recognition of their significance (Clifford *et al.*, 1996; Rosenzweig and Hillel, 1998). For instance, a doubling of CO_2 has been conclusively shown to increase C3 and C4 crop yield by about 33 per cent and 10 per cent respectively in over 1000 studies (Kimball, 1985; Cure and Acock, 1986), but whether these benefits will be realized in the presence of pest, disease and weeds is yet to be inferred.

Impact on Diseases

Plant diseases are significant constraint to global productivity of crops and worldwide losses from diseases range from 9 to 16 per cent in wheat, rice, barley, maize, potato, soybean and coffee. Only in USA alone, fungicides worth over 5 billion US $ are used to control disease (Oerke *et al,* 1994). The economic impact of disease stems from losses in productivity, the cost of disease management and the economic penalty paid for having to grow less profitable alternative crops. The Irish potato late blight (1845–46) and the great Bengal famine (1943) are grim reminders of the fact that the socio political repercussions of major epidemics go far beyond simple economic impact (Padmanabhan, 1973). In the recent times, diseases such as Panama wilt have resulted in the abandonment of entire banana plantation in central America (Chakraborty *et al.*, 2000).

The classic disease triangle emphasize the role of physical environment in the plant diseases as no virulent pathogen can induce disease on a highly susceptible host if weather conditions are not favourable. Weather influences all stages of host and pathogen life cycles as well as the development of disease. Correlation between weather and disease are routinely used for forecasting and managing epidemics and disease severity over a number of years and can fluctuate according to variations in climate (Coakley, 1979; Scherm and Yang, 1995)

Most research on how climate change influence plant diseases has concentrated on the effects of a single atmospheric constituent or meteorological variable on the host, pathogen or the interaction of two under controlled conditions (Chakrobarty, 1997; Coakley, 1988; 1995; Magan and Baxter, 1996; Manning and Tiedemann, 1995). In a dynamic environment, interactions are more complex as multiple biological and climatological factors are involved. Climate change has potential to modify host physiology and resistance, alter stages and rates of development of the pathogen, shift in the geographical distribution of host and pathogen, change the physiology of host–pathogen interaction and changes in the management strategies.

Geographical Distribution of Host and Pathogen

New disease complexes may arise and some disease will lose their economic importance if warming causes a poleward shift in agroclimatic zone and host plants migrate into new regions. Carter *et al.* (1996) predicted that maize could be cultivated in southern Finland by 2050 and warming will extend the northern limit for cereal cultivation. At the same time risk of damage from late blight of potato would increase in all regions. Pathogens would follow the migrating hosts and their dispersal and survival between seasons and changes in the host physiology and ecology in the new environment would largely determine how rapidly the pathogens establish in the new environment. Facultative parasites with broad host range would mostly fall in this category, although obligate parasites may also expand their host range to infect plant in their proximity (Eshed and Dinoor, 1981; Savile and Urban, 1982). If only climatic effects are considered, changes in geographical distribution itself may not have a major economic impact on crop production. Impacts only become obvious when associated changes such as terrain, remnant, vegetation, soil characteristics etc. are considered.

Using three temperature change scenarios and a 20 per cent increase in rainfall, Brasier and Scott (1994) predicted that global warming will make the pathogen, *Phytopthora cinnamomi* more severe in regions of Europe where it is currently present and spread it in northward and eastward direction. Temperature and rainfall were considered in simulation studies to predict the impact of climate change on rice blast caused by *Pyricularia oryzae* (Luo *et al.*, 1995). It was found that rainfall had no signficant effect on blast epidemics but changes in temperature had significant effect on five Asian rice–growing countries, included in the simulations. However, there was variation of impact with the agro–ecological zone. In cool subtropical rice growing zones, such as Japan there was higher risk of rice blast disease where as in the humid tropics and subtropics such as Phillippines there was lower risk of disease with increased temperature.

Pathogenic genera with broad host range such as *Rhizoctonia*, *Sclerotium*, *Sclerotinia* etc. will have a tendency to migrate from agricultural crops to natural plant communities. Similarly, pathogens, which are normally less aggressive in natural plant community, could devastate crop monocultures growing in close proximity. Pathogens may extend their host range to cause new disease problems in migrating crops. For example, *Erwinia amylovora*, the fire blight bacterium was pathogenic to some indigenous plant of the family Rosaceae without causing much loss but when European settlers grew apples and pears in some region, the bacterium caused severe losses. Similarly, the coffee rust disease was facilitated by growing an introduced susceptible host *Coffea arabica* in a region where the native pathogen, *Hemileia vastatrix* was already present on alternative hosts in the forests outlining the coffee plantation (Carefoot and Sprott, 1967). If the frostline moves north in he northern hemisphere, higher winter temperatures could be accompanied by increased survival of insects

(Porter *et al.*, 1991; Sutherst *et al.*, 1995). For virus–vector aphids, this could lead to higher incidence of virus disease, especially in that region where the timing of virus arrival is linked to winter survival and spring flight of aphids (Harrington *et al.*, 1995). Barley yellow dwarf polyvirus (BYDV) and several viruses of potato and sugar beet show significant increase with warmer winters (Carter and Harrington, 1991; Mackerron *et al.*, 1993; Thomas, 1989).

Physiology of Host-Pathogen Interaction

Increased CO_2

Recent findings on the physiological, biochemical and molecular mechanisms of host–pathogen interaction under high CO_2 and/or other elements of global change are very encouraging, although reduced stomatal opening, increases in leaf area and duration, leaf thickness, branching, tillering, stem and root length and dry weight are well known effects of increased CO_2 on many plants (Mc Elrone *et al.*, 2005; Hibberd *et al.*, 1996; Titone *et al.*, 2008; Wolfe, 1995; Bowes, 1993). Faster development of certain diseases under elevated CO_2 was reported in the early 1930. In a comprehensive review of literature covering the period 1970 to 1993, Manning and Tiedemann (1995) concluded that while not much is known about the influence of elevated CO_2 on plant diseases, potential impact on disease could be predicted from known effects of CO_2 on host plants and their pathogens. They suggested that elevated CO_2 would produce increased plant size and canopy density with high nutritional quantity foliage and microclimates more conducive to development of rust, mildews, leaf spots and blights. Necrotrophic pathogens, which will overwinter under warmer conditions, may increase in their severity as a result of enhanced inoculums survival on the greater amount of crop residues.

More recently a mixed report of enhancement and reduction of plant disease severity under elevated CO_2 have brought to the fore two important facts (Lupton *et al.*, 1995; Thompson *et al.*, 1993). Firstly, the initial establishment of the pathogen may be delayed because of modification in pathogen aggressiveness and/or host susceptibility. *Colletotrichum gloeosporioides* showed delayed or reduced conidial germination, germ tube growth and appressorium production when inoculated onto susceptible *Stylosanthes scabra* plant under increased CO_2 (Chakraborty *et al.*, 2000). In barley, infection of *Erysiphe graminis hordei* caused larger but delayed reduction in host growth at elevated CO_2. In these examples, host resistance may have increased because of changes in host morphology, physiology, nutrient and water balance. A decrease in stomatal density (Bowes, 1993; Wolfe, 1995) increases resistance to pathogen that penetrates through stomata. In barley, although the thickness of epicuticular wax did not play a role in resistance to *E. graminis hordei*, plants in elevated CO_2 were able to mobilize assimilates into defense structures including the formation of papillae and accumulation of silicon at sites of apppressorial penetration (Hibberd *et al.*, 1996).

Secondly, an increase in the fecundity of the pathogen under elevated CO_2 occurs (Klironomos, 1997). Established colonies of *E. graminis hordei* (Hibberd *et al.*, 1996) and *C. gloesporioides* (Chakraborty *et al.*, 2000; Lupton *et al.*, 1995) grew faster when the CO_2 concentration is doubled. However, the latent period was extended under high CO_2 conditions, in all pathogen studies so far, because of delay and/or reduction in spore germination and initial establishment of the pathogen. For pathogens with broad genetic diversity, increased population size and the number of generation in conducive microclimate may lead to the development and proliferation of well adopted and possibly more destructive sub population (Sutherst *et al.*, 1996).

Increased Temperature

Elevation in temperature and duration of exposure can modify host physiology and resistance (Wolfe, 1995). An increase of yield by 10 –15 per cent in wheat and soyabean and 8 per cent in rice and maize have been reported by an increase of 2°C temperature coupled with ± 20 per cent precipitation and doubling of CO_2 (Rosenzweig and Parry, 1994). However, global yields of all four crops were reduced with 4°C increase indicating a threshold of compensation for the direct effect of CO_2.

Information regarding heat induced susceptibility and temperature sensitive genes are available (Dyck and Johnson, 1983; Gerechter–Amitai *et al.*, 1984; Sanden and Moore, 1978). In some forage species, at higher temperature lignification of cell walls increases (Wilson *et al.*, 1991) and this can enhance the level of resistance to pathogen (Strange, 1993). On the contrary, an increase in temperature above 20°C can inactive pg3 and pg4 genes conferring resistance to oats against stem rust caused by *Puccinia graminis avenae* (Marten *et al.*, 1967). Severe drought condition might cause severe diseases by *Armillaria* spp. that are normally not very pathogenic (Rishbeth, 1991).

Increased Ozone and Ultraviolet B

The effect of elevated ozone on plant metabolism, crop yield and productivity (Laurence and Weinstein, 1981; Heagle, 1989) plant health (Sandermann, 1996), host–pathogen interaction (Heath, 1980) and host defence mechanism have been extensively studied. Of the 49 bacterial and fungal pathogens examined, exposure to elevated ozone concentration enhanced disease in 25, reduced 14 and did not affect ten (Manning and Tiedemann, 1995).

Ultraviolet B (UV–B) also cast a major effect on crop and growth and life cycle of pathogens (Krupa, 1997). The effect of UV–B radiation on diseases is mainly through altered host physiology and morphology. Continued exposure to enhanced UV–B radiation lowers the level of antifungal compound in foliar plant parts (Asthana *et al.*, 1993).

UV–B has been reported to reduce tolerance of rice to blast caused by *Pyricularia grisea* and although higher UV–B reduced plant biomass and leaf area, there was no increase in blast severity (Finckh *et al.*, 1995). Again, out of seventeen reported instances between 1970 and 1993, enhanced UV–B increased diseases in ten, reduced five and had no effect in two (Manning and Tiedemann, 1995).

Disease Management

As the information level regarding the impact of climate change on plant pathosystem in scanty, it is difficult to assess implication for disease management. The management strategies may require restructuring under the changed context. It may be assumed that the effect would occur chiefly through influence on host resistance or chemical and biological control agent.

Host Resistance

Cultivar resistance to pathogen may become more effective because of increased static and dynamic defenses from changes in physiology, nutritional status and water availability. Durability of resistance may be jeopardized, if the number of infection cycle escalate within a growing season, due to either increased fecundity, more pathogen generation per season or more suitable microclimate for disease development. This may lead to more rapid evolution of aggressive pathogen races. *Colletotricum gloeosporioides* exhibited a gradual increase in fecundity under elevated CO_2 conditions (Chakraborty *et al.*, 2000).

Chemical Control

Fungicides will be affected by the climate change in either of the two ways. Firstly, changes in crop plants, morphologically or physiologically due to increased CO_2 could affect uptake, translocation and metabolism of fungicides. Due to increased thickness of the epicutiular wax layer on leaves (Bowes, 1993; Wolfe, 1995), caused by increased CO_2, uptake of fungicides will be slow and/or reduced by the plants. Increased canopy size will reduce spray coverage, and dilute the active ingredient in the host tissues. On the contrary, increased metabolic rates due to higher temperatures would result in faster uptake by and greater toxicity to the target organism (Coakley and Scherm, 1999). Similar observation were reported in herbicides, chlorotoluron (Edis *et al.*, 1996).

Secondly, the dynamics of fungicides residues may be changed due to changes in temperature and precipitation. Increase in the frequency of intense rainfall (Fowler and Hennessy, 1995) will lead to increased wash off of fungicides, and hence reduced control. Again for certain fungicides, precipitation following application may result in enhanced disease control because of a redistribution of the active ingredient on the foliage (Schepers, 1996).

Biological Control

One of the major problems with application of biological control for plant disease management in the field has been the vulnerability of biocontrol agent population to environmental variations and extremes (Wong *et al.*, 2002). If appropriate temperature and moisture are not consistently available, biocontrol agent population may not recover as rapidly as pathogen population when conducive conditions recur. Changes in temperature may have highly non-linear effect on tritrophic interaction of host, pathogen and biocontrol agent as observed in wheat.

Quarantine and Exclusion

Climate change will put extra pressure on agencies responsible for exclusion as a plant disease control strategy. In some regions, certain diseases of economic importance do not occur because the climatic conditions are not congenial for the pathogen. Use of Geographical Information System (GIS) and climate matching tools may assist quarantine agencies in determining the threat posed by a given pathogen under current and future climates.

Impact Models

In order to understand the complex interaction of atmospheric, climatic and biological factors with technological and socioeconomic changes, quantitative or modeling approaches are more pertinent for impact assessment (Coakley and Scherm, 1996). Guidelines for such model based assessments are the need of the day and the preliminary frame work of the same has been reported (Sutherst *et al.*, 1996; Teng and Yang, 1993). Modelling approaches are based on climate matching which involves the calculation of a "match index" to quantify the similarity in climate between two locations. The match index is based on the two variables such as monthly minimum and maximum temperatures, precipitation and evaporation. Several software packages namely BIOCLIM (Busby, 1991), CLIMEX (Sutherst and Maywald, 1985), HABITAT (Walker and Cocks, 1991), WORLD (Booth, 1996) etc. often come with additional useful features such as internal algorithms for generating "climatic surfaces" through interpolation between locations. Brasier and Scott (1994) used CLIMEX program to map region in Europe favourable or unfavourable for *Phytophthora cinnamomi*, a soil born oomycete, responsible for oak decline disease. Empirical models have been used to study the effect of climate change on diseases of rice and wheat (Coakley *et al.*, 1998) where as population models have divulged considerable information on late blight of potato (Kaukoranta, 1996). Simulation models have been used extensively to predict yield of various crops in different agroecological zones under climate change (Goudriaan, 1996; Riha *et al.*, 1996). Luo *et al.* (1995; 1998a; 1998b) have used simulation models to study the combined effects on yield of increased temperature, elevated UV–B radiation and rice blast disease.

Impact on Weeds

Effect of Increased CO_2

Weed flora has a genetic diversity, which exceeds that of the cultivated crops. If a resource component (light, water, nutrients or carbon dioxide) is altered within the environment, it is more likely that weeds will show a greater growth and reproductive response. Weed species have C3 and C4 photosynthetic pathways, the later being more than the former. So, weed species having the C4 photosynthetic pathway will show a smaller response to atmospheric CO_2 relative to C3 crops. On the contrary, it can be opined that the range of available C3 and C4 weeds present in any agronomic environment is usually not taken into consideration in assessing these changes. It has been reported that the number of obnoxious weed species associated is almost ten times than that of cultivated crops (Bridges, 1992). Hence, if a C4 weed species does not respond, it is likely that a C3 weed species will. In addition, many growers experienced that the worst weeds for a given crop are almost homologous in growth habit or photosynthetic pathway to the main crop. They are often the same uncultivated or "wild" species of the crop, *e.g.* oat and wild oat, sorghum and shatter cane, rice and red rice etc.

In a weed–crop competition study, where the photosynthetic pathway is the same, with the increase in CO_2 concentration in the environment, usually weed growth is favoured. In addition to agronomic weeds, there is an additional category of plants that are considered "noxious" or "invasive" weeds. These are usually non-native plants, whose introduction results in widespread economic or environmental consequences (*e.g.* water hyacinth). Most of these weeds reproduce by vegetatively *i.e.* by roots, stolons, etc. and recent studies reveal that as a group, these weeds may show a strong response to increases in atmospheric CO_2 (Ziska and George, 2004). The mechanism is however, yet to be clearly understood about the role played by the escalating concentration of CO_2 in the *in situ* proliferation of these weeds.

The data that are available on the response of weeds and alterations in weed ecology due to the changing climate are limited. Information regarding the weed interaction with other abiotic factors like temperature, nutrient availability and precipitation is wanting and needs to be generated. Effect of other biotic factors like fungi, nematodes and bacteria on the weed flora in the changing environment needs to be assessed.

Effect of Increased Temperature

Escalation in temperature implies an expansion or bio–invasion of weeds into higher latitudes or higher altitudes. Very aggressive weeds that are currently found in the warm equatorial regions are limited in the temperate regions by virtue of low temperature. In other words, temperature forms an ecological barrier in the spread of the weed species. For example, many C4 grass weeds are serious problems in the southern U.S. but their incidence is not reported at problem levels in the U.S. Corn Belt. Studies have shown that itch grass, a profusely tillering, robust grass weed could invade the central Midwest and California with only a slight warming trend (Patterson, 1995). A root parasite of corn, witch weed, which is limited to the coastal plain of north and south Carolina is speculated to be established in the Corn Belt with disastrous consequences with an increase of temperature. The current distribution of both Japanese honeysuckle and kudzu is limited by low winter temperatures but increase in temperature could extend their northern limits by several hundred miles.

Effect of Changing Precipitation

Both excess and scanty rainfall can predispose the plants to diseases and pest attack. Response to drought in agronomic conditions is dependent on species and cultural conditions. Any factor which increases environmental stress on crops may make them more vulnerable to attack by insects and plant pathogens and less competitive with weeds (Patterson, 1995).

Controlling of CO_2, Climate and Weeds

Chemical management of weeds *i.e.* herbicide intervention will have a direct effect if there is any alteration in the environmental factors. Changes in temperature, wind speed, soil moisture and atmospheric humidity can influence the effectiveness of herbicide applications. Thicker cuticle development or increased leaf pubescence due to drought conditions can result in reduced herbicide entry into the leaf. These same variables can also interfere with crop growth and recovery following herbicide application. Herbicides will be most effective when they are applied to plants at an active growth stage *i.e.* their metabolic activity is very high. This is only possible when the plants are free from environmental stress.

There are several reports that highlight that the efficacy of herbicides is greatly reduced with the rising CO_2 (Ziska and George, 2004, Ziska and Caulfield, 2000). The mechanism for this reduction is yet to be deciphered. A greater root to shoot ratio and subsequent dilution effect of glyphosate was observed in case of Canada thistle when grown in field at elevated CO_2 conditions (Ziska and George, 2004) However, ubiquitousness of the response is yet to be defined. The observed phenomenon that CO_2 does reduce herbicide efficacy tantamount to additional work that is to be done to revise herbicide specificity, concentration and application rates as possible means of adaptation.

Biological control of weeds by natural or manipulated means is likely to be affected by increasing atmospheric CO_2 and climatic change. Climate, as well as CO_2, could alter the efficacy of weed bio–control agents by potentially altering the development, morphology and reproduction of the target weed. Changes in the C:N ratio and modified feeding habit and growth rate of herbivores are directly affected by CO_2. As per the reports of Patterson (1995), warming could also result in increased overwintering of insect populations and changes in their potential range. Although this could increase both the biological control of some weeds, it could also increase the incidence of specific crop pests, with subsequent indirect effects on crop–weed competition. Synchronization between development and reproduction of bio–control agents and their selected targets is a mandate for successful biocontrol. In a changed environmental condition of climatic change and climatic extremes, this synchrony is difficult to maintain. Whether it will be a boon or a bane needs to be seen.

The most popular and accepted means of controlling weeds in developing countries is mechanical removal. Tillage either by animals or mechanical means is regarded as a global method of weed control in agronomic systems. Below ground carbon storage with subsequent increases in the growth of roots or rhizomes, particularly in perennial weeds is an evident expression of elevated CO_2 in the environment (Rogers *et al.*, 1994). Thus, mechanical tillage may lead to additional plant propagation in a higher CO_2 environment, with increased asexual reproduction from below ground structures and negative effects on weed control *e.g.* Canada thistle (Ziska and George, 2004). New strategies are available to combat the weed menace in the changing climatic conditions, but the cost of implementing such strategies (*e.g.* new herbicides, higher chemical concentrations, new biocontrol agents) is high and require considerable amount of time.

Weed Biology and its Impact on Human Health

Weeds affect human health through allergenic reactions, skin irritations, mechanical injury or internal poisoning (Ziska, 2001) is a reported fact. As on date science is at a nascent stage to quantify how climate changes, especially changes in CO_2, may affect specific weed population associated with human health. Changes in pollen production and allergenicity in common ragweed (a recognized cause of allergic rhinitis) with changing CO_2 and temperature in both indoor (Ziska and Caulfield 2000, Wayne *et al.*, 2002) and *in situ* experiments have been reported and further research on how escalating CO_2 concentration can affect both the growth and toxicity of poison ivy is ongoing.

Impact on Insects

Temperature is presumably the environmental criterion, which has a profound influence on insect behavior, distribution, development, survival and reproduction. The main reason being, the insect are poikilothermal *i.e.* their body temperature is approximately the same as that of the environment. Insect life stage predictions are most often calculated using accumulated degree–days from a base temperature and biofix point. It has been reported that the effect of temperature on insects largely masks the effects of other environmental factors (Bale *et al.*, 2002). It has been estimated that with a 2°C increase in temperature, insects might experience one to five additional life cycle per seasons (Yamamura and Kiritani, 1998). Hamilton (2005) and Hunter (2001) found that moisture and CO_2 effects on insects are also of importance in a global climate change setting.

Elevated Temperature could Increase Insect Pest Populations

Increased temperature can potentially affect insect survival, development, geographical range and population size. Depending on the development strategy of an insect species, temperature can exert different effects. Cicadas and arctic moths, which take several years to complete one life cycle, will tend to moderate temperature variability over the course of their life history. Some insect like cabbage head borer, onion borers develop more rapidly during periods of time with suitable temperatures. As scientists use degree–day or pathology based models to predict the emergence and the potential to damage of these insects increased temperature will accelerate development of these types of insects resulting in more generation and subsequently more crop damage in a year. Temperature may change gender ratios of thrips (Lewis, 1997) potentially affecting reproduction rates. Insect that spend major parts of their life history in the soil is slowly affected by temperature fluctuation than the ones that are above the ground because soil provides an insulating medium that will tend to buffer temperature changes more than the air (Bale *et al.*, 2002)

Insect species diversity per area tends to decrease with higher latitude and altitude (Andrew and Hughes, 2005), which means that rising temperature could result in more insects species attacking more host in temperate climate. Based on evidence developed by studying the fossil record, it has been conducted that the diversity of insect species and the intensity of their feeding have increased manifold with increasing temperature (Gaston and William, 1996; Bale *et al.*, 2002)

Natural enemies and host insect population may respond differently to changes in temperature. Parasitism could be reduced if host populations emerge and pass through vulnerable life stages before parasitoids emerge. Host may pass through vulnerable life stages more quickly at higher temperature, narrowing the opportunity for parasitism.

Elevated Temperature could Decrease Pest Population

Monophagous or oligophagous insects are confined to a specific single or few set of host crops. Due to rise in temperature, farmers may be discouraged to grow these specific crops resulting in the dwindling population of pest due to starvation and dearth of food. The same environmental factors that affect the insect pests can affect their predators and parasites resulting in an escalated, aggressive mode of hyperparasitism. At higher temperature, aphids have been shown to be less responsive to the aphid alarm pheromone, they release when attacked by predators and parasitoids resulting in potential increase of pest control (Awmack *et al.*, 1997)

Effect of Changing Precipitation on Insects

Just like temperature, alteration in precipitation can have an impact on insect pests, predators and parasites resulting in a complex dynamics. Some insects are sensitive to precipitation and are killed by heavy rains *e.g.* onion thrips (Reiners and Petzoldt, 2005). Flooding the soil has been used as a control measure for those insects that over winter in soil (Vincent *et al.*, 2003). Fungal hyperparasites of insects are favoured by high humidity and their incidence will be increased with the lengthening period of high humidity and vice versa.

Effect of Rising CO_2 Levels

Usually CO_2 impacts on insects are thought to be indirect as impact on insect damage result from changes in the host crop. It has been observed that the soyabean grown in elevated CO_2 atmosphere had 57 per cent more damage from insects than those grown in present day atmosphere. It is assumed that measured increases in the level of simple sugars in the soyabean level may have stimulated the additional insects feeding (Hamilton *et al.*, 2005). Recently, free air gas concentration enrichment (FACE) technology was used to create an atmosphere with CO_2 and O_2 concentration similar to what climate change models predict for the middle of the 21st century. Researchers have observed that insects sometimes feed more in leaves that have a lowered nitrogen content in order to obtain sufficient nitrogen for their metabolism (Hunter, 2001). Increased carbon to nitrogen ratios in plant tissue resulting from increased CO_2 levels may slow insect development and increase the length of life stages vulnerable to attack by parasitoids (Coviella and Trumble, 1999)

Effect on Farmers

As entomologists expect that insects will expand their geographic range and increase reproduction rates along with over wintering success, farmers will confront more types and higher number of insects to manage. This in turn means more round of insecticide spray as compared to the present time. Moreover, some classes of pesticide *e.g.* pyrethroids and spinosad have been shown to be less effective in controlling insects at higher temperatures (Musser and Shelton, 2005). With more insecticide applications being predicted, the probability of applying a given insecticide in a season will increase and hence the risk of developing insecticide resistance is catapulted. Cultural practices adopted by the farmers will also undergo certain changes. For example, using crop rotation as an insect management strategy could be less effective with earlier insect arrival or increased over wintering of insects. Row covers used for insect exclusion might have to be removed earlier to prevent crop damage by excessive temperature under the covers.

Remarks

Climate change effects are challenging to investigate but are potentially of great importance. Hence, the topic has been reviewed and researched upon, continuously. Increased focus has been diverted towards how a changing environment affects evolution of pest and several question have cropped up – which pest characteristics (such as, frequency of generation and proportion of sexual reproduction) affect the rate of adaptation in both host and pest population? Are invasive plant species better able to adapt to climate change and move to new areas rapidly, leaving the pest behind? Keeping these questions at the back drop, it is known that climate change will alter the suitability of crops and other plants for certain locations. However, information gathered so far has been fragmented and a comprehensive analysis of climate change impact on biotic stress is limiting with the present knowledge base. Apart from the technical constraints, the most significant bottleneck to climate change impact assessment is the inability to predict how technological and socioeconomic forces will interact with atmospheric, climate and biological factors to shape the future agriculture. The land use pattern change in response to market demands, transgenic technology and new chemicals for pest management are perhaps more priority area in agricultural research, but climate change and climate variability add another layer of complexity and uncertainty onto a system that is already exceedingly difficult to manage.

References

Andrew, N.R. and Hughes, L. 2005. Diversity and assemblage structure of phytophagous Hemiptera along a latitudinal gradient: Predicting the potential impacts of climate change. *Global Ecol. Biogeogr.*, 14: 249–262.

Asthana, A., McCloud, E.S., Berenbaum, M.R. *et al.* 1993. Phytotoxity of *Citrus jambhiri* to fungi under enhanced UV–B radiation: role of furanocaumarins. *Journal of Chemical Ecology*, 19: 2813–2830.

Awmack, C.S., Woodcock, C.M. and Harrington, R. 1997. Climate change may increase vulnerability of aphids to natural enemies. *Ecological Entomology*, 22: 366–368.

Bale J.S., Masters G.J., Hodkinson I.D. *et al.* 2002. Herbivory in global climate change research: Direct effects of rising temperature on insect herbivores. *Glob. Change Biol.*, 8:1–16.

Booth, T.H. 1996. Mapping region climatically suitable for particular tree species at the global scale. *For. Ecol. Manage.*, 36: 47–60.

Bowes, G. 1993. Facing the inevitable: plants and increasing atmospheric CO_2. *Annu. Rev. Plant Physiol. Pla. Mol.Biol.*, 44: 309–32.

Brasier, C.M. and Scott, J.K. 1994. European oak declines and global warming: a theoretical assessment with special reference to the activity of Phytophthora cinnamomi. *EPPO Bull.*, 24: 221–32.

Bridges, D.C. 1992. *Crop Losses Due to Weeds in the United States*. Weed Science Society of America, Champaign, IL, USA. Pp. 403.

Busby, J.R. 1991. BIOCLIM– a bioclimate analysis and prediction system. *Plant Protection Quaterly*, 6: 8–9.

Carefoot, G.L. and Sprott, E.R., 1967. Famine on the Wind: Plant Diseases and Human History. Angus and Robertson, Sydney.

Carter, T.R., Nurro, M. and Torkko, S. 1996. Global climate change and agriculture in the North. *Agric. Food Sci. Finl.*, 5: 223–385.

Carter, N. and Harrington, R. 1991. Factors influencing aphid population dynamics and behavior and the consequences for virus spread. In: *Advances in Disease Vector Research*, New York: Springer. (Harris, K.F. ed.) 7: 19–51.

Chakraborty, S. 1997. How will plant diseases impact on pasture production under climate change: a case study of Stylosanthes anthracnose. *Proc. 18th Int. Grasslands Conf.*, Calgary, Canada. Pp. 96 (Abstr.).

Chakraborty, S., Pangga, I.B., Lupton, J. *et al.* 2000. Production and dispersal of Colletotrichum gloeosporioides spores on Stylosanthes scabra under elevated CO_2. *Environmental Pollution Change*, 32:293–311.

Clifford, B.C., Davies, A., Grith, G., *et al.* 1996. UK.climate change models to predict crop disease and pests threats. *Asp. App. Biol.*, 45: 269–276.

Coakley, S.M. 1979. Climatic variability in the Pacific Northwest and its effect on stripe rust of winter wheat. *Climate Change*, 2: 33–51.

Coakley, S.M. 1988. Variation in climate and prediction of disease in plants. *Annu. Rev. Phytopathol.*, 26: 163–81.

Coakley, S.M. 1995. Biospheric change: will it matter in plant pathology? *Canadian Journal of Plant Pathology*, 17: 147–153.

Coakley, S.M. and Scherm, H. 1996. Plant disease in a changing global environment. *Asp. Appl. Biol.*, 45: 227–38.

Coviella, C. and Trumble, J. 1999. Effects of elevated atmospheric carbon dioxide on insectplant interactions. *Conserv. Biol.*, 13:700–712.

Cure, J.D. and Acock, B., 1986. Crops response to carbon dioxide doubling; a literature survey. *Agricultural and Forest Meteorology*, 38: 127–145.

Dyck, P.L. and Johnson, R. 1983. Temperature sensitivity of genes for resistance in wheat to *Puccinia recondita*. *Can. J. Plant Pathol.*, 5: 229–234.

Edis, D., Hull, M.R., Cobb, A.H. *et al.*1996. A study of herbicide action and resistance at elevated levels of carbon dioxide. *Asp. Appl. Biol.*, 45: 205–210.

Eshed, N. and Dinoor, A. 1981. Genetics of pathogenicity in *Puccinia coronata*: the host range among grasses. *Phytopathology*, 71: 156–163.

Finckh, M.R., Chavez, A.Q., Dain, Q. *et al.* 1995. Effect of UV–B radiation on the growth of rice and its susceptibility to rice blast under glass house condition. *Agric. Ecosyst.Environ.*, 52: 223–233.

Gaston, K.J. and Williams, P.H. 1996. Spatial patterns in taxonomic diversity. In: *Biodiversity*, Blackwell Science, Oxford. Pp. 202–229.

Gerechter–Amitai, Z.K., Sharp, E.L. and Reinhold, M. 1984. Temperature–sensitive genes for resistance to *Puccinia striiformis* in *Triticum dicoccoides*. *Euphytica*, 33: 665–672.

Goudriaan, J. 1996. Predicting crop yields under global change. In: *Global changes and Terrestrial Ecosystems*. Cambridge: Cambridge Univ.Press. Pp. 260–274.

Hamilton, J.G., Dermody, O., Aldea, M. *et al*. 2005. Anthropogenic Changes in Tropospheric Composition Increase Susceptibility of Soyabean to Insect Herbivory. *Envirn. Entomol.*, 34(2): 479–485.

Harrington, R. and Stork, N.E. (eds.) 1995. *Insects in a Changing Environment*. Academic Press, London.

Heagle, A.S. 1989. Ozone and crop yield. *Annu. Rev. Phytopathol.*, 27:397–423.

Heath, R.L. 1980. Initial events in injury to plants by air pollutants. *Annual Review of Plant Pathology*, 31: 395–431.

Hibberd, J.M., Whitbread, R. and Farrar, J.F. 1996. Effect of elevated concentrations of CO_2 on infection of barley by *Erysiphe graminis*. *Physiol. Mol. Plant Pathol.*, 48: 37–53. Available at: http://www.nysaes.cornell.edu/recommends/

Hunter, M.D. 2001. Effects of elevated atmospheric carbon dioxide on insect–plant interactions. *Ag. Forest. Entomol.*, 3: 153–159.

Intergovernmental Panel on Climate Change (IPCC). 2001. *Climate Change 2001: The Scientific Basis.* (Houghton, J.T., Ding, Y., Griggs, D.J. *et al.*, eds.), Cambridge University Press, Cambridge, UK. Pp. 881.

Kaukoranta, T. 1996. Impact of global warming on potato late blight: Risk, yield loss and control. *Agricultural and Food Science in Finland*, 5: 311–327.

Kimball, B.A., 1985. Adaptation of vegetation and management practices to a higher carbon dioxide world. In: *Direct Effects of Increasing Carbon Dioxide on Vegetation.* (Strain, B.R. and Cure, J.D. eds.), US Department of Energy, Washington. Pp. 185–204.

Klironomos, J.N., Rillig, M.C., Allen, M.F. *et al*. 1997. Soil fungal arthropod responses to *Populus tremuloides* grown under enriched atmospheric CO_2 under field conditions. *Glob. Change Biol.*, 3: 473–478.

Krupa, S.V. 1997. *Air Pollution, People, and Plants*. APS Press, St. Paul.

Laurence, J.A. and Weinstein, L.H. 1981. Effects of air pollutants on plant productivity. *Annu. Rev. Phytopathol.*, 19: 257–271.

Lewis, T. 1997. *Thrips as crop pests*. CAB International, Cambridge University Press, Cambridge, UK. Pp. 740.

Luo, Y., TeBeest, D.O., Teng, P.S. *et al*. 1995. Simulation studies on risk analysis of rice leaf blast epidemics associated with global climate change in several Asian countries. *J. Biogeogr.*, 22: 673–678.

Luo, Y., Teng, P.S., Fabellar, N.G. *et al*. 1998a. The effects of global temperature change on rice leaf blast epidemics: a simulation study in three agroecological zones. *Agric. Ecosyst. Environ.*, 68: 187–196.

Luo, Y., Teng, P.S., Fabellar, N.G. *et al.* 1998b. Risk analysis of yield losses caused by rice leaf blast associated with temperature changes above and below for five Asian countries. *Agric. Ecosyst. Environ.*, 68: 197–205.

Lupton, J., Chakraborty, S., Dale, M. *et al.* 1995. Assessment of the enhanced greenhouse effect on plant diseases: a case study of Stylosanthes anthracnose. In: *Proceedings of the 10th Biennial Australasian Plant Pathology Conference*, Lincoln University, New Zealand, 28–30 August.

Mackerron, D., Boag, B., Duncan, J.M., *et al.*1993. The prospect of climate change and its implications for crop pests and diseases. In. Plant Health and the European Single Market (Ebbels, D. ed) Br.Crop Prot.Counc., Farnham. Pp.181–193.

Magan, N. and Baxter, E.S. 1996. Effect of increased CO_2 concentration and temperature on the phyllosphere mycoflora of winter wheat flag leaves during ripening. *Ann. Appl. Biol.*, 129: 189–195.

Mahlman, J.D. 1997. Uncertainties in projections of human–caused climate warming. *Science*, 278: 1416–1417.

Manning, W.J. and Tiedemann, A.V. 1995. Climate change: potential effects of increased atmospheric Carbon dioxide (CO_2), Ozone (O_3), and Ultraviolet–B (UVB) radiation on plant diseases. *Environ. Pollut.*, 88: 219–245.

Martens, J.W., McKenzie, R.I.H. and Green, G.J. 1967. Thermal stability of stem rust resistance in oat seedlings. *Can. J. Bot.*, 45: 451–458.

McElrone, A.J., Reid, C.D., Hoye, K.A. *et al.* 2005. Elevated CO_2 reduces disease incidence and severity of a red maple fungal pathogen via changes in host physiology and leaf chemistry. *Glob. Change Biol.*, 11: 1828–1836.

Musser, F.P. and Shelton, A.M. 2005. The influence of post–exposure temperature on the activity of peroxidases. *Phytopathology*, 68: 1164–1167.

Oerke, E.C., Dehne, H.W., Schonbeck, F., *et al.* 1994. *Crop Production and Crop Protection, Estimated Losses in Major Food and Cash Crops*. Elsevier Sc. Pub., Amsterdam.

Padmanabhan, S.Y. 1973. The great Bengal famine. *Ann. Rev. Phytopathology*, 11: 11–26.

Patterson, D.T. 1995. Weeds in a changing climate. *Weed Science*, 43: 685–701.

Porter, J.H., Parry, M.L. and Carter, T.R. 1991. The potential effects of climatic change on agricultural insect pests. *Agric. For. Meteorol.*, 57: 221–240.

Reiners, S. and Petzoldt, C. (eds). 2005. *Integrated Crop and Pest Management Guidelines for Commercial Vegetable Production*. Cornell Cooperative Extension publication. Available at: http://www.nysacs.cornell.edu/recommends/

Riha, S.J., Wilks, D.S. and Simoens, P. 1996. Impact of temperature and precipitation variability on crop model predictions. *Climate change*, 32: 293–311.

Rishbeth J. 1991. Armillaria in an ancient broad leaved woodland. *Eur. J. For.Pathol.*, 21: 239–49.

Rogers, B.R. and Krupa, S.V. 1994. Plant response to atmospheric CO_2 enrichment with emphasis on roots and the rhizosphere. *Env. pollution.*, 83 (1–2): 155–189.

Rosenzweig, C. and Parry, M.L. 1994. Potential impact of climate change on world food supply. *Nature*, 367: 133–138.

Rosenzweig, C. and Hillel, D. 1998. *Climate Change and the Global Harvest*. Oxford University Press, New York.

Rosenzweig, C. and Parry, M.L. 1994. Potential impact of climate change on world food supply. *Nature*, 367: 133–138.

Sanden, G.E. and Moore, L.D. 1978. Effect of heat–induced susceptibility to tobacco black shank on protein content and on activity of peroxidases. *Phytopathology*, 68: 1164–1167.

Sandermann, H. Jr. 1996. Ozone and plant health. *Annu. Rev. Phytopathol.*, 34: 347–366.

Savile, D.B.O. and Urban, Z. 1982. Evolution and ecology of *Puccinia graminis*. *Preslia*, 54: 97–104.

Scherm, H. and Yang, X.B. 1995. Interannual variations in wheat rust development in China and the United States in relation to the El Nino/Southern Oscillation. *Phytopathology*, 85: 970–976.

Strange, R.N. 1993. *Plant Disease Control: Towards Environmentally Acceptable Methods*. Chapman and Hall, London.

Sutherst, R.W., Maywald, G.F. and Skarratt, D.B. 1995. Predicting insect distributions in a changed climate. In. *Insets in changing environment*, Academic Press, London, Pp. 59–91.

Sutherst, R.W. and Maywald, G.F. 1985. A computerized system for matching climates in ecology. *Agric. Ecosyst. Environ.* 13: 281–299.

Sutherst, R.W., Yonow, T., Chakraborty, S. *et al.* 1996. A generic approach to defining impacts of climate temperate forage species. *Neth. J. Agric.Sci.*, 39: 31–48.

Teng PS, Yang XB. 1993. Biological impact and risk assessment in plant pathology. *Annu. Rev. Phytopathol.*, 31: 495–521.

Thomas, T. 1989. Sugar beet in the green house–a global warming warning. *Br. Sugar.*, 57: 24–26.

Thompson, G.B., Brown, J.K.M. and Woodward, F.I. 1993. The effects of host carbon dioxide, nitrogen and water supply on the infection of wheat by powdery mildew and aphids. *Plant, Cell and Environment*, 16: 687–694.

Titone, P., Gullino, M.L. and Garibaldi, A. 2008. Effect of global change on powdery mildew–grapevine pathosystems: study under controlled environment. *Pest Manag. Sci.*, 61: 508–510.

Vincent, C., Hallman, G., Panneton, B. *et al.* 2003. Management of agricultural insects with physical control methods. *Ann. Rev. Entomol.*, 48: 261–281.

Walker, P.A. and Cocks, K.D. 1991. Habitat: a procedure for modelling a disjoint environmental envelope for a plant or animal species. *Global Ecology and Biogeography Letters,* 1: 108–118.

Wayne, P.S., Foster, S., Connolly, J., *et at.* 2002. Production of allergenic pollen by ragweed (*Ambrosia artemissiifolia* L.) is increased in CO_2–enriched atmospheres. *Annal of Allergy, Asthma and Immunology,* 80: 669–679.

Wilson, J.R., Deinum, B. and Engels, F.M. 1991. Temperature effects on anatomy and digestibility of leaf and stem of tropical and temperate forage species. *Netherland Journal of Agricultural Sciences,* 39: 31–48.

Wittwer, S.H. 1995. *Food, Climate and Carbon Dioxide: The Global Environment and World Food Production.* CRC Press, Boca Raton.

Wolfe, D.W. 1995. Physiological and growth responses to atmospheric carbon dioxide concentration. In: *Handbook of Plant and Crop Physiology,* (Pessarakli M. ed.) Marcel Dekker Inc., Dekker, New York. Pp. 223–242.

Wong, P.T.W., Mead, J.A. and Croft, M.C. 2002. Effect of temperature, moisture, soil type and *Trichoderma* species on the survival of *Fusarium pseudograminearum* in wheat straw. *Aust. Plant Pathol.,* 31: 253–257.

Yamamura, K. and Kiritani, K. 1998. A simple method to estimate the potential increase in the number of generations under global warming in temperate zones. *Appl. Ent. and Zool.,* 33: 289–298.

Ziska, L.H. and George, K. 2004. Rising carbon dioxide and invasive, noxious plants: potential threats and consequences. *World Resource Review,* 16: 427–447.

Ziska, L.H. 2001. My View: Weed science and public health. *Weed Science,* 49: 437–438.

Ziska, L.H. and Caulfield, F.A. 2000. Rising carbon dioxide and pollen production of common ragweed, a known allergy–inducing species: Implications for public health. *Australian Journal of Plant Physiology,* 27: 893–898.

Climate Change: Man and Environment (2012)
Editor: **Goutam Kumar Saha**
Published by: **DAYA PUBLISHING HOUSE, NEW DELHI**

Pages **138–165**

10

Global Climate Change vis-à-vis Agriculture and Food Security

☆ *Rathindra Narayan Basu**

Introduction

Climate change is undoubtedly a matter of utmost concern to the humankind as well as many other members of the biosphere and their well–being and even survival till the end of the present century may depend on how effectively the formidable challenge posed by anthropogenic greenhouse gas induced global warming can be countered.

Lack of general awareness on the issue among the general public, even many in the more enlightened section of the people and particularly among many decision makers, is widespread and perhaps underestimated. A clear understanding of the causative factors behind climate change is necessary to visualize its impact on agriculture especially in view of the fact that present day industrial–agricultural system including deforestation is responsible for generation of nearly one third of the anthropogenic greenhouse gas emissions that are responsible for global warming. Thus, contrary to the general belief of global climate change disrupting agricultural systems, mitigation of overall agriculture related emissions of greenhouse gases is urgently necessary not only to save agriculture itself but also to protect many other vital spheres of human activities. Fortunately, there are overwhelming and most convincing documented evidences of ecosustainable agricultural practices that are capable of greatly reducing greenhouse gas emissions and at the same time ensure food security of the people, specifically of the poor small and marginal farmers of the

* Formerly of University of Calcutta. E-mail: basurathindranarayan@gmail.com

developing countries like India which incidentally is a country with the largest number of hungry people in the world.

The present article deals with the climate change–agriculture interface and brings out the salient features of the climate change scenario as well as relevant features of conventional (industrial) and holistic ecosustainable farming systems. It is primarily targeted towards readers with a broad understanding of the general environmental and agricultural issues and their socioeconomic implications on the livelihood security of small–holder farmers of the developing countries, with emphasis on the Indian situation, in particular.

Climate Change

The earth had to wait for around a thousand million years for the origin of life since it came into being some 4,500 million years ago as conditions in the newly formed earth were too hot and hostile for the creation of life. The primitive earth's atmosphere lacked free oxygen and it was the advent of chlorophyllous green plants that made possible the release of oxygen from water molecules through chlorophyll–mediated photolysis of water as a part of the photosynthetic carbon assimilation process. Besides harvesting solar energy for organic food production it opened up and boosted the origin and evolution of all oxygen dependent higher organisms.

Primitive earth had no organized atmosphere, though water vapour, carbon dioxide, various compounds of nitrogen, sulphur, etc. in gaseous forms were intermittently released in large quantities through volcanic activities and cooling of the earth over time to serve as constituents of the evolving atmosphere. While in due course life evolved under water to escape the killing effect of hazardous ultraviolet rays, the evolution of more and more free oxygen by chlorophyllous green algae and primitive aquatic plants also enhanced the formation of ozone molecules that provided protection to living organisms from impinging UV–rays enabling them to move to land areas. Over millions of years a structured and organized atmosphere was evolved. Perhaps from a partially developed prototype, over further millions of years the present day lower atmosphere (troposphere), middle atmosphere (mesosphere) and upper atmosphere (stratosphere) were gradually organized. In due course of time (may be simultaneously) various climate systems, all dynamic in nature, developed and the biosphere with ever increasing complexity coevolved accordingly for adaptation to the changing situations. The most important factor in the modulation of climate systems and development of biosphere has been the prevailing ambient temperature regimes. Interestingly and most significantly, the composition of air, specially the presence of a few gases in very low concentrations would greatly influence atmospheric temperature. Unlike the two major constituents of the present–day air, nitrogen and oxygen that together account for 99 per cent of dry air (nitrogen 78 per cent and oxygen 21 per cent) with little effect on ambient temperature, these trace gases (carbon dioxide, methane, nitrous oxide, etc.), collectively constituting less than 0.05 per cent of air composition, would very disproportionately enhance air temperature.

Global Temperature Changes Over the Ages

There have been major periodic changes in temperatures in different continents during the past two million years through phenomena known as primary and secondary oscillations resulting in around ten glacial interspaced by requisite numbers of interglacial periods attributable primarily to a gradual change in the shape of the earth's orbit oscillating from elliptical to nearly circular and back over a period of 100,000 years; there are secondary oscillations such as the earth's tilt fluctuating from 21.5° to 24.5° with a period of 41,000 years and the wobbling of the earth's spin axis over a 23,000–year period – all of which would influence the amount of sunlight that would hit the earth with resultant differential effect on atmospheric temperature. Shifts in ocean currents attributable to various factors would also affect ambient temperature and vice versa. Besides ocean currents, other variables such as winds, precipitation and run–offs, snow cover, polar ice caps and glaciers, all of which are basically temperature dependent events would also significantly influence temperature itself.

Earth's Ambient Base Temperature and Greenhouse Effect

Physicists have treated earth as a black body, which is a useful theoretical abstraction defining an object that absorbs all radiation impinging upon it, as well as one that radiates at the maximum possible rate for any object with the same dimensions and temperature (Masters, 1995). Equating the rate at which energy from the sun is absorbed with the rate at which energy is radiated back to space (assuming a steady–state condition whereby the earth is isothermal and its temperature is not changing), application of the Stefan–Boltzmann equation would give the earth's temperature as 279 Kelvin (K), a value very close to the actual global average surface temperature of about 288 K (approx. 15°C). This is, however, largely coincidental as in the aforesaid calculation earth's albedo effect (the reflecting back of a fraction of incoming radiation) has not been taken into consideration (the global mean value for which is estimated to be about 30 per cent). Incorporation of the albedo factor into the equation would give an effective temperature of the earth as 255 K (–18°C or approx. 0°F) which is an extremely cold situation for sustenance of life; further it is 33 K (288–255 = 33 K) less than the actual average surface temperature of the earth. Obviously, something in the atmosphere has been acting as a blanket and imparting a warming effect equivalent to 33°C. This effect is attributed to the presence of gases such as carbon dioxide, water vapour, methane, nitrous oxide, tropospheric ozone, etc. that are collectively known as greenhouse gases because similar phenomenon of warming up happens in conventional greenhouses or glasshouses for growing plants particularly in winter months in the temperate countries. Glass cover would readily allow entry of short wave length solar energy into the greenhouse but would prevent longer wave length radiation (heat generating) to escape from the inside of the glasshouse (synonymous with greenhouse). The aforesaid greenhouse gases (GHGs) in the same manner would trap long wave length energy radiated back from the earth's surface, warming up the earth's lower atmosphere (air) and surface areas including land and water.

Greenhouse Gases and Global Warming

The role of carbon dioxide in raising air temperature was first suggested by the famous Swedish physicist–chemist Svante August Arrhenius (1859–1927) in 1896. Subsequently many scientists supported the concept of a major role of carbon dioxide in global warming and evidences have accumulated over the years confirming its overwhelming global warming role as the major greenhouse gas with the potential devastating power to destabilize the earth's climatic scenario. Despite the fact that carbon dioxide through its normal (non-anthropogenic) greenhouse effect (it has been primarily responsible for the non-anthropogenic 33°C temperature rise as a benevolent greenhouse effect, enabling continued sustenance of life on earth), its current increasing emission through human activities is a matter of utmost concern to the sustainability of the biosphere.

It is worthwhile in this connection to recall the most convincing evidence regarding the role of carbon dioxide as a modulator of air temperature over thousands and thousands of years. Through a brilliant and exhaustive research endeavor (Jouzel *et al.* 1987), the Soviet scientists in Vostok in east Antarctica recovered a 2083 m long ice core (Vostok ice core) that gave a continuous 160,000–year record of Antarctic surface temperature along with the concentration of gases, specifically carbon dioxide, trapped and sealed in the air bubbles within the ice core. There were major as well as minor ups and downs in the two sets of data over the long 160,000–year period but the remarkable correlation between atmospheric carbon dioxide concentration and Antarctic surface temperature over a vast expanse of time should convince even the most obstinate skeptics about the utmost necessity of curtailing the emission of carbon dioxide. The publication of the results of the Vostok ice core study in the famous science journal *"Nature"* in 1987 has virtually closed the ranks of scientists on the issue but the decision makers of the world, the politicians, are still to follow suit and as yet evasive in approach.

Unlike the past ages, in which greenhouse gas generation was mostly natural and non-anthropogenic, in the recent times human involvement in greenhouse gas emission has been quite high. In 2005, the atmospheric concentrations (in ppm) of the GHGs were as follows (the concentrations in 1750 are given in parenthesis): CO_2, 379 (280); CH_4, 1.77 (1.15); N_2O, 0.32 (0.28); Chlorofluorocarbons/F–gases, 0.003 (0). Excessive use of fossil fuels, such as coal, oil, natural gas (which are all products of past photosynthesis in the ancient times) for industry, transport and agriculture including deforestation and changes in land use) are the major reasons of the recent anthropogenic emission of carbon dioxide as the major greenhouse gas (besides a wide range air pollutants such as carbon monoxide, nitrogen dioxide, ozone, sulphur dioxide, suspended particulate matters, etc. – all of which are proven health hazards).

Some Recent Observations of the IPCC

A few salient observations from the Fourth Assessment Report of the Intergovernmental Panel on Climate Change (IPCC) are worth mentioning here to understand the complex greenhouse gas scenario primarily attributable to human activities (IPCC, 2007). These are outlined below:

Global Anthropogenic GHG Emissions Allowed an Increase of 70 per cent between 1970 and 2004

Over the 10–year period of 1995–2004, the rate of greenhouse gas emission has been 0.92 Gt CO_2–eq/year compared to an emission rate of 0.43 Gt/year during the period 1970–1994. The values for annual emissions in gigatons (1 gigaton = 1 billion ton) carbon dioxide equivalents (Gt CO_2–eq.) have been as follows: year 1970, 28.7; year 1980, 35.6; year 1990, 39.4; year 2000, 47.7; year 2004, 49.0.

The Share of Different Anthropogenic GHGs in Total Emissions in Terms of CO_2

Eq. Values in 2004 has been as follows (total emission taken as 100)

1. Carbon dioxide (CO_2 from three sources); (a) CO_2 based on fossil fuel use, 56.6 per cent, (b) CO_2 from deforestation and decay of biomass, etc., 17.3 per cent, (c) CO_2 from other sources, 2.8 per cent ;(total CO_2 all sources = 76.7 per cent).
2. Methane (CH_4) from agriculture and other sources, 14.3 per cent.
3. Nitrous oxide (N_2O) from agriculture and others, 7.9 per cent.
4. F–gases (fluorinated gases HFCs, PFCs, SF_6), 1.1 per cent.

The Share of Different Sectors in Total Anthropogenic GHG Emissions in 2004 in Terms of CO_2

Eq. values (total emission = 100) :

1. Energy supply, 25.9 per cent.
2. Industry, 19.4 per cent.
3. Forestry (including deforestation), 17.4 per cent.
4. Agriculture (all sectors), 13.5 per cent (agriculture+deforestation, etc. = 30.9 per cent).
5. Transport, 13.1 per cent.
6. Residential and commercial buildings, 7.9 per cent.
7. Waste and waste water, 2.8 per cent.

Agriculture which is closely related to deforestation and changed land use would collectively account for the highest emission of GHGs with the maximum global warming potential (GWP) (IAASTD, 2008). Broadly, a doubling of CO_2 (or CO_2–eq.) concentration in ppm over a span of time would raise the air temperature in a logarithmic relationship by approximately 3°C with the possibility of large deviations (±) due to thermal inertia and various other reasons.

Agriculture and Climate

Natural 'wild' plants and animals as well as lower organisms and microbes coevolved with the changing climatic scenario over millions and millions of years. Geographically climate itself varied widely, as such, there have been large variations, in plants, animals and microbes in macro and innumerable microclimatic niches as reflected in their diverse kinds, orders, families, genera, species and types. The coevolution of animals that are basically dependent on plants as well as intra–plant and intra–animal coevolutions (also similar coevolutions within lower organisms and microbes), would continuously add to the expanding realm of biodiversity. The present day cultivated plants, specifically the short–lived annual crops on the other hand, have been developed through human efforts of conscious and purposeful selection of plant types successfully adapting to the microclimate prevailing in a zone or locality and that has been done rather recently dating back to mid–agrarian civilization of several thousand years. Conscious breeding of plants and animals followed by selection for climatic adaptation (rather edapho–climatic adaptation) as well as productivity and quality related characters has been a much more recent event.

The coevolution of higher and lower organisms particularly with microbes, that are integral parts of the natural cycles, is intimately associated with the functioning of all sustainable agricultural systems. The manifold interactions between the different constituents of the biosphere involving natural and anthropologically altered complex ecosystems would indeed be a highly complicated issue defying clear understanding because of complex interactions. Agroecologists the world over are aware of the fact that of all the ecosystems, biodiversity loss is the highest in the agroecosystem and the same would be further exacerbated by climate change. They therefore advocate adoption of agricultural systems that would as far as practicable be in harmony with nature.

The launching of the monumental report of the United Nations and World Bank sponsored International Assessment of Agricultural Knowledge, Science and Technology for Development (IAASTD) in April, 2008 sounding a stark warning note, "Business as usual is not an option", would imply the urgent need to shift from the present conventional (industrial) agriculture and allied activities to natural resource management and societal welfare based sustainable agricultural systems that would particularly benefit the underprivileged small–holder farmers in the developing countries.

It is worthwhile to note that major observations of IAASTD on agriculture and climate change would broadly reflect the views of the Intergovernmental Panel on Climate Change (IPCC). Incidentally, the chairman of IAASTD, Professor Robert Watson has been the former chairman of the IPCC (and a co–sharer of the Nobel Prize for Peace won by the IPCC team in 2007). The broad observations presented in the synthesis report of IAASTD are briefly outlined hereunder.

IAASTD on Climate Change vis-à-vis Agriculture

Complexity of the Climate Systems

Climate change is a highly complex subject and it is difficult to elucidate and fully understand the different issues involved. Nevertheless, the IPCC has provided convincing evidences, implicating the greenhouse gases in global warming and climate change and all countries responsible for significant emissions must agree to urgently implement emission reduction proposals as per ratified international protocols. Even

—

if emission reduction proposals are readily implemented the greenhouse gases already released would continue to increase global warming for many years because of the time lag between emission and its actual manifestation on global warming depending on the physical and chemical properties of the individual greenhouse gases, their atmospheric residence times and reactions with other constituents of the atmosphere and sunlight.

Observed Climate Change and Impacts

In general, longer and more intense droughts have been observed since 1970s. Extreme events such as floods, droughts and tropical cyclones are presently more intense than before. Usually, such extreme events are associated with significant adverse consequences on food and forestry production and very often associated with food insecurity. During the summer of 2003, the European heat wave that was accompanied by drought, caused a 20 per cent reduction of maize yields. Recurrence of such events is quite likely, along with the incidence and severity of forest fires in the coming years exacerbated by unmitigated climate change. Throughout North America and Europe (NAE) there has been significant increase in serious forest fires attributable to dense biomass and human access. As was anticipated, climate change has considerably aggravated the adverse effects of El Nino Southern Oscillation (ENSO).

Future Climate Change and Projected Impacts

1. With a temperature increase of 1–2°C, crop productivity in the countries in low latitudes would decrease but with that much temperature increase, countries in the mid and high latitudes with more carbon dioxide in the air, coupled with longer growing seasons (the thermal growing season has already lengthened by about 10 days in many temperate zones) may show some increase in productivity but even in those countries production would go down with still higher temperature rise. From an ecosystem perspective the rate of change is more important and by 2030 a rate of temperature rise by 0.2°C per decade is projected. Experts are of the opinion that rates of rise in excess of that could be dangerous.

2. The current scenario in respect of changes in precipitation is as yet not consistent enough to make confident future projections although water scarcity for many crops and consequent adverse effects on productivity is likely. A rise in temperature would most likely increase the rate of evaporation as well as transpiration from plants. As such, more precipitation in the form of rainfall or snowfall would be expected but it is the pattern of distribution and extent of precipitation in different regions that would be very important from the agricultural viewpoint. Also the relative contribution of rainfall and snowfall to overall precipitation needs to be elucidated and projections for the same would be necessary. By the year 2050, internal renewable water may increase in several developed countries but the same would possibly decrease in all developing countries.

3. Overall increase in summer monsoon in Asia is projected in most climate models. But the likelihood of prolonged droughts that would be particularly damaging for rice has been projected during the same period. The semiarid regions of Africa would possibly get less rainfall. Further seasonal and annual differences in rainfall would increase, bringing uncertainty in crop production. Climate change induced rise in temperature when coupled with drought would adversely impact the "breadbasket" regions of China, India and USA. A rise of 2°C temperature may decrease the production of rice by 5–12 per cent in the rainfed paddy cultivation regions of China.

4. Cattle and other livestock are very sensitive to temperature rise and suffer a lot because of drought and lack of drinking water. Climate change induced changes in marine ecosystems as well as inland water bodies would cause greater hardship to fishermen who depend on adequate fish catch from these sources. Sea level rise, changes in water current are affecting the movement of fish and other sea animals. Enhanced dissolution of carbon dioxide that is being increasingly added to the atmospheric pool would greatly affect the survival of shellfish and other animals with calcareous shells due to greater acidification of seawater. The entry of seawater to coastal areas because of sea level rise as well as by tidal floods (also increased incidence of cyclones) would threaten the livelihood of farmers inhabiting the coastal areas.

5. Climate change, besides direct effect on agricultural production would indirectly reduce productivity by adversely affecting the health of farmers of developing countries where farming is so very dependent on human labour and draft animals. Working in the field at a high temperature for long hours would be really tough for women farmers and labours in particular. Already annually 2–5 million farmers in the world become ill from use of toxic pesticides of which 0.22 million ultimately die (besides 170,000 occupational deaths in agriculture every year). Climate change would further add to the misery with higher incidence of human diseases as projected by IAASTD and as a result crop productivity is bound to go down. The failing health of farm animals would further add to the misery of small–holder farmers.

6. The greater incidence of crop pests and diseases in a warmer climate would reduce crop production causing economic distress of the farmers. The possibility of greater soil erosion because of prolonged drought followed by heavy rainfall and flood, and consequent washing away of the topsoil and nutrients and ultimately leading to desertification is quite strong. Tropical soils with low organic matter would have the greatest impact of erosion on productivity. There would be greater desertification because of low average annual rainfall and increased evapotranspiration, and poor aggregate stability of the soil. The circumstances would force many farmers to move to urban areas for livelihood. Rural areas devoid of able–bodied male members would have to depend more on women workers and elderly people for farming and allied activities creating difficult sociological

problems besides production loss. The loss of health of women farmers in particular could be disastrous as they are not only producers but also caretakers of the families bearing the burden of nursing children, cooking and many other activities including farming. The conflict for natural resources would increase and according to IAASTD by the end of 2010, 50 million people (mostly of the developing countries) would become climate refugees and by the end of 2050 that number may go up to 200 million.

IAASTD on Mutual Dependency of Climate Change and Agriculture

Dependency of Agriculture on Climate

In the history of human civilization, climate extremes and climatic aberrations have wreaked havoc in agriculture and livelihood. All over the world people, rural people in particular, suffered immensely from such events. Through all such tragedies, people learnt how to adapt to such situation especially those who could not move to alternative places or livelihoods. An example of such a phenomenon is El Nino Southern Oscillation (ENSO) which is a recurring climatic event that occurs at intervals of 2–5 years. El Nino originates in the southern Pacific ocean off South America and involves a huge warm water current that flows along the west coast of Latin America bringing with it huge tidal waves and very high rainfall greatly damaging marine ecosystems along the coast and devastating the livelihood of fishing communities throughout the western coast of the continent. Climatologists are of the opinion that El Nino has been responsible for droughts in south India a few years back, as also the prolonged drought in Australia that has drastically reduced Australian wheat production. It is further argued that global climate change has been exacerbating the damaging effects of El Nino, and agricultural systems in many countries are badly hit.

Dependency of Climate on Agriculture

The effect of climate on agriculture is an issue that has received much attention over the years; IAASTD has put forward the other side of the story, overall agriculture and allied landuse activities including deforestation are responsible for nearly 31 per cent of the anthropogenic greenhouse gas emission.

The major sources of GHG emission in agriculture are as follows:

1. Land conversion and ploughing would release a large amount of carbon dioxide stored in vegetation and soil. Half of the surface land of the world has been converted to crop cultivation and grazing resulting in a loss of half of the forests through deforestation and degradation releasing stored carbon to the atmospheric pool as CO_2.

2. Fossil fuels used to power a range of farm machineries release large amount of CO_2 (also nitrogen oxides, particulate matters, etc. as pollutants), use of fossil fuels in production of fertilizers and pesticides also serves as a significant source of CO_2.

3. Transport of agricultural produce over long distances from farmers to consumers (attributed to regional monoculture and speciality crops in

specific belts) is currently making a significant contribution to the carbon dioxide pool.

4. Emissions of nitrous oxide (N_2O) take place due to application of fertilizers and manures as well as during decomposition of agricultural wastes.

5. Increasing level of industrial scale animal husbandry to meet the growing demand for livestock products (specifically ruminant–based meat and milk products) as well as wetland paddy cultivation are significant sources of methane gas emissions from agriculture.

Crop and animal husbandry together would account for 13.5 per cent of global emissions mostly through methane (47 per cent of total anthropogenic emission of methane) and nitrous oxide (58 per cent of total anthropogenic N_2O emissions). Against 13.5 per cent estimate of IPCC for cropping and livestock there are reports from other sources, which suggest emissions from livestock alone to be as high as 18 per cent of total emissions (including entire commodity chain for livestock). Land use, land use changes and forestry contribute 17.4 per cent mostly as carbon dioxide, thus agriculture (cropping and livestock together with land use, land use change and forestry) would contribute nearly 31 per cent (13.5+17.4=30.9 per cent) of anthropogenic greenhouse gases. As diets change and demand for animal products rise (ruminant meat in particular), GHG emission may increase by another 7–20 per cent in the near future.

Climate Change Impacts on Various Subsectors of Agriculture and Allied Practices in Different Regions

1. Although the highest emissions of greenhouse gases are associated with the most intensive farming systems and the least for subsistence type rainfed agriculture as in sub–Saharan Africa yet the latter is amongst the most vulnerable regions to the impact of climate change due to multiple stresses.

2. Changes in land use have adversely affected the net ability of ecosystems to sequester carbon from the atmosphere. The replacement of carbon–rich grasslands and forests by crops in the temperate zone would result in much lower carbon sequestration. Although in the northern hemisphere there is a slow increase in forests, the overall advantage is being lost due to increased deforestation in the tropics and subtropics. Forest replacement by perennial oil palm plantations, as in southeast Asia, may be advantageous for carbon sequestration but cannot compensate for the loss of biodiversity and its potential advantages. Some vulnerable natural pools of carbon such as peat lands are likely to become sources of CO_2 emission due to climate change.

3. Livestock holders would suffer in various ways as animals are very sensitive to heat stress and drought; they require reliable resource of water and pasture. Further incidences of infectious and vector-borne diseases, to which livestock are vulnerable, would considerably increase globally with climate change. Incidentally, 75 per cent of emerging diseases are zoonotic (diseases that are transmitted between animals and humans) and the indirect impact

on agriculture would be considerable especially in the human labour intensive developing countries.

4. Regarding the future impacts of climate change on crop and tree yields, fisheries, forestry and livestock, a wide region–to–region variation would be expected. Local biomes and terrestrial ecosystems will be adversely impacted. As there is little doubt that climate change will impact regional patterns of temperature and precipitation, climate projections are as yet unable to elucidate exactly when, where and how the changes will be experienced.

5. Global climate change is expected to alter marine and fresh water ecosystems and habitats. Rising sea levels will alter coastal habitats and their future productivity, threatening some of the most productive fishing areas in the world. Altered ocean currents consequent upon temperature rise will bring about changes in distribution and ranges of marine animals including fish populations. Sea level rise leading to salt water intrusion will result in reduced agricultural productivity in many coastal areas. Secretion of calcareous shells or skeletons by corals, molluscs etc., so essential for their survival, will be disrupted because of acidification of sea water due to excess carbon dioxide dissolving in water. All these will have a serious impact on marine ecosystems and vital food webs, even affecting the diversity of fish species in marine as well as fresh water lakes and rivers.

6. Climate change is already affecting and will continue to affect the geographic range of many animal and plant pests, disease vectors and a wide variety of invasive species that will inhabit new ecological niches with negative impact on agricultural activities through their effect on the health of farmers and functioning of ecosystems, especially in the developing countries. An increase in temperature and precipitation would be conducive to the expansion of a wide range of vector transmitted diseases with the strong possibility of those being established outside their present range and also at higher elevations. Increased irrigation to counter water scarcity may increase the incidence of water related diseases including malaria. Seasonal weather changes (short–term version of climate change) as well as long-term climate change will strongly influence the incidence of pests and diseases. Conditions favorable for pests and diseases such as higher winter temperature (implying reduced winter–kill) and higher rainfall (and humidity) may encourage new pest introductions that would alter pest–predator–parasite dynamics through changes in growth and development rates, number of generations produced per year, the severity and density of populations, the pest virulence to host plants or the extent of susceptibility of the host to the pest. Changing weather patterns also increase crop vulnerability to pests, weeds and invasive plants, thus decreasing yields and increasing pesticide applications. Increased temperatures are likely to facilitate range expansion of some highly damaging weeds, which are usually limited by cool temperatures.

7. IAASTD has cautioned about the serious potential for future conflict and possible violent clashes over habitable land and natural resources, such as fresh water, as a result of climate change that would seriously jeopardize food security and poverty reduction. Annually an estimated 25 million people flee from weather related disasters, this number is projected to increase to about 200 million by 2050 with semi–arid ecosystems expected to be the most vulnerable to the adverse impacts of climate change. IAASTD further points out that climate change, combined with other socioeconomic stresses could alter the regional distribution of hunger and malnutrition particularly in the poor developing countries of the world.

Options for Action to Meet Climate Change Impacts

The two conclusive statements of the IPCC – "warming of the climate system is unequivocal" and "most of the observed increase in globally averaged temperatures since the mid 20th century is very likely due to observed increase in anthropogenic greenhouse gas concentrations" (IPCC,2007) are now almost universally accepted. The focus, as the IPCC (also IAASTD) rightly mentioned, should therefore shift from defining the threat to seek appropriate solutions.

A coordinated global approach to tackle the root cause of climate change namely GHG emission into the atmosphere is essential for the purpose and that requires a meticulously planned set of strategies by appropriate international bodies and above all with the firm determination to implement emission reductions. The whole issue is of utmost urgency as further negative changes are inevitable requiring interim adaptation and mitigation strategies imperative in the coming decades.

Climate change is not merely an environmental issue; it encompasses broader issues such as sustainable development and livelihood security that involves equitable access to resources and appropriate technologies as well as support systems and mechanisms to cope with risks.

Agricultural policies need to be developed for emission reductions as well as adaptation to climate change; these should be closer to carbon neutral, minimize trace gas emissions and reduce natural capital degradation. The focus should be on emission reduction in agriculture and forestry, production of food with greater input efficiency and less GHG emissions, and how best agriculture, agroforesty and forestry can adapt to given local conditions. For all these a revised and retooled Agricultural Knowledge, Science and Technology (AKST) would be required for meeting the challenges of energy efficient farming systems development as well as more comprehensive cost–benefit analysis than those now available. Further interconnected issues such as the negative effects of land use changes on biodiversity and on land degradation need to be addressed in order to exploit the synergies between the goals of UN conventions on biodiversity, and desertification and climate change.

Complementary Nature of Mitigation and Adaptation Strategies

In view of the fact that the effects of reduced emission (mitigation) to avoid temperature rise will not be manifested soon, and may even be delayed for several decades, due to the inertia of climate system (time lag between emission of GHGs and temperature rise), adaptation will be important in coping with the early impacts. In fact adaptation will be essential to meet the challenge of impacts on agriculture to which it is already committed in near or distant future. Mitigation would be the best recourse at all times and it will be the only option when unmitigated climate change exceeds the adaptive capacity of the existing agricultural systems. Mitigation options include a range of approaches such as lower rates of agricultural expansion into natural habitats, afforestation, reforestation, agroforestry and restoration of underutilized or degraded land, carbon sequestration in agricultural soils, appropriate application of nitrogenous inputs, effective manure management and use of feed that increases livestock digestive efficiency.

Policy options include financial incentives to maintain and increase forest area through reduced deforestation and degradation and improved management to increase the production of renewable energy sources.

Local, national and regional agricultural development frameworks will have to take into account the trade–offs between the need for promoting higher yields and the need for the maintenance and enhancement of environmental services that support agriculture.

Climate Change Adaptation Options

Two adaptation options have been recognized, namely autonomous adaptation and planned adaptation. Autonomous adaptation does not constitute a conscious response to climatic stimuli but is triggered by ecological changes in natural systems and by market and welfare changes in human systems. Planned adaptation is the result of a deliberate policy decision and involves implementation of existing knowledge and technology and increasing the adaptive capacity through improved policies and investments in new technologies and infrastructure for more effective adaptation activities.

Autonomous Adaptation Options

These are largely extensions or intensifications of existing risk management or production enhancement activities as given below:

1. Changing varieties/species to fit more appropriately to the changed climatic conditions.
2. Changing time of irrigation and adjusting nutrient management.
3. Applying water conserving technologies and promoting agrobiodiversity for increased resilience of the agricultural systems.
4. Altering timing or location of cropping activities, and diversification of agriculture.

Planned Adaptations

These include the following:

1. Implementation of specific policies aimed at reducing poverty and increasing livelihood security.

2. Provision of infrastructure to support integrated spatial planning and to generate and disseminate new knowledge and technologies.

3. Development of management practices tailored to anticipated changes in climate.

According to IAASTD the aforesaid strategies are in fact, examples of the mainstreaming of climate change adaptation into policies intended to enhance broad resilience.

The effectiveness of the adaptation strategies is likely to vary significantly between and within regions depending on the impacts and adaptive capacity of regions or areas that may differ very much infrastructurally and socioeconomically.

For climate predictions to meet the needs of agriculture it would be necessary to increase observational networks in the most vulnerable areas along with further improvements in forecast accuracy along with the integration of seasonal prediction with information at shorter as well as longer time scales.

Climate Change Mitigation Options (as suggested by IAASTD)

1. Reducing emissions of carbon dioxide, deforestation and forest degradation that would include policy measures to prevent deforestation, improving forest management, forest fire management, improving silvicultural practices, promoting afforestation and reforestation to increase carbon storage in forests.

2. Improving soil carbon retention by promoting global biodiversity and associated ecosystem services (valued at US $ 1,542 billion annually by IAASTD), as a strategy for climate change mitigation and adaptation, and improving management of residues using zero or reduced tillage, growing legumes in crop rotation, reducing fallow periods and converting marginal lands into woodlots.

3. Reducing levels of methane emission in livestock production by improving animal diets and using feed additives to increase food conversion efficiency by reducing enteric fermentation and consequent methane emissions, aerating manure before composting, recycling agricultural and forestry residues to produce biogas or biofuels.

4. Giving support to low input farming in agriculture that relies on renewable sources of energy.

5. Policy options should include giving financial incentives to increase forest area, reduce deforestation and for enhancing the production of renewable energy source.

IAASTD cautions that efforts towards emission reduction of a particular GHG should not encourage GHG emission from other potential sources; often reduction in methane emission may lead to greater N_2O emission through changes in soil nitrogen dynamics; conservation tillage for carbon sequestration may result in elevated N_2O emissions through increased use of agrochemicals and accelerated denitrification in soils.

Climate Change Regimes

Although IAASTD (2008) regarded the Kyoto Protocol as one representing the highest level of International consensus to address climate change, its report pointed out that mitigation options for agriculture has, however, not been well covered under the Protocol. Incidentally, Kyoto Protocol itself has been effectively shelved in the Copenhagen summit meeting of the heads of states in December, 2009 because of intervention by USA and several other countries.

IAASTD is of the opinion that to take full advantage of the opportunities offered by agriculture and forestry sectors, a long term (30–50 years) comprehensive and regulatory framework would be necessary. Within such a framework, a modified Clean Development Mechanism (CDM) with a comprehensive set of eligible agricultural mitigation activities could help in meeting the development and sustainability goals. Such activities should include afforestation and reforestation, avoiding deforestation, using a national sectoral approach rather than a project approach allowing for policy interventions, and a wide range of agricultural practices (that includes organic agriculture) for development and long term sustainability goals. The future of conservation agriculture in the current format with lot of emphasis on agrochemicals, particularly toxic herbicides is, however, doubtful.

GHG Mitigation and Energy Saving in Organic Farming

Although IPCC (2007) pointed out the significant contribution of conventional industrial agriculture along with deforestation on GHG emission (together 30.9 per cent) and global warming, it was IAASTD (2008) that indicated the possibility of countering the same by ecologically sustainable natural resource management based agriculture.

Mae–Wan Ho (2008) in a press release by the Institute of Science in Society, UK (ISIS) reestimated the contribution of industrial agriculture and food systems to global anthropogenic greenhouse gas emissions and produced an estimate of 34 per cent GHG emission as the overall agriculture and food related contribution as shown in Table 10.1.

Table 10.1: Sub-sectorwise contributions of industrial agriculture to GHG emissions

Subsectors	GHGs (per cent)	Explanation
Agriculture	11	Mostly methane (3.39 gigaton/year) and nitrous oxide (2.84 gigaton/year).
Landuse change	9	Deforestation for agriculture.
Industry	3	Fertilizer manufacture, machinery, food industry.
Energy	2	Direct use on farm, machinery, heating, cooling, irrigation.
Transport	4	Food transport, distribution.
Processing and packaging	2	Food processing, packaging, etc.
Buildings and infrastructure	2	Storage and allied activities.
Waste	1	Food and packaging wastes.
Total	34	

Further, there is a large scope for reversing the damages of the current agriculture and faulty food systems by organic agriculture and localized food systems that besides effectively mitigating the whole of the agricultural GHG emission would save energy equal (or nearly equal) to all that is being used at present in industrial agriculture and associated activities (ISIS/Mae–Wan Ho, 2008), as shown in Table 10.2.

Table 10.2: Global potential of organic agriculture and sustainable food systems for mitigating greenhouse gas emission and saving fossil fuel energy

Sources/Sectors, etc.	GHG Emission Mitigation (per cent)	Energy Saving (per cent)
Carbon sequestration in organic soil	11	–
No deforestation	9	5
Localizing food systems		
Reduced transport	3	5
Reduced building and infrastructure	1	1
Reduced processing and packaging	1.5	3.5
Reduced wastes	0.5	0.5
Phasing out N synthetic fertilizers		
Reduced N$_2$O emissions	3.8	–
No fossil fuels used in manufacture	0.7	2.3
Total	**30.5**	**17.3**

Reduction of factory–scale beef cattle production in the west by half would increase the mitigation potential by 1.6 per cent (making the total 32.1 per cent), and installation of biogas plants in crop livestock integrated farming systems would serve as net providers of energy to the farms.

David Pimentel and coauthors have shown that in the USA fossil fuel energy inputs for organic crop production were about 30 per cent lower than for conventionally produced corn (Pimentel *et al.*, 2005), that is indeed a considerable energy saving.

There is no valid reason to be afraid of the ultimate outcome of organic farming in terms of productivity loss in our country. Using a large data set, Catherine Badgley and coworkers (2007), working in the University of Michigan, have shown that at a stable state of soil, either under conventional or under organic or near–organic management (not during conversion from high intensive input to organic or near–organic systems), the yield ratios of organic versus conventionally grown food crops that included grains, starchy roots, sugar crops, legumes (pulses), oil yielding crops, vegetables and fruits (total number of studies had been 293) gave an average (average of all different types of plants) yield ratio value (organic versus conventional) of 0.914 (S.E.±0.02) for developed countries of the world, showing a minor yield reduction in those countries. In developing countries, on the other hand, the yield ratio value was 1.74 (S.E.± 0.09) indicating considerable overall superiority of organic systems vis–a–vis conventional agriculture. Badgley *et al.* (2007) therefore suggested that organic agriculture could feed the world while doing away with the adverse impacts of conventional agriculture on environment and ecosystems.

Indian Climate Scenario

Concerns and National Action Plan

According to the Fourth Assessment Report (2007) of the IPCC, the Indian subcontinent will have to cope up with challenges of the adverse effects of climate change that may be moderate to severe. Over the last 100 years, average surface temperature of the country (all India level) has risen by 0.4°C with the adverse effects more apparent in coastal areas, central India and Deccan plateau. Monsoon rains have decreased in the eastern part of Madhya Pradesh, northeast India, Gujarat and Kerala and over the last 100 years about 6–8 per cent decrease in monsoon rainfall has taken place.

Sea level rise (SLR) measured at high tide over the last 40 years has been 1.06– 1.75 mm per year and according to National Communications, India (NATCOM I), by 2050 the total sea level rise will be 15–18 cm that will go up to 46–59 cm by the end of the century resulting in inundation of many low lying areas with salt water and increased salinity due to intrusion of sea water over large tracts of land along the coast line. NATCOM I (2004) also projects that extreme weather events such as cyclones will increase by 15 per cent damaging the coastal regions in particular with lot of suffering of people inhabiting such vulnerable areas.

Regarding mitigation of greenhouse gas emissions, a NATCOM I study has shown that GHG emission from agricultural sources is next to that for fossil fuel consumption in transport. According to Parikh (2010) agriculture in India presently uses 12 per cent of total diesel consumed in India (trucks 37 per cent, passenger cars 15 per cent, buses 12 per cent, industry 10 per cent, power generators 8 per cent, and railways 6 per cent, are the other important consumers).

The greenhouse gases, methane and nitrous oxide owe their emission mostly from agricultural activities. It also noted that crossbred dairy cattle would show more methane emission per head. Suitable food additives would be required to check enteric emission. A very useful suggestion is to stop the burning of crop residues in mechanized farming and divert the same as cattle feed (or using the same for making compost or vermicompost). Reduced dependency on synthetic N–fertilizers would significantly decrease emission of nitrous oxide, a highly potent GHG.

Increasing ambient temperature will accelerate melting of glaciers that may initially increase water availability but will eventually recede and supply less water to the snow–fed Himalayan rivers of northern India in the lean season of the year, when water will be necessary for irrigation. Recharging of aquifers will be affected because of erratic rainfall pattern and intermittent droughts. By the end of the 21st century water availability may go down by 20–30 per cent with large negative effects on agriculture (Kumar, 2006); rise in average temperature further exacerbating the situation. Predictions regarding health sector indicate greater incidence of vector borne diseases like dengue and malaria as also enteric diseases, which have indirect effects on agricultural productivity by adversely affecting the health of farmers in the labour intensive farming practices in the country. Farming in the coastal areas will suffer from more frequent floods, storm surges and salinitization of land.

Recently, Chakraborty *et al.* (2010) of West Bengal State Pollution Control Board, Department of Environment have deliberated on the multiple adverse effects of climate change on various sectors including health, agriculture, industry, energy, etc. along with ways and means of tackling the situation. They have pointed out that negative impacts of water shortage will be more in the intensively cultivated Ganga basin particularly in the lower part (in West Bengal) where the decreased flow will encourage sea water intrusion and increased salinity. Water shortage and temperature rise in general will be especially harmful for *rabi* (winter–summer) crops that require the supply of stored surface water from reservoirs or underground aquifers.

In fact NATCOM I projections indicate that water shortage will occur in all major rivers including Ganga, Brahmaputra, Sind and their tributaries affecting agriculture and allied activities in their respective basins.

As pest and disease infestations will be more, crop production will also go down. Animals being more sensitive to high temperature and water shortage (and also disease susceptible) will be less productive. Marine, coastal and inland fisheries are being and will be adversely impacted by climate change with consequent economic distress to the fisher–folk.

National Initiatives to Face the Challenge of Climate Change

The Government of India has recently published the National Action Plan on Climate Change (NAPCC) through the Prime Minister's Council. Among the NAPCC's guiding principles, the following points are considered to have significant bearing on food security and agriculture.

1. Protecting the poor and vulnerable sections of the society through an inclusive and sustainable development strategy, sensitive to climate change.
2. Achieving national growth objectives through a qualitative change in direction that enhances ecological sustainability leading to further mitigation of greenhouse gas emission.
3. Deploying appropriate technologies for both adaptation and mitigation of greenhouse gas emission extensively as well as at an accelerated pace.
4. Effecting implementation of programme through unique linkages, including civil society and local government institutions and through public–private partnership.

For fulfilling developmental objectives by limiting the per capita GHG emission below that of the developed countries, eight national missions have been identified for necessary directional shifts towards implementation of growth strategies, namely (i) National Solar Mission, (ii) National Mission for Enhanced Energy Efficiency, (iii) National Mission on Sustainable Habitat, (iv) National Water Mission, (v) National Mission for Sustaining the Himalayan Ecosystem, (iv) National Mission for Sustainable Agriculture, (vii) National Mission for a "Green India", and (viii) National Mission on Strategic Knowledge for Climate Change.

The widely varied geoclimatic features of a vast country like India implies that climate change impact would be different in geoclimatelogically different regions. In fact IPCC (2007) as well as NATCOM I findings confirm the need for differential and regional approach. As such, the different states should formulate state policies in keeping with regional needs and national policies. It is heartening to note that most of the aforesaid eight national missions directly and indirectly focus on the need for ecologically sustainable agriculture which is so very relevant to the small and marginal farmer dominated rural India; in fact they would be the worst sufferers of the adverse impacts of climate change.

The Urgent Need for Rescheduling the Farming Practices of Small-holder Indian Farmers

Directly and indirectly over 60 per cent of the Indian population are dependent on agriculture and allied activities. The small–holder farmers are socioeconomically extremely hard pressed, climate change will inevitably aggravate their food and livelihood security.

The report of the Suresh Tendulkar Committee, submitted to the Planning Commission in December 2009, reveals that at the national level 37 per cent of Indians are below the poverty line (BPL) (about 10 per cent more than the previous estimate). If only rural areas are considered, the figure goes up to 42 per cent implying that nearly half of rural India is starving. It is a pity that most of them are the people who produce food to feed the nation. To nearly half a billion very poor Indians "India Shining" is an over–hyped myth.

Of the various reasons behind the precarious condition of the farming community, a few relevant ones may be cited as follows:

1. Agricultural growth in the country is falling behind population growth, excessive fragmentation of holdings, soil degradation due to lack of soil management and wrong agricultural policies and practices; high input costs – as such, farming is becoming uneconomic and nonremunerative..

2. Loss of biodiversity, crop biodiversity in particular, has been a major fall-out of the so–called green revolution technology (GRT) in view of the fact that thousands of varieties of different crops (especially cereals, pulses, oilseeds and vegetables) have been irretrievably lost because farmers who have been maintaining them (and also adding to their numbers) for specific agro–ecological niches and special purposes shifted over to a handful of GRT varieties. Faunal diversity (particularly the population of beneficial and farmer–friendly insects, amphibians and birds) has been greatly reduced by the pesticide menace.

3. Globalization of agricultural trade is encouraging cash crops over staple foods; export orientation in the face of unequal competition from highly subsidized imports is also adversely affecting farmers.

4. Climate change is negatively impacting agricultural production specifically in low latitude countries like India. Of these, most are beyond the immediate

control of the average small–holder farmer except short term changes (and in due course medium and long term changes) in farming practices.

5. Corporatization of agriculture and food systems would pose a very serious threat to the livelihood and survival of our small–holder farmers.

The sole focus on production increase at any cost through conventional (industrial) chemical–intensive agriculture and its all pervasive formula of "fertilizer (NPK)–irrigation water–pesticides", with initial short term success, has in 3–4 decades ruined the natural fertility of the soil, greatly diminished soil organic matter, the abode of soil microbes, extensively depleted the aquifers, almost irreversibly polluted air and water and incurring huge external costs with no attempt to minimize the cost of production. The tolerance of the insect pests to pesticides has reached a point where even the expensive third and fourth generation toxic pesticides are failing to control them. Increased applications of NPK fertilizers are becoming counterproductive, the deficiency of secondary nutrients and micronutrients is widespread. Further cost of fertilizers is increasing in the world market and as the country now imports over 40 per cent of nitrogen, 97 per cent of phosphorus and 100 per cent of potash at international price and makes that available at around 15–20 per cent of actual costs, the situation is destined to turn serious in the near future. According to Roberts and Stewart (2002) economically recoverable phosphorus in the world would be available may be for another three decades only (global potassium resource, however, may not deplete before another 100 years or so). The depleting fossil fuel reserves will certainly affect mechanized agriculture and will also limit their use in production of fertilizers, pesticides etc. Recycling of crop residues and all organic wastes in a closed agricultural system (earth itself with limited resources is also a closed system but for the almost unlimited and inexhaustible supply of solar energy) would perhaps be the only viable option left to millions and millions of small–holder farmers not only in India but also in all developing countries more so in the perspective of climate change intensifying its negative impacts on most human activities including agriculture.

In natural resource management based agricultural systems the dependency on synthetic fertilizers, pesticides, fossil fuels, etc. should ideally be nil but from a practical viewpoint, as low as possible to tide over initial set–backs in the absence of adequate experience and training in genuine ecologically sustainable farming practices. High yielding varieties (HYVs) are in fact high input requiring varieties and in the changed situation they must be avoided as far as possible. The presently available expensive genetically modified (GM) seeds which are also tailor–made for specific inputs, besides their still unknown short and long term effects on health of the ecosystem including that of humans and livestock and allied plants of the same family must not be included in the cropping schedule of the farmers whose seed requirements should be basically made from locally adapted crop cultivars. Hybrids are also high input requiring seeds and cannot be saved for next sowing. Incidentally all GM cotton cultivars commercialized in India are hybrids as also all other GM crops in the pipeline. So the farmer would be unable to save the seed for next sowing. Under no circumstances, the farmers as individuals or communitywise should be deprived from their traditional right to save the seeds of crops raised by them. As such, setting up seed villages with emphasis on locally adapted varieties also those in the public domain, and provision of adequate support systems and training must be encouraged throughout the country.

As against the top–down approach in conventional industrial agriculture, all natural resource management based ecosustainable agricultural systems would be essentially bottom–up in nature. All decisions in the individual farms and at the community level must be taken after a threadbare discussion with experienced farmers, *panchayat* and block level functionaries.

Biodiversity and Climate Change

Adequate exploitation of existing crop biodiversity will be a key factor in successful adaptation to the changing climate scenario. Increased emphasis on polyculture and mixed farming, if necessary sowing different compatible crops at intervals or even different varieties of the same crops, would serve as natural insurance to cover the risk of monoculture of a single species or variety failing to withstand unknown or known biotic and abiotic stresses; with more crops or varieties, the chances of compensation of the loss incurred by susceptible ones by the resistant or tolerant ones would be of great help to the farmers. In the climate change projections intermittent as well as prolonged droughts followed by heavy showers and floods have been made; we do not have as yet commercial cultivars of crops, which would successfully withstand drought as well as flood simultaneously. Mixed cropping employing appropriate cultivars of the same crop, or if necessary altogether different crops depending on the situation, would be a more practical approach. Creation of varieties genetically equipped with such opposite traits (tolerant to drought as well as flood) may be a very difficult proposition.

In a recent review article, Cotter and Tirado (2010) have also strongly vindicated the concept that the most effective strategy to adapt to global climatic changes particularly to temperature rise, unpredictable rainfall pattern and weather extremes would be to put due emphasis on increased biodiversity.

Judicious use of different but compatible crops as well as different varieties within a crop species for cultivation in the same field or in adjacent small subplots, or as mixed stand in alternate lines or cluster of rows would be a useful strategy. Crops of similar durations sown by broadcasting seed mixtures also would serve as effective compensatory mechanisms against climate change induced biotic and abiotic stresses. Care should be taken to avoid the problem of non-synchronous harvesting schedule of a mixed stand of varieties of annual crops by selecting as far as possible cultivars of similar durations specifically when broadcasting is done to sow the crops with seed mixtures. Sowing in lines or groups of lines of different varieties would, however, allow hand cutting and separately harvesting the different varieties without much difficulty. Such mixed cropping would cause a dilution of the availability of pest and disease susceptible host plants, indirectly but effectively protecting them from crop pest and disease attacks.

Inclusion of drought tolerant and water logging resistant varieties in the same or adjacent crop stands may enable farmers to tide over uncertain climatic conditions, particularly irregular rainfall patterns. Similarly in coastal semisaline tracts, salt tolerant varieties should be used along with better yielding salt–susceptible ones, and only salt–tolerant varieties, preferably a number of them, should be grown in a mixed stand in more saline soils.

Planting of suitable hardy perennial plants of economic importance on farm bunds and other suitable locations would provide further economic security to the farmer. Working with a rice–based mixed farming in a drought–prone area in eastern Bihar, Roy (1982) noted that the genetic diversity, within a crop species and that of mixed cropping systems would provide overall yield advantage and also serve as a safety measure against disease–pest epidemics.

Crop–livestock integrated farming has been a traditional practice with the Indian small–holder farmers concerned with the nutritional security of the family as well as the great importance given to animal manure for maintaining soil fertility but with a shift to conventional chemical–intensive agriculture, it is only sporadically practiced. There are, however, overwhelming evidences of much larger economic benefits of crop–livestock (fishery included) integrated farming practices in India, south–east Asia and elsewhere with 2–3 times net income along with food and nutrition security of the practicing farmers (Basu, 2007, 2008, 2010). Planned incorporation of animal husbandry and fishery into the crop sector is not simply a question of added benefit; it is highly synergistic as the combined effect on integration is more than the additive value of the individual beneficial effects of the different sectors.

For effectively tackling the adverse impacts of global climate change on agriculture, it is necessary to undertake research on regional and agro–climatic zone basis in a state and also on specific problem areas within a zone. It is now widely recognized that ecologically sustainable farming will greatly reduce greenhouse gas emissions serving as an effective GHG mitigation technology (ISIS/Mae–Wan Ho, 2008). Research on adaptation needs immediate strengthening.

Adaptation of plants and animals introduced from different countries to other climatologically different countries has given a lot of experience that would need recapitulation. Animal breeding studies have made possible the incorporation of useful genes from Indian livestock to their counterparts in Australia, Africa, etc. Such experiences with plants and animals need to be shared at all possible levels. Firstly, the characters of plants and animals that will enable them to adapt to projected altered climatic scenario need to be identified and sources within the country critically investigated. May be within acknowledged varieties and breeds, commercial as well as indigenous ones (including local types), desired characters are present. If so, they need to be tested under simulated conditions before outdoor trials, which would be easier with plants. In case of plants, conventional (traditional) breeding combined with the biotechnological (non-GM) technique of MAB (Marker Assisted Breeding) and MAS (Marker Assisted Selection) has enabled the creation of submergence tolerant rice (Xu *et al.*, 2006) that may prove useful in flood prone areas. Such studies would be necessary for other crops as well but the onus for critical analytical studies on planned adaptations would be on institutions of high level research at national and state levels.

Options for Small and Marginal Farmers

For the small–holder farmers at the village level several options are given hereunder, the choosing of appropriate ones depending on the situation would enable them to considerably meet many of the impending challenges of climate change.

1. Digging rainwater–harvesting structures such as ponds, dug wells of different sizes, shallow rectangular water harvesting structures for lands with gentle slopes in drought prone areas, contour bunds across the slope to arrest the run off and consequent erosion of the soil. The fertile topsoil along with invaluable plant nutrients would otherwise be lost irretrievably to the sea via creeks, canals and rivers.

2. Land shaping to raise low lands to accommodate a pond and raise a part of the land (surrounding the pond along with the bunds around the plot) to enable growing of fruits and vegetables in the uplands, paddy with fish in low land, paddy (HYV) in the medium land; the pond serving as rainwater harvesting structure would supply irrigation water in the lean period (*rabi* season) besides raising fish and rearing ducks for nutritional security and food security. Rainwater harvesting and crop–livestock integrated farming system (IFS) would assure year round employment generation and a 2–3 fold net rise in income (Swami Sadananda, 2005; Ghosh, 2006; Ravisankar and Pramanik, 2007). Cattle keeping for draft, milk and cowdung, the latter in many instances has enabled installation of small biogas plants to supply cooking fuel and slurry as valuable manure (also used for vermicompost preparation and as fish feed).

3. Wherever possible adoption of System of Rice Intensification (SRI) for paddy cultivation is suggested; it would save nearly half of the water used in conventional cultivation, seed rate will be less than one fifth; with organic manure only, an increase in yield to the extent of one ton per hectare may be expected. SRI in *rabi* season on 20 per cent of paddy area would enable a significant diversion of land to pulses, oilseeds and wheat that are in short supply. The acreage should go up by 10 per cent annually. Necessary motivation of farmers along with arrangements for training will go a long way in popularizing SRI. As in SRI much less water is used, production of the greenhouse gas methane would be less. SRI would impart more tolerance to climatic stresses such as dry spells and erratic rainfalls. Data from northern India in the drought year of 2009, showed that conventional paddy yield declined by 31 per cent while with SRI yield declined by 13 per cent only (Adhikari *et al*, 2010).

4. The Non-Pesticidal Management (NPM) of insect pests, presently covered under Community Managed Sustainable Agriculture (CMSA) has by 2009 spread to 1.39 million acres (estimated at 2.8 million acres *i.e.* 1.12 million hectares in 2010) in Andhra Pradesh involving 318,000 farmers of 3,171 villages in the state (Vijay Kumar *et al.*, 2009). NPM has been similar to Integrated Pest Management (IPM) but one step ahead in that it altogether discarded chemical pesticides (the last resort for pest control in IPM).

Summer ploughing, bonfires, and pheromone traps, use of trap crops, bird perches, light traps, spraying diluted fermented cow dung–cow urine along with range of botanical preparations would substantially reduce pest attack, significantly lowering the cost of cultivation. Although initially CMSA has been continuing the use of chemical fertilizers, those are being gradually phased out with emphasis on soil fertility management through application of tank silt, composts and vermicomposts (augmented by biomass plantation on farm bunds, azolla culture in rice) and inoculation with nitrogen fixing bacteria, etc. With total shift to organic farming within several years, the emphasis will be on intercropping and biodiversity-based multi–cropping. The Govt. of Kerala has already adopted a state policy of a total shift to organic agriculture within 5 to 10 years to save the farmers from the current multidimensional agrarian distress and to protect the environment from further degradation and bring back Kerala's pristine ecosystems (Vijayan, 2010).

It is worthwhile to note in this connection that within a short period the certified area under organic agriculture in India has gone up to over one million hectares with Madhya Pradesh heading the list of the practicing states. Another feature is the sharp reversal of trend of consumption, the earlier export oriented organic agriculture has given way to domestic consumption of nearly 96 per cent of the total produce of 976,000 tons (valued at 2.63 billion US dollars) in 2007–2008; the 3.8 per cent export of organic produce gave a foreign exchange earning of 100 million US dollars (Export–Import Bank of India, 2010). The demand of metropolitan cities is estimated to be around Rs.1.4 billion annually (Menon, 2007). Greater adoption of organic farming systems would be a key factor in mitigating GHG emission besides enhanced nutritive values and health benefits of organic food (Lu *et al.*, 2006; Mitchell *et al.*, 2007).

1. The small–holder farmer is sensible enough to give enough priority to food and nutritional security of the family but very often unable to ensure that because of acute poverty. A community based approach through which the farmers would first meet the basic needs of the community by cultivating (through mutual arrangements) a wide range of crops for nutritional security, and after meeting local needs, sell the rest to the nearby semiurban or urban markets. Transport cost is steadily rising and along with that the "food miles" related GHG emission; as such, localized food production and consumption (collectively known as localized food system) are widely advocated by environmentalists all over the world (ISIS/Mae–Wan Ho, 2008). Unfortunately the corporate controlled food and agricultural sectors, as in the developed countries in the North, follow totally opposite policies and strategies on the issue and to them neither food and nutritional security nor environment would mean the same thing as they do to the poverty stricken small and marginal farmers fighting for mere subsistence.

Remarks

Ignoring the extreme seriousness of the GHG induced climate change will be utter foolishness on the part of the humankind who are directly responsible for the gradually unfolding impacts of climate change and its future consequences on the sustenance of the biosphere. Both adaptation to the present and future more severe impacts of climate change and urgent steps to mitigate GHG emission would be essential features of our defense mechanisms for sustenance of the biosphere and ultimate survival. Adaptation may be possible upto a certain limit; adaptive capacity will fail thereafter.

The sense of complacency that comes out of a "nothing much to worry" attitude is one of the reasons for the lack of seriousness of many developed countries in emission cuts. It is because of low impacts of climate change at initial stages in the countries at high latitudes. A small increase of 1–2°C temperature will lengthen the growing season by several days increasing agricultural production; warming of cooler regions will be advantageous for livestock rearing and for many other human activities. Further, the economic prosperity of the developed countries will enable them to more effectively counter the adverse impacts of future climate change through advanced adaptation technologies. As such, they can wait for a longer period than the much more vulnerable developing countries in the low latitudes. Immediate emission cuts would mean lower resource and energy consumption and a partial compromise in living standards which is rather difficult to reconcile for people long accustomed to a high standard of living.

The hope that was generated especially among the poor developing countries before the Copenhagen World Summit of the heads of states in December, 2009 was ultimately shattered through deliberate intervention by the USA, the biggest emitter of greenhouse gases to effectively stall the implementation of the Kyoto Protocol for emission reductions to mitigate climate change impacts thereby reconfirming the continuation of the US policy of utter disregard for any strategy that would compromise their standard of living. In the Earth Summit in Rio de Janeiro in June, 1992 the–then US President George Bush (Sr.) forewarned, "Our standard of living is not negotiable"; President Barack Obama like his predecessors may not have any other option but to continue with the legacy.

The mindless consumerism (a product of corporate culture) and wasteful living, and individual and collective selfishness resulting in disproportionately draining out and exploiting earth's reserves and natural resources (and in the process vitiating the earth's environment and ecosystems), can hardly be justified as acts of progressive human civilization.

The Latin name of "modern" human being, which evolved some 120,000 years ago is, *Homo sapiens sapiens*. The word *"sapiens"* means 'wise', as such, the present day human is doubly wise. Unfortunately, this overintelligent sole surviving species of the genus *Homo* is fully responsible for the present calamity being perpetuated on earth through climate change.

Perhaps the last hope of a global reconciliation on GHG emission reduction (to eventually become C–neutral), rests on the realization that the end is not very far and ultimately nobody would be spared from the inevitable onslaught of climate change. May be that would be too late for fruitful mitigation of the adverse impacts but let us hope for the best!

References

Adhikari, P., Sen, D. and Uphoff, N. 2010. System of rice intensification as a resource–conserving methodology : Contributing to food security in an era of climate change. *Satsa Mukhapatra*, Annual Technical Issue 14: 26–44.

Arrhenius, S. 1896. On the influence of carbonic acid in the air upon the temperature of the ground. *Philosophical Magazine and Journal of Science*, S5, 41(251):237–276.

Badgley, C., Moghtader, J., Quintero, E., *et al*. 2007. Organic agriculture and the global food supply. *Renewable Agriculture and Food Systems*, 22: 86–108.

Basu, R. N. 2007. The rationale behind an immediate return to crop–livestock integrated organic agriculture by small and marginal farmers. Proceedings of the National Workshop on "Organic Horticulture–It's Production, Processing, Marketing and Export for Sustainability," June 8–10, 2007 (Chattopadhyay, P. K. and Mitra, S. K. eds.), BCKV, West Bengal. Pp. 16–35.

Basu, R. N. 2008. Indian Agriculture at a crossroads: An imperative need to turn to crop–livestock integrated sustainable organic farming by small holders. *Indian Agriculturist*, Golden Jubilee Number. Pp. 32–49.

Basu, R. N. 2010. Ecologically sustainable organic agriculture for mitigating the adverse effects of global climate change. Professor J. C. Sengupta Endowment Lecture, 2010 (in press); also see WBSAC Final Report, Pp. 1–732.
Available at: http://www.banglarmukh.com

Chakraborty, D., Maulik, D. and Gupta, D. 2010. Climate Change : Suggested Action Plan. Environmental Factsheets. West Bengal State Pollution Control Board, pp. 33–70.

Cotter, J. and Tirado. R. 2010. Food security and climate change. *Satsa Mukhopatra*, Annual Technical Issue 14: 18–25.

Export–Import Bank of India, 2010. Agri Export Advantage: Organic Agriculture Worldwide, Global Market, New EU organic regulation and organic labels for food in the EU, Organic production in India. 14(1): 1–12.

Ghosh, M. 2006. Landshaping for better livelihood for the Sundarbans. *Sustainable Agriculture: A pathway out of poverty for India's rural poor*. GTZ, Sustainet, Eschborn, Germany. Pp. 94–99.

Institute of Science in Society (ISIS), UK. 2008. Press Release October 10, 2008 (Mae–Wan Ho) : organic and localised food and energy systems for mitigating climate change.
Available at: http://www/i–iss.org.uk/index.php

International Assessment of Agricultural Knowledge, Science and Technology for Development, (IAASTD). 2008 *Synthesis Report*.
Available at: http://www.agassessment.org

Intergovernmental Panel on Climate Change (IPCC) 2007. *Fourth Assessment Report*,
Available at: http://www.ipcc.ch/

Jouzel, J. C., Lorius, C., Petit, J. R. *et al.*, 1987. Vostok ice core: A continuous isotope temperature record over the last climatic cycle (160,000 years), *Nature*, 329: 403–407.

Kumar, H. D. 2006. Global Climate Change: Insights, Impacts and Concerns, Vitasta Publishing Pvt. Ltd., New Delhi.

Lu, C., Toepl, K., Irish, R., *et al.* 2006. Organic diets significantly lower children's dietary exposure to organophosphorus pesticides. *Environmental Health Perspectives*, 114(2): 260 – 263.

Masters, G. M. 1995. Introduction to Environmental Engineering and Science. Prentice-Hall of India Pvt. Ltd. New Delhi, Chapter 8, Global Atmospheric Change, pp. 375–446.

Menon, M. K. 2007. The market potentials for organic foods in India. *Organic Farming News Letter*, 3(2) : 22–26.

Mitchell, A., Hong, Y., Koh, E. *et al.* 2007. Ten–year comparison of the influence of organic and conventional crop management practices on the content of flavonoids in tomatoes. *Journal of Agricultural and Food Chemistry*, 55: 6154–6159.

NATCOM I. 2004. National Communication (India). Ministry of Environment and Forests,Govt.of India.
Available at: http://www.natcomindia.org/natcomreport.htm

Parikh, K. S. 2010. Food for Thought. Times of India, Kolkata, February 18, 2010, p.16.

Pimentel, D., Hepperly, P., Hanson, J. *et al.* 2005. Environmental, energetic and economic comparisons of organic and conventional farming systems. *Bioscience*, 55(7): 573–582.

Ravisankar, N. and Pramanik, S.C. 2007. Integrated farming system: Mode for hilly uplands of Bay Islands. *Indian Farming*, November :11–16.

Roberts, T.L. and Stewart, W.M. 2002. Inorganic phosphorus and potassium reserves. *Better Crops*, 86(2): 6–7.

Roy, S.K. 1982. Intervarietal and interspecies interactions: Indian experience. In: Roy, S.K. (ed.) Frontiers of Research in Agriculture. Indian Statistical Institute Golden Jubilee International Conference, 1982, Calcutta. Pp. 321–354.

Swami Sadananda, 2005. Land Shaping: An effective water utilization module for the Sundarbans. *Best Practices in Water Management: Case studies from Rural India.* German Agro Action (India). Pp. 65–77.

Vijayan, V. S. 2010. Kerala's planning for sustainable agriculture. Presentation in National Seminar on Sustainable Agriculture, Bhubaneswar; *also see* Kerala State Organic Farming Policy: Strategy and Action Plan, Government of Kerala 2010. Pp. 1–21.

Vijay Kumar, T., Raidu, D. V., Killi, J., Pillai, M., Shah, P., Kalavadonda, V. and Lakhey, S. 2009. Ecologically sound, economically viable: Community managed sustainable agriculture in Andhra Pradesh, India. The World Bank, Washington DC. Pp. 1–29.

West Bengal State Agriculture Commission. 2009. Final Report submitted to the Govt. of West Bengal in March 2009. Pp. 1–732.
Available at: http://www.banglarmukh.com

Xu, K., Xu, X., Fukao, T. *et al.* 2006. Sub 1A is an ethylene responsive–factor–like gene that confers submergence tolerance to rice. *Nature*, 442: 705–708.

Climate Change: Man and Environment (2012)   *Pages* **166–173**
Editor: **Goutam Kumar Saha**
Published by: **DAYA PUBLISHING HOUSE, NEW DELHI**

11

Climate Change and its Impact on Antarctic Life

☆ *Asok Kanti Sanyal**

Introduction

Antarctica is the fifth largest continent of the seven continents of the world. The continent itself makes up about 9.4 per cent of the land surface of the ice sheets and ice shelves being about 14 x 10^6 km² (Stonehouse, 2002). It is roughly circular in outline and forms eccentric hub around the south pole. The continent comprises the area south of 66°33´S latitude with two deep indentations the Weddell Sea, south of the Atlantic Ocean and the Ross Sea, south of the Pacific Ocean breaking the almost circular landmass. It contains open water in summer to above Lat. 78°S and floating ice–shelves all the year round extending to about 82°S and 85°30´S respectively at the heads of these seas. The tail of the continent, the Peninsula extending northwards from lesser Antarctica to south American continent at Lat. 63°15´S. The nearest country is Chile (950 km, Lat. 55°50´S), next come New Zealand (2000 km, Lat 46°40´S), Australia (2500 km, Lat 42°50´S) and South Africa (3800 km, Lat 34°24´S). It is over 10,000 km away from India.

Geomorphology

Geomorphologically, the Antarctica is twin continents, the Greater and the Lesser Antarctica separated by some deep under ice channels. Antarctica is divided into three main geographic regions: West Antarctica, East Antarctica and the Antarctic Peninsula. The West Antarctica is the youngest of the three geographic regions due to the buckling and folding of the earth's crust, and is characterized by huge mountain

* Zoological Survey of India, Kolkata. E-mail: asokzsi@yahoo.co.in

ranges both above and below sea level. The Antarctic Peninsula is the northern most section of Antarctica as well as the most moderate in climate due to its latitude. It is an extension of the Trans–Antarctic Mountain Range and is a fertile breeding ground of many life forms.

About two per cent or 8000 km² of Antarctic rock and soil are exposed from the ice. The oases in general are continuous ice–free rocky desert–like areas of low precipitation and freezing temperatures with some snow cover in winter in Antarctica and collectively form only a small part of the total land surface of the continent.

Antarctic Ice and Icebergs

Ninety eight per cent of Antarctica is covered by thick ice sheets that contains about 90 per cent of world's 'permanent' ice and 70 per cent of its fresh water (Fox and Cooper, 1994). Sea ice is made of sea water freezing at –1.5°C to –1.9°C. It may be called as fast or pack sea ice as per the location. There are numerous floating slabs of extended land ice sheet, the ice shelves. The Antarctic ice sheets are nearly 2.5 to 3 million years (Ma) old and have an average thickness of 2,400 m with maximum thickness of 4,786 m (Stonehouse, 2002) at Terre Adelie, about 400 km from the coast. The thinnest ice sheets are near the centre of the continent. The Greater Antarctica fronts on the Atlantic and Indian Oceans and has an ice dome upto 4,200 m thick ice and the Lesser Antarctica fronts on the Pacific Ocean with ice dome upto 2,000 m thick ice.

Icebergs are large floating mass of ice calved from ice shelves. It is of varying shapes and sizes having a covered area of about 31,000 km² with about 335 km in length and 97 km in width–the size of Belgium. The icebergs generally drift northwards by the ocean currents at a speed of about 10–20 km/day and normally remain within Lat. 50°S–62°50′S.

Life in Antarctica

The harshest climate in Antarctica obviously became the greatest hurdle for survival of life there. However, the southern sea surrounding the Antarctic continent is enormously rich in marine life. How the life will survive in a continent like Antarctica 98 per cent of which is covered with 2–3 km thick ice sheet? So the life is very scare on continental Antarctica. Whatever richness of plant and animal is found that is in Antarctic Peninsula and surrounding island where climate is to some extent tolerable.

Plants

Only two species of native flowering climbers *viz.*, the Antarctic hair grass *Deschampsia antarctica* and a cushion–forming pearlwort, *Colobanthus quitensis* are found to grow in western coast of Antarctic Peninsula (56°S) and nearby islands like South Orkney Islands and the South Shetland Islands. Besides these, in other areas of Antarctica, small forms of plants such as algae, lichens and mosses can be found beneath insulating layers of rocks or any other shelter to avoid cold mostly near the coasts and Trans–Antarctic mountains where rocky lands are present. The most abundant of these three plants is algae. They grow on rocks, ice and in sea. These blue–green, yellow, orange, brown and grey coloured algae are very old life–form in

Antarctica, the age of some algae is 6000 years or more. There are 300–400 species of algae, 200 species of lichen having beautiful colours like yellow, red, white, black, etc., 100 species of mosses and 25 species of liverworts. Mosses mostly grow surrounding the glacier lakes. They may be deep green or brown in colour and 1 cm to 30 cm thick.

Animals

In Antarctica faunal diversity is poor but very rich in population. The terrestrial fauna of Antarctica consists mostly of invertebrates, mostly microscopic living in soil and vegetation.The invertebrates in the Continental Antarctica include moss and soil inhabiting wingless collembola (Insecta), acarine (Arachnida), nematoda, protozoa, tardigrada, rotifera and wingless midges (*Belgica antarctica*). Besides these animals, around 65 species of winged parasitic insects are found in Peninsular Antarctica and neighbouring islands. These animals are capable of tolerating very low temperature.

Unlike the Continental and Peninsular Antarctica, the water around Antarctica or southern ocean are teeming with life of plant and animal species. The most fertile land in the world will loss to southern ocean in respect of production of protein, carbohydrate, fat, sugar, etc. This is due to extreme cold of sea for which the ocean can contain much more oxygen and carbon dioxide. The sea water in winter contains about 95 per cent oxygen. Due to highly fertile sea the phytoplankton shows enormous growth and very rich in population. As the zooplanktons live on phytoplanktons, they are also very rich both in quality and quantity. The most important and unique zooplankton is a Crustacea–the krill (*Euphausia superba*). These pinkish prawn–like animals of 2" to 3" long are present in sea in such a number (5–6 billion ton or more) that the colour of water turns reddish. These krills are full of protein and one of the body pigments may be used as medicine of ulcer. The whale, seal, penguin totally live on krill. A whale eats about three ton krill per day.

The other wildlife of Antarctica mostly thrive in the sea are squids, sea anemones, jelly fishes, arrow worms, combe jellies, snails, salps, starfishes, seals, whales, penguins and other sea birds like skuas, terns, petrels and albatrosses.

Fish population in Antarctic sea is very rich and over 200 species are known to live in the water. About 75 per cent of fishes belonging to five families under the order Notothenioidea, a relative of cod fish, make up the fish population. The blood of some species of fish is white in color due to absent of red blood cells. There are some species like *Dissotichus mawsani* and *Trematomus* sp. which can withstand extreme cold for presence of an antifreeze substance glycopeptide which prohibits coagulation of blood.

Antarctic seals include crabeater seal (*Lobodon carcinophagus*), Antarctic fur seal (*Arctocephalus gazella*), leopard seal (*Hydrurga leptonyx*), Weddell seals (*Leptonychotes weddellii*) and elephant seal (*Mirounga angustirostris*). Only the Weddell Seals can be found in Antarctica year–round. Other seals migrate to the north by following the pack ice.

Whale the largest mammal can be found in Antarctic waters and only during the summer time. The species of whale found around the Antarctic continent are blue whale, minke whale, humpback whale, fin whale, sperm whale, killer whale (*Orca* sp.), southern right whale, Arnoux's beake whale, southern bottlenose whale and hourglass dolphin.

Around 35 species of birds visit Antarctica and nearby islands to breed or live. They range from magnificent wandering albatross capable of flying thousands of kilometers for feeding and mating to gulls, cormorants and terns. Most of them return to the same site each year. Eighteen species of penguins live south of equator including southern coast of South America, south Australia, New Zealand, tip of Africa and all around Antarctica. One species, the Galapagos penguin, lives on the equator near Peru. Seven species are the regular visitor of Antarctica and only two emperor and Adelie penguins breed exclusively on the Antarctic continent.

Recent Climate Change in Antarctica

Since the International Geophysical year of 1957–58, Antarctic surface temperatures have remained stable over much of the continent. The majority of stations in East Antarctica showed no significant warming or cooling trends (Turner *et al.*, 2005). On the contrary, large and sufficiently significant warming trends are seen at the Antarctic Peninsula. The most rapidly warming part during the past 50 years period is the west coast of the peninsula. The annual mean temperatures at the peninsula have risen by nearly 3°C, with the largest warming occurring in winter (Turner *et al.*, 2005; Vaughan *et al.*, 2001; King *et al.*, 2004). This rate of temperature increase is approximately 10 times the mean rate of global warming as reported by the Intergovernmental Panel on Climate Change (IPCC, 2001). The east coast of the peninsula warmed slowly showing largest warming in summer and autumn (King *et al.*, 2004).

The Southern Ocean showed significant warming. Upper ocean temperatures to the west of the Antarctic Peninsula showed an increase by over 1°C since 1955 (Meredith and King, 2005). It is well established that the waters of the Antarctic circumpolar current are warming more rapidly than the global ocean as a whole. This rate of warming is of around 0.2°C at 700–1000 m depth.

The result of study of vertical warming over the last 30 years has shown that Antarctic atmosphere has warmed below 8 km and cooled above this height. This trend of warming in the troposphere and cooling in the stratosphere indicates increase in greenhouse gasses, such as carbon dioxide. Year round satellite observations on Antarctic sea ice extent has indicated that sea ice cover has declined substantially in the seas to the west of Antarctic Peninsula while it has increased in other parts of the Antarctica (Zwally *et al.*, 2002). The reports on atmospheric circulation around Antarctica mentioned that atmospheric pressure has dropped over Antarctica and risen in the mid–latitudes of the southern hemisphere (Marshall, 2003).

The ice cover of the peninsula has largely changed, many glaciers have retreated (Cook *et al.*, 2005) and around 10 ice shelves have been observed to retreat in recent years (Vaughan and Doake, 1996) and some have collapsed totally. Eighty seven per cent of glaciers along the west coast of the peninsula have greatly retreated resulting sea–level rise at about the same rate as Alaska glaciers. The climate change has weakened the carbon sink of the Southern Ocean since 1981. This is due to increased upwelling of carbon–rich waters associated with strengthening of the westerly winds (IeQuere *et al.*, 2007).

The analysis of the result of works of different climate models by IPCC showed greatest warming in the peninsula region and little change elsewhere (Chapman and Walsh, 2007).

Impact of Climate Change on Antarctic Life

The recent climate change in Antarctica has made significant changes in the physical and biological environment. The environmental changes are most remarkable in the Antarctic Peninsula, where climate change has been most significant. The study conducted by Doran *et al.* (2002) in the McMurdo Dry Valleys had evidenced rapid terrestrial ecosystem response to climate cooling in Antarctica. Their spatial analysis of Antarctic meteorological data showed a net cooling on the Antarctic continent between 1966 and 2000, particularly during summer and autumn. They reported decreased primary productivity of lakes (6–9 per cent per year) and declining numbers of soil invertebrates (more than 10 per cent per year). Fourteen years of continuous weather recording from the shore of Lake Hoare by them revealed that seasonally average surface air temperature has decreased by 0.7 per cent per decade. The temperature decrease was more pronounced in summer and autumn. Winter (June–August) and spring (September–November) showed smaller temperature increase. They opined that climate cooling has significantly impacted ecosystem properties. They also reported decrease in irradiance during November–December which affected the rate of primary production in the lakes.

Soil invertebrate communities showed changes in diversity and abundance from 1993 to 1998. The abundance of tardigrades and nematodes were specially declined even after climate manipulation experiment. This indicates that change in climate in Antarctica has declined the diversity, life cycle, trophic relationships and functioning of dry valley soils.

It is difficult to assess the response of soil organisms to the climate change. However any change in the climate will consequently affect the soil organisms in their density and biomass. Soil organisms play a significant role in soil nutrient cycling, decomposition of organic wastes and biodegradation of the major hydrocarbons as soil spills. The fauna is moderately abundant in Antarctic soil and vegetation. It's a fact that soil is a thermally buffered environment that responds slowly to temperature changes in their column above. Therefore, the biota living within the soil is less likely than the surface vegetation to show large responses to changes in atmospheric temperature. Although climate does not have direct effects on soil biota, soil biota are more strongly affected by vegetation than they are by atmospheric conditions directly. This is for two reasons: (1) the physical structure of the vegetation affects the soil temperature and moisture, and (2) vegetation provides vast amount of organic matter (*i.e.* energy supply) and nutrients which may affect the species composition and structure of the soil biotic communities. This phenomenon has been observed correct in Antarctic Peninsula, where melting of perennial snow and ice covers encouraged the growth and spreading of established plants and increased establishment of seedlings. The increase in abundance in vegetation has resulted increased abundance of soil and other terrestrial fauna in some sites of Antarctic Peninsula.

Plant responses to atmospheric CO_2 increase, to global warming and to increased ultraviolet–B radiation as a result of stratospheric ozone depletion. Changes in temperature and precipitation have increased biological production in lakes, mainly due to decreases in the duration and extent of lake ice cover. Some lakes have become more saline due to drier condition, and resulted depletion of lake biota.

The review of studies done during 1985–2000 by several Indian scientists including the author in the Schirmacher Oasis, East Antarctica revealed that the faunal density in soil and lake waters was much higher in the year 1987 (Ingole and Parulekar, 1993). The result also indicated that the population of microscopic protozoa and nematoda was much more in 1996 and decreased later. Mitra (1999) observed dominancy of immature forms of protozoa, nematoda and acarina and stated that it was due to suitable environment for reproduction. Sanyal (2004) reported that mite population was more in quantity and quality that the earlier records in the Oasis.

Compared to many marine organisms the terrestrial biota in Antarctica often has a wide environmental tolerance. It includes some of the most robust life forms on earth, the cyanobacteria, which can survive extremes of low temperature, water availability, light and high UV radiation (Hodgson *et al.*, 2004).

Environmental change is most apparent in the Antarctic Peninsula, where climate change has been largest (Turner *et al.*, 2009). The changes have declined in number of Adelie penguins, a species well adapted to sea ice conditions, and been replaced by open–water species such as chinstrap penguins (Frazer *et al.*, 1992). The number of emperor penguins has dropped from 300 breeding pairs to just nine in the western Antarctic Peninsula.

The Antarctic marine ecosystem is also severely affected by climate change, especially on the western side of peninsula, with its warm water and reduction in sea ice. Among the significant cases of decline in population of different marine fauna, decline in Antarctic krill stocks, decrease in phytoplankton and southward shift in the population of gelatinous salps are noteworthy (Atkinson *et al.*, 2004). The decline in phytoplankton may reflect a decrease in iron input form the continental margin that is in turn related to a reduction in the formation of sea ice in this region and hence to climate change. There are records of decline in population of fish especially *Pleuragramma antarcticum*, whose reproduction is associated with sea ice. Antarctic marine animals are of typical stenothermal in nature which would be highly sensitive to significant warming. Most species have upper lethal temperature below 10°C and some can survive just a 5°C change. The behavior of organisms can be affected at lower temperature long before lethal levels are reached. The question whether the rise may depend upon their ability to carry out critical activities like feeding. The planktonic snails, key species in food chain, could suffer from ocean acidification. This decline in food regime is likely to affect the Antarctic sea bed biodiversity.

The increase in temperature in terrestrial ecosystem promote growth and production of biota, but also cause drought and associated effects. Water availability would have greater effect on vegetation and faunal composition than temperature. An increase in the frequency and intensity of freezing events could readily exceed the tolerance limits of many arthropods. With increase in temperature many terrestrial species may exhibit faster metabolic rates, shorter life cycles and local expansion of population, which is presently observed in Peninsular Antarctica. Warming also favours alien species to migrate and settle into the region, competing with and replacing original Antarctic inhabitants.

Remarks

The analysis of instrumental data from some limited research stations in Antarctica revealed that Antarctica have undergone complex and significant temperature changes in recent decades, with greatest warming in the peninsula region and little change elsewhere. Recent climate observations indicate that changes in the strength of the wasterlies strongly influence temperature variations on the east coast of the Antarctic Peninsula (Marshall *et al.*, 2006). Contrary to this, the interior parts of East Antarctica have experienced less warming, with large spatial variability. It is evidently difficult to separate natural climate variability and anthropogenic influences. However, the effect of greenhouse gases are already evident and if they continue to rise at the current rate, some catastrophic effects on physical and living environment of Antarctica are inevitable. We can make reasonably a broad estimate of how the parameters of physical environment like temperature, precipitation and sea ice extent might change and consider the possible impact on terrestrial and marine biota. Antarctica is earth's most powerful natural laboratory, a pivotal part of the earth's climate system and a sensitive barometer of environmental change. For peaceful sustenance of life on earth, the Antarctic continent must be saved along with its fragile environment.

References

Atkinson, A., Siegel, V., Pakhomov, E. *et al.* 2004. Long-term decline in krill stock and increase in salps within the southern ocean. *Nature*, 432:100–103.

Chapman, W.L. and Walsh, J.E. 2007. A synthesis of Antarctic temperatures. *Journal of Climate*, 20: 4096–4117.

Cook, A.J., Fox, A.J., Vaughan, D.G. *et al.* 2005. Retreating glacier fronts on the Antarctic Peninsula over the past half–century. *Science*, 308: 541–544.

Doran, P., Priseu, T., John, C. *et al.* 2002. Antarctic climate and terrestrial ecosystem response. *Nature*, 415: 517–520.

Fox, A.J. and Cooper, A.P.R. 1994. Measured properties of the Antarctic Ice Sheet derived from the SCAR Antarctic Digital Database. *Polar Records*, 30 (174): 201–206.

Frazer, W.R., Trivelpiece, W.Z., Ainley, D.G. *et al.* 1992. Increases in Antarctic Penguin populations: reduced competition with whales or a loss of sea ice due to environmental warming? *Polar Biology*, 11: 525–531.

Hodgson, D.A., Vyverman, W., Verleyen, E. *et al.* 2004. Environmental factors influencing the pigment composition of *in situ* benthic microbial communities in east Antarctic lakes. *Aquatic Microbial Ecology*, 37: 247–263.

Ie Quere, C., Rodenbeck, C., Buitenhuis, E.T. *et al.* 2007. Saturation of the Southern Ocean CO_2 sink due to recent climate change. *Science*, 316: 1735–1738.

Ingole, B.S. and Parulekar, A.H. 1993. Limnology of freshwater lakes at Schirmacher Oasis, East Antarctica. *Proc. Indian natn. Sci. Acad.*, B 59 (6): 589–600.

Intergovernmental Panel on Climate Change (IPCC). 2001. *Climate Change: Impacts, Adaptation, and Vulnerability* (McCarthy, J.J., Canziani, O.F., Leary, N.A. *et al.* eds.). Cambridge University Press, Cambridge, UK. Pp. 1031.

King, J.C., Turner, J., Marshall, G.J. *et al.* 2004. Antarctic Peninsula a Climate Variability and its Causes as Revealed by Analysis of Instrumental Records. In: *Antarctic Peninsula Climate Variability: A Historical and Paleoenvironmental Perspective* (Domack, E., Burnett, A., Convey, P. *et al.* eds.), American Geophysical Union. 17–30.

Marshall, G.J. 2003. Trends in the southern annular mode from observations and reanalyses. *Journal of Climate*, 16: 4134–4143.

Marshall, G.J., van Lipzig, O.P.M. and King, J.C. 2006. The impact of a changing Southern hemisphere annular mode on Antarctic Peninsula summer temperatures. *Journal of Climate*, 19: 5388–5409.

Meredith, M.P. and King, J.C. 2005. Rapid climate change in the ocean west of the Antarctic Peninsula during the second half of the 20th century. *Geophysical Research Letters*, 32.

Mitra, B. 1999. Studies on moss–inhabiting invertebrate fauna of Schirmacher Oasis. *XIIIth Indian Expedition to Antarctica, Scientific Report, Department of Ocean Development, Tech. Publ.*, 13: 93–108.

Sanyal, A. K. 2004. Notes on the ecology of soil mites (Acari) in two contrasting sites of Scirmacher Oasis, East Antarctica. *Acarina*, 12(2): 151–157.

Stonehouse, B. 2002. *Encyclopedia of Antarctica and the Southern oceans*. John Wiley and Sons Ltd, Chichester.

Turner, J., Bindschadler, R., Convey, P. *et al.* 2009. *Antarctic climate change and the Environment: a contribution to the International Polar Year*. Scientific Committee on Antarctic Research, Cambridge.

Turner, J., Colwell, S.R., Marshall, G.J. *et al.* 2005. Antarctic climate change during the last 50 years. *International Journal of Climatology*, 25: 279–294.

Vaughan, D.G. and Doake, C.S.M. 1996. Recent atmospheric warming and retreat of the ice shelves on the Antarctic Peninsula. *Nature*, 379: 328–330.

Vaughan, D.G., Marshall, G.J., Connolley, W.M. *et al.* 2001. Devil in the detail. *Science*, 293: 1777–1779.

Zwally, H.J., Comiso, J.C., Parkinson, C.L. *et al.* 2002. Variability of Antarctic sea ice 1979–1998. *Journal of Geophysical Research*, 107: 1–19.

Climate Change: Man and Environment (2012)   *Pages* **174–187**
Editor: **Goutam Kumar Saha**
Published by: **DAYA PUBLISHING HOUSE, NEW DELHI**

12

Climate Change and its Impact on the Distribution of Birds in Southern Indian Ocean and Antarctica

☆ *K. Sivakumar and S. Sathyakumar**

Introduction

Antarctica and the surrounding ocean have a unique assemblage of flora and fauna that have been threatened in the recent past due to global warming and also due to increasing human activities. Climate change impacts ocean biota in different ways and biological feedbacks in course of time may become amplified during any climate change signal (Camille, 2006). These feedbacks arise from changes in primary production or shifts in the phytoplankton community composition. Several ocean models have suggested that the Southern Ocean surface waters are going to become warmer and fresher with increased vertical stratification, reduced sea–ice, higher oceanic CO_2 concentrations and ultimately increased upwelling. Alteration of these ocean properties may affect the plankton dynamics and community composition in these waters. This is in turn could have a long lasting impact on oceanic birds and mammals, which are directly or indirectly linked to the availability of certain species of phytoplankton. Penguins, oceanic birds and whales are expected to be affected badly among vertebrates due to global warming in circumpolar region around Antarctica.

* Wildlife Institute of India, Dehradun. E-mail: ksivakumar@wii.gov.in, ssk@wii.gov.in

Monitoring the wildlife populations provides crucial information about the viability of the population being monitored and about the quality of the habitat or landscape in which the population occurs. Although the broad distribution, abundance and ecology of major mammal and bird species that occur in the southern Indian Ocean and Antarctica are known, information on population trends is essential for understanding the long-term conservation status of these species with respect to climate change. The impacts of regional climate change and extreme weather on wild species has been studied for several decades (Camille, 2006). Paleoclimatic studies have also shown that species have adjusted to climate changes at times in the past without mass extinctions (Isabel, 2007). Yet, it is uncertain if projected climate change widely foreseen today would mimic climate change events in the geological record and if human–ecosystem relationships would help or hurt adaptation. In the last ten years, scientists have documented the effects of climate change on species and populations on every continent and in most taxonomic groups. Some studies provide correlations between climate change and species changes, others predict changes with climate and species models and some demonstrate mechanistic connections between species changes and climate change (Root, 2003). Many studies report that climate change acts in concert with other factors to affect species and their habitat. For this reason and because of differences across biomes and species, it is difficult to generalize about the overall impacts of climate change on biodiversity and ecosystems.

The IUCN, the Scientific Committee on Antarctic Research (SCAR) and the Convention for the Conservation of Antarctic Marine Living Resources (CCAMLR, 2007) stress the importance of ecosystem monitoring that includes monitoring of wildlife species such as the penguins and seals that are indicators of Antarctic ecosystem. Having better understanding about the impact of climate change on wildlife in a ecosystem where anthropogenic pressures are comparatively less would help to prepare better conservation plan for the marine biodiversity of India and elsewhere, which are threatened by both anthropogenic pressures and climate change.

There are few studies that report the effects of climate change on species distribution, partly because of the difficulties of gathering data throughout a range during a period long enough to produce significant results (Camille, 2006). Such kind of studies in the southern Indian Ocean was even less (Jayasankar *et al.*, 2007). Therefore, this study was initiated by the Wildlife Institute of India aimed to observe the effect of climate change on birds of southern Indian Ocean.

Study Area

The sea voyage route of 28th Indian Scientific Expedition to Antarctica covering a stretch of the southern Indian Ocean from 40°S to 69°S was used as the study area. We carried out ship based surveys along the sea voyage route that started from Cape Town in Republic of South Africa to Larsemann Hills of Antarctica (68°54'92.1" S, 75°30'40.2" E) in almost a straight line; then to the Princess Astrid Coast (5–20°E and 69–72°N), also known as the India Bay along the coastal area of Antarctica; and from India Bay to Cape Town in a straight line. The unique biodiversity assemblage of southern Indian Ocean and Antarctica is home to 21 mammalian species and about 45 species of birds which include seven species of penguins. About 75 species of oceanic birds have been recorded in the southern Indian Ocean especially between South Africa and Antarctica.

Methods

Abundance and Distribution of Oceanic Birds

The ship M.V. Emerald Sea was chartered by the Govt. of India for the 29th Indian Scientific Expedition to Antarctica (January–March 2009). Oceanic birds all along the sea voyage were sampled at regular intervals to estimate their abundance and to understand their distribution pattern from defined co–ordinates. Standard techniques for censusing seabirds at sea were used following Tasker *et al.* (1984). Briefly, all seabirds that entered a 90 arc from the bow to the beam and out to 150 m on the one side with best visibility (*e.g.* lowest sun glare) were enumerated and their behaviour recorded by two observers stationed on the bridge. A hand–held binoculars equipped with reticules was used to ground–truth the width of the 300–m survey strip. Belt transects were laid along the voyage line at the interval of two hours in a day for two hour observation each. Likewise, three observations covering six hours in a day were made. Start and end points of each transect was marked using a GPS. Speed of the ship during the transect sampling was consistent throughout voyage at the speed of 12 nm/hour. The average length of each belt transect in sea was 24 nm. The ocean between Cape Town and Antarctica was divided into following three zones (Bhatnagar and Sathyakumar, 1999) for data analysis as each zones differed in various environmental settings (Figure 12.1).

Zone 1: Temperate (36°S to 49°S)

This zone is characterised by an abrupt reduction in sea temperature (of up to 4°C) and salinity along the north–south axis due to an oceanic frontal system, the *sub–tropical convergence* (STC), at approx. 40°S. Here, the cold sub–Antarctic and the warmer sub–tropical surface water meet and sink deep, leaving a distinct line of oceanic disturbance, popularly called the 'Roaring Forties'. Temperatures just north of the STC are ca. 14°C during winter and 18°C during summer, and usually remain ca. 4°C lower to the south of the STC. Salinity is also high in this zone 34.9 ppt (Watson, 1975; Eastman, 1993).

Zone 2: Sub-Antarctic/Sub-Polar (50°S to 59°S)

In this zone, a decline of another 2° to 3°C occurs in sea water temperature across another frontal system, the *Antarctic convergence* (AC). This convergence marks the subduction of the cold and denser Antarctic surface water under the slightly warmer, sub–Antarctic surface water. Water temperatures are ca. 5°C to 10°C during winter and 8°C to 14°C during summer months in the sub–Antarctic zone, north of the AC. Ambient temperatures dip to 4°C to 8°C and the ocean usually remains rough. Salinity is reported to be ca. 34.3 ppt (Eastman, 1993).

Zone 3 : Antarctic/Polar (60°S to 69°S)

Water temperature near the shelf usually remains sub–zero round the year, but further north, on an average is ca. 1°C to 2°C in winter and 4°C to 5°C during summer (Watson, 1975). An ocean front, the *Antarctic divergence*, occurs at ca. 65°S. The

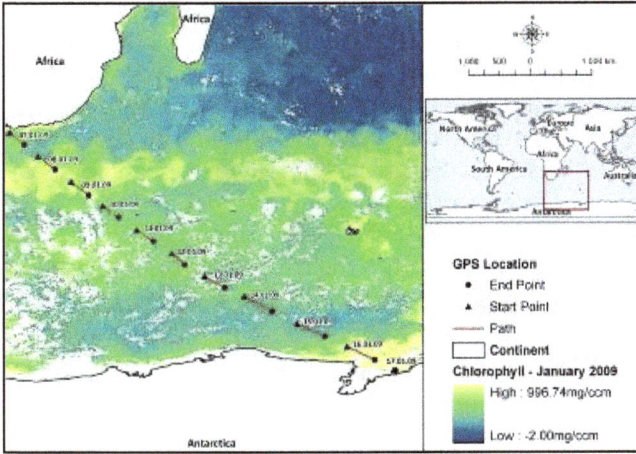

Cape Town in South Africa to Larsemann Hills in East Antarctica, January 2009

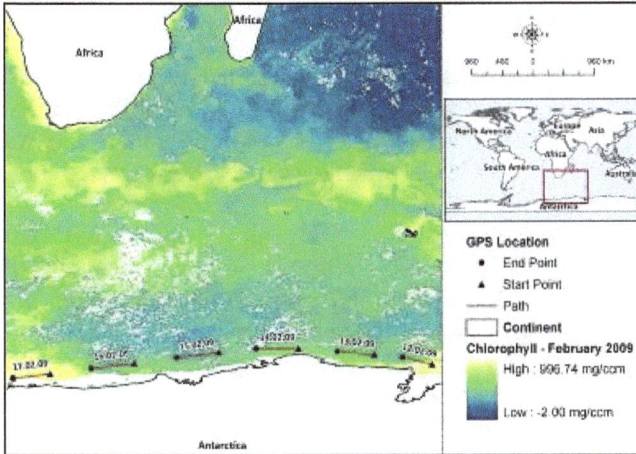

(A) Larsemann Hills to Princess Astrid Coast (India Bay, near Maitri), February, 2009

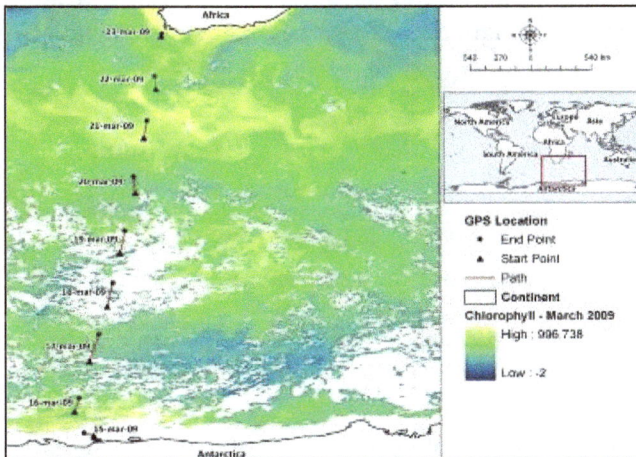

(B) India Bay (near Maitri) to Cape Town, March, 2009

Figure 12.1: A map of the southern Indian Ocean showing the sea voyage route undertaken during the study and the average chlorophyll content during the study period, January–March 2009.

Antarctic water, cooled by ice and wind, off the ice shelf becomes dense and sinks to the bottom. The 'hole' thus formed is filled by the circumpolar deep water which rises and diverges south towards Antarctica and north towards the AC. This deep water is profuse in nutrition and consequently, this zone is known to be a biologically rich zone. The Antarctic sea ice has considerable influence on the phytoplankton productivity. The ice edge is particularly rich because of the 'phyto–planktonic bloom' (Sakshaug and Skjoldal, 1989). The salinity of the sea water in the immediate vicinity of the shelf varies between 30.5 to 34.5 ppt seasonally, with a mean of 33.9 ppt. The summer melt dilutes the sea water and in winter, the freezing sea leaves out salt which makes the water more saline (Watson, 1975; Eastman, 1993).

A total of 66 belt transects of 24 nm each were laid in ocean between January and March 2009. Total observation time was 136 hours and sampled 1613 nm. Total area of bird count was 261 nm^{-2}.

Abundance and Distribution of Birds in the Coastal Habitat of Antarctica

Aerial strip–line transect method was used. Building on this method, the density of birds of Antarctica had been estimated with relation to different habitats which occurred there all along the coast of Larsemann Hills and India Bay (near Maitri). All identified major habitats in Antarctica *i.e.* the maritime, polynea–pack ice, ice shelf and mainland habitats were surveyed using a helicopter. For the aerial transects, the helicopter was flown at a ground speed of 100 km h^{-1} at an altitude of around 70 to 100 m above sea surface all along the coastal line. Number of groups or flocks of birds, size of groups or flocks and their demographic pattern was noted from the aircraft. Aerial photography was also done to reconfirm the aerial census data. On the mainland habitat of Antarctica, few 'Variable width ice–line transects or total count or flock count' was used to monitor the populations of major avian species. A total of three aerial surveys had been carried out along the Larsemann Hills and four surveys along the India Bay. Total distance surveyed at Larsemann Hills was about 60 nm with three replications. Total length of coasts surveyed along India Bay was about 140 nm with four replications.

Results

Species Richness

In the southern Indian Ocean, between Cape Town and Antarctica, a total of 46 oceanic bird species were recorded during the summer months of January, February and March 2009. The list of birds along with their common and scientific names recorded during the expedition are given in Table 12.1. The species composition of birds changed significantly along latitudinal gradient (R^2=0.531) but number of species sighted throughout the region was not changed significantly (R^2=0.003, Figure 12.2).

At Larsemann Hills where India is establishing her third Antarctic Research Station, six species of birds were recorded. These include: Adelie penguin, emperor penguin, south polar skua, snow petrel, Wilson's storm petrel and light mantled

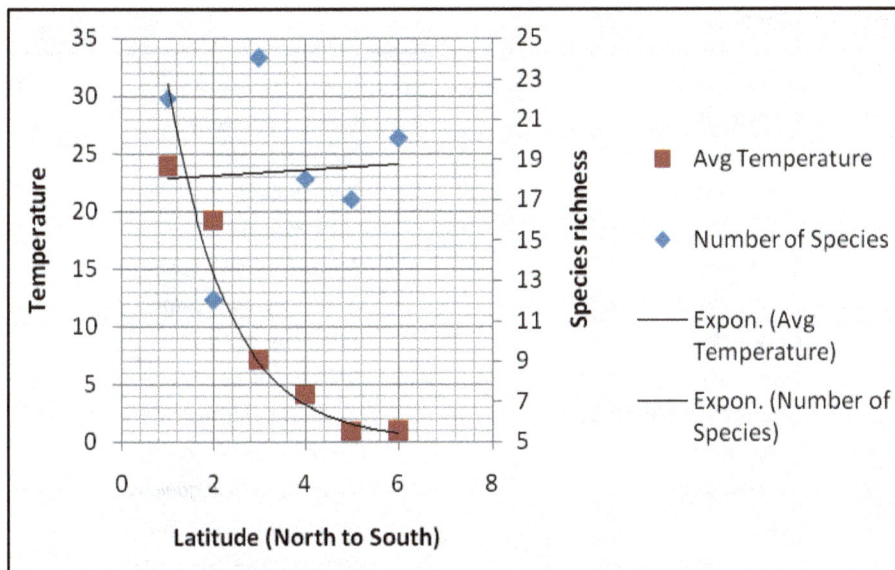

Figure 12.2: Species richness of oceanic birds in the southern Indian Ocean (from Cape Town to Antarctica) January–March 2010.

sooty albatross. Of these, south polar skua and storm petrel were observed breeding in this group of islands. It was also observed that Adelie and emperor penguins were using some of these islands for moulting especially on Fisher and Stornes islands.

Table 12.1: Checklist of birds recorded in the southern Indian Ocean and off-Antarctica, January – March, 2009.

S.No.	Species	Scientific Names
1.	Adelie Penguin	*Pygoscelis adeliae*
2.	Antarctic Fulmar	*Fulmarus glacialoides*
3.	Antarctic Petrel	*Thalassoica antarctica*
4.	Antarctic Skua	*Catharacta antarctica*
5.	Antarctic Tern	*Sterna vittata*
6.	Atlantic petrel	*Pterodroma incerta*
7.	Black–bellied Storm Petrel	*Fregetta tropica*
8.	Black–browed Albatross	*Diodedea melanophris*
9.	Blue Petrel	*Halobaena caerulea*
10.	Broad–billed Prion	*Pachyptila vittata*
11.	Cape Gannet	*Sula capensis*
12.	Cape Petrel	*Daption capense*
13.	Coommon Diving Petrel	*Pelecanoides urinatrix*
14.	Cory's Shearwater	*Calonectris diomedea*

Contd...

Table 12.1–*Contd...*

S.No.	Species	Scientific Names
15.	Emperor Penguin	*Aptenodytes forsteri*
16.	Fairy prion	*Pachyptila turtur*
17.	Flesh footed shearwater	*Puffinus pacificus*
18.	Great Black backed Gull	*Larus marinus*
19.	Great–winged Petrel	*Pterodroma macroptera*
20.	Grey headed albatross	*Deomedea chrysostoma*
21.	Grey Petrel	*Procellaria cinerea*
22.	Grey–backed Storm Petrel	*Garrodia nereis*
23.	Grey–headed Albatross	*Diomedea chrysostoma*
24.	Kerguelen Petrel	*Pterodroma brevirostris*
25.	King Penguin	*Aptenodytes patagonicus*
26.	Laysen albatross	*Diomedea immutabilis*
27.	Leach's Storm Petrel	*Oceanodroma leucorhoa*
28.	Light mantled sooty albatross	*Phoebetria palpebrata*
29.	Pomarine skua	*Stercorarius pomarinus*
30.	Royal albatross	*Diomedea epomophora*
31.	Salvin's Albatross	*Diomedea cauta salvini*
32.	Snow Petrel	*Pagodroma nivea*
33.	Soft–plumaged Petrel	*Pterodroma mollis*
34.	Sooty Albatross	*Phoebetria fusca*
35.	Sooty Shearwater	*Puffinus griseus*
36.	South Polar Skua	*Catharacta maccormicki*
37.	Southern Giant Petrel	*Macronectes giganteus*
38.	Thin–billed Prion	*Pachyptila belcheri*
39.	Wandering Albatross	*Diomedea exulans*
40.	White capped albatross	*Deomedea cauta cauta*
41.	White chinned Petrel	*Procellaria aequinoctialis*
42.	White–bellied Storm Petrel	*Fregetta grallaria*
43.	White–capped Albatross	*Deomedea cauta*
44.	White–headed Petrel	*Pterodroma lessonii*
45.	Wilson's Storm Petrel	*Oceanites oceanicus*
46.	Yellow nosed albatross	*Diomedea chlororhynchos*

At India Bay and *Maitri*, six species of birds were recorded. These include: Adelie penguin, emperor penguin, south polar skua, snow petrel, Wilson's storm petrel and southern giant petrel. Of these, skua and snow petrel were observed breeding in the Schirmacher Oasis. More number of species recorded along the coast. South polar skua is a common bird in interior Antarctica and they are largely seen around the scientific research stations looking for food and also known to hunt snow petrels. South polar skua and snow petrels along the Adelie were seen with chicks during survey period.

Abundance and Distribution Pattern

In the 1,613 nm long total transects, a total of 5,727 birds belonging to 46 species were recorded and the density estimated at 21.9 birds/nm² in the southern Indian Ocean. Although, the number of bird species recorded in the different latitudinal zones were more or less similar (R^2=0.003, Figure 12.1), higher number of oceanic birds were recorded between 60°S and 70°S. Density and sightings of oceanic birds towards south was higher than north in the southern Indian Ocean (Figure 12.3).

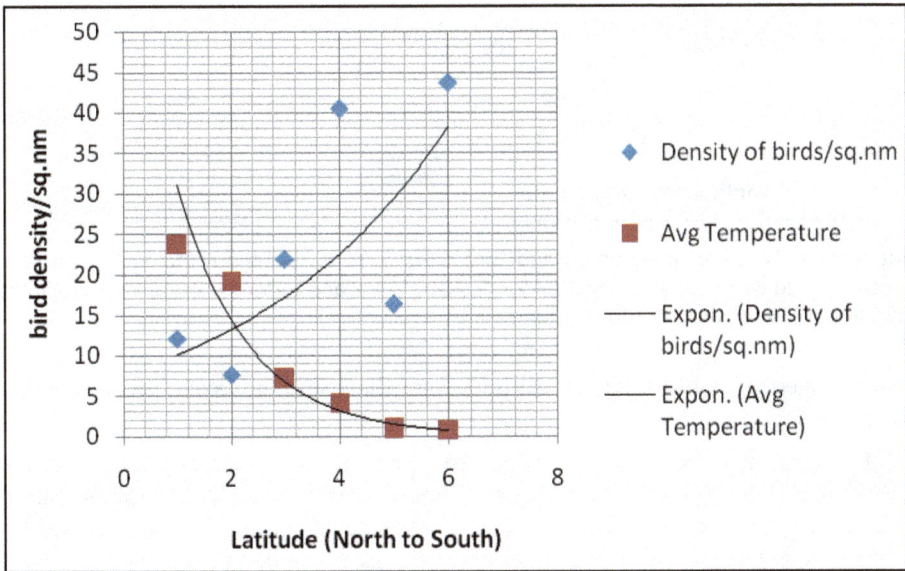

Figure 12.3: Density of oceanic birds in the southern Indian Ocean (from Cape Town to Antarctica), January–March 2010.

In the southern Indian ocean, Antarctic petrel, broad–billed prion, Antarctic skua, Kerguelen petrel, light–mantled sooty albatross, wandering albatross, cape petrel, great–winged petrel, Antarctic fulmer and thin–billed prion were recorded as common birds and had more sightings. Black bellied storm petrel, Atlantic petrel, Antarctic tern, common diving petrel, grey petrel, Salvin's albatross, Layson albatross, Leach's storm petrel, black browed albatross and king penguin were observed rarely in the study area with few sightings. There was a significant negative correlation between ambient temperature and bird density as well as sightings in the southern Indian Ocean. Number of sightings and density of birds increased when the ambient temperature reduced (Figure 12.4).

Range extension of Cory's shearwater, Antarctic skua, cape petrel, white–capped albatross, grey–headed albatross, blue petrel, sooty albatross, kerguelen petrel and

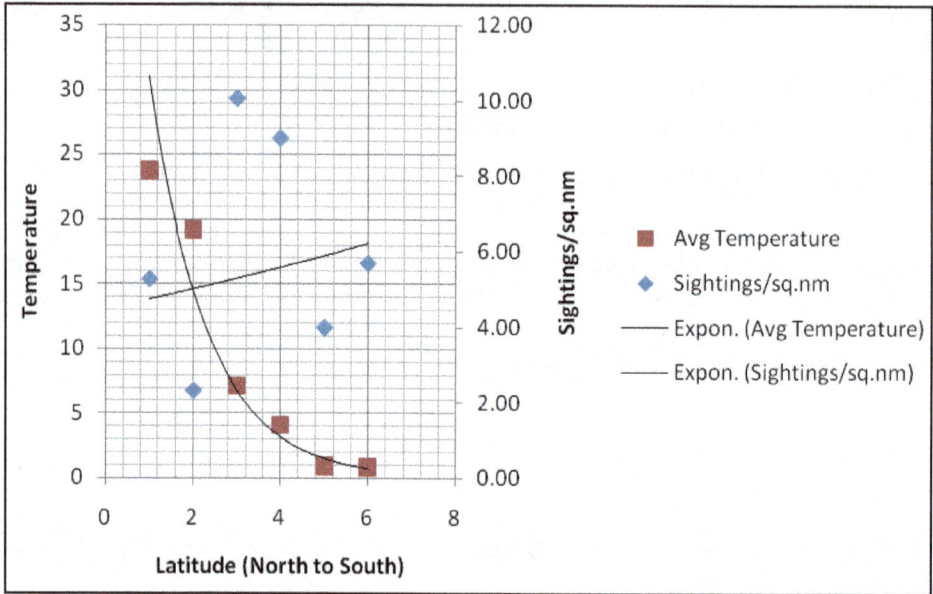

Figure 12.4: Number of sightings and Bird density in the Southern Indian Ocean (from Cape Town to Antarctica), January–March 2010. Increase in temperature due to climate change may affect this population trends.

great winged petrel towards south in the southern Indian Ocean were observed during this study.

In Antarctica, the encounter rate of two penguin species, along the coast from Clements Bay to Stornes Island in Larsemann Hills, were 24±21/aerial sortie for Adelie penguin and 2±1/aerial sortie for emperor penguin. The total coastal line surveyed was 60 nm. Although, king penguin were sighted in the Laresemann Hills, they were not recorded during the aerial survey as they were very rare. Observations on Adelie penguins in the rookery at Hop Island was also made and we estimated the population to about 5,000 birds which were mostly chicks as their parents were in sea, searching food for their young ones.

The skua population at and around *Maitri*, the second research station of India, was also estimated as six breeding pairs. The mean number of individuals (±SD) recorded during aerial surveys along the Princess Astrid Coast were as follows: Adelie penguin 13±10, emperor penguin 23±19, each sortie covered around 140 nm.

Remarks

In the southern Indian Ocean, bird species richness seems to be more or less similar from north to south during the southern summer. However, the bird density increased significantly from north to south due to low temperature towards Antarctica (Figure 12.4). In the circumpolar region, life gets activated during summer due to favourable temperature that facilitates the production of primary producers and consumers. This ecological phenomenon has been attracting several major vertebrates

Figure 12.5: Observations on birds and mammals made during the sea journeys.

Antarctic Petrel

Cape Petrel

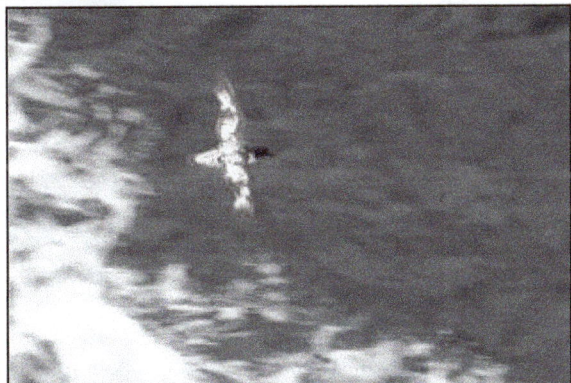

Contd...

Figure 12.5–*Contd...*

Humpbacked whales

including ocean birds to forage and breed here. Any change in the temperature of southern Indian Ocean may have adverse impacts on the species composition as well as abundance of ocean birds as climate change impacts ocean biota in different ways and biological feedbacks in course of time may become amplified during climate change (Fung *et al.*, 2005). Because of alteration of ocean properties such as temperature and salinity, these may affect the phytoplankton dynamics and could result in a shift in phytoplankton community composition in these waters. This in turn could have a long lasting impact on zooplankton and other faunal populations such as oceanic birds, which are directly or indirectly linked to the availability of certain species of phytoplankton (Moline *et al.*, 2004).

Figure 12.6: Wildlife species encountered during aerial surveys.

Emperor Penguins and Weddel Seals at Thala Fjord, Larsemann Hills

Contd...

Figure 12.6–*Contd...*

Adelie Penguins

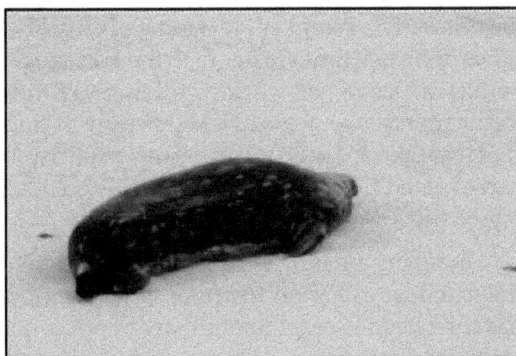

Leopard Seal

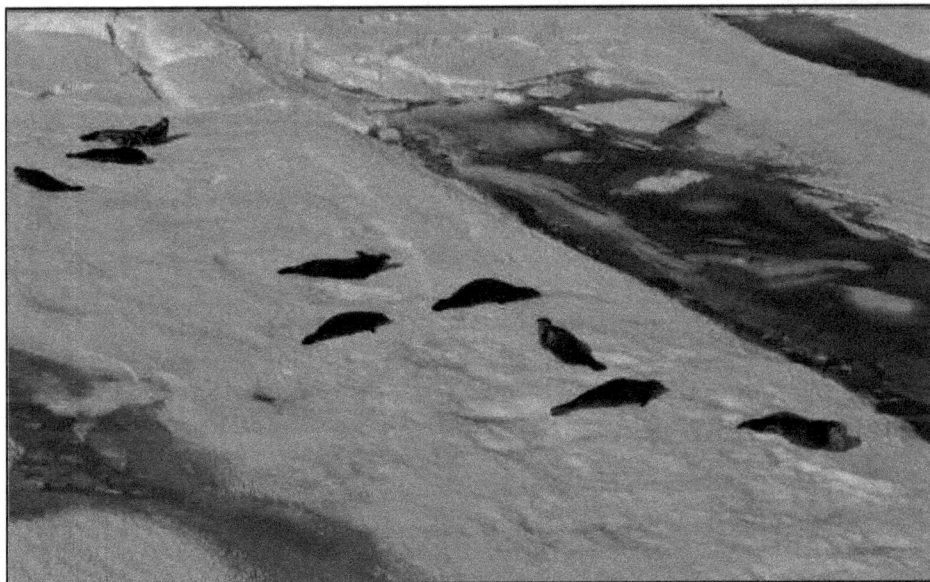

A group of Weddel Seals

Range extension of Cory's shearwater, Antarctic skua, cape petrel, white–capped albatross, grey–headed albatross, blue petrel, sooty albatross, kerguelen petrel and great winged petrel towards south in the southern Indian Ocean were observed during the summer 2009, this may be due to increase in temperature in south because of climate change.

The Fourth Assessment Report of the Intergovernmental Panel on Climate Change (IPCC) concluded that the earth's dramatic warming is "unequivocal". Across the globe, the atmosphere and the ocean are warming, and snow and ice have been melting at faster rates. Many plants and animal species have had to adapt, migrate or change the timing of their growth stages in order to avoid extinction. Even oceanic birds and penguins towards the southern pole have not escaped from these changes.

The polar regions at the northern and southern ends of the planet have been seriously affected by global warming. In fact, they are among the regions warming the fastest. The World Wide Fund for Nature has assessed that the Antarctic Peninsula is warming five times faster than the average rate of earth's overall warming. The vast Southern Ocean has warmed all the way down to a depth of 3,000 m. Sea ice that forms from sea water and a key feature of polar oceans covers an area that is 40 per cent less than it did 26 years ago off the west Antarctic Peninsula. Many species that had evolved the capacity to live in the cold, icy and harsh conditions of these polar regions, are now losing their only home.

Breeding areas of emperor penguin have suffered dramatic changes in Antarctica due to climate change. Warmer winter temperatures have led to thinner ice which has then been broken up and swept out to sea by frequently stronger winds. As a result, emperor penguin eggs and chicks have been blown away and before the chicks being able to survive on their own. Of all the Antarctic bird and mammal species, the emperor penguin has become the most vulnerable to the rapidly changing climate. It needs stable, land–locked sea ice on which to breed (it is too clumsy to climb over icy, coastal slopes), but wind–swept, ice–free ocean areas in which to feed. Ironically, climate change has made it easier to feed at the expense of strong thick ice needed for nesting.

We need a long term database on the abundance and distribution patterns of oceanic birds, along with data on the changing environmental settings of the southern Indian Ocean to relate the effects of climate change on biodiversity, which is largely lacking at present. However, it is certain that birds in Antarctica and in the southern Indian Ocean have been affected by the climate change as many oceanic birds appear to be extending their range towards south. Range extension of oceanic birds either towards south or north will affect the normal distribution pattern and species composition of the oceanic birds. Change in species composition may increase the competition for food and habitat, which will ultimately affect the biodiversity in the southern polar region. Actions initiated by the global community to reduce the gas emission therefore minimise the effect of climate change needs to be supported to safeguard the biodiversity of southern Indian Ocean and Antarctica.

Acknowledgements

We are grateful to the National Centre for Antarctica and Ocean Research, Goa, and the Wildlife Institute of India, Dehradun, for financial and logistic support to this study. We thank the Director and Dean of WII, and the Director, Logistic and Programme Directors of NCAOR, for their support and encouragement. We are thankful to our family members for supporting us and managing home during our long absence during this expedition. We are also grateful to our team members who participated in the 28th Indian Scientific Expedition to Antarctica for their help and encouragement. We would like to dedicate this chapter to late Dr. Ravi Sankaran, former Director of Salim Ali Centre for Ornithology and Natural History, Coimbator, for his encouragement.

References

Bhatnagar, Y.V. and Sathyakumar, S. 1999. *Fifteenth Indian Expedition to Antarctica, Scientific Report, 1999*. Department of Ocean Development, Technical Publication No. 13, Pp. 131–164.

Camille, P. 2006. Ecological and Evolutionary Responses to Recent Climate Change. *Annual Review of Ecology, Evolution and Systematics*, 37: 637–669.

CCAMLR, 2007. *Scientific Observers Manual – observing fish, birds and mammals in Indian Ocean and Antarctica*. Convention on Conservation of Antarctic Marine Living Resources, Tasmania, Australia.

Eastman, J.T. 1993. *Antarctic Fish Biology: Evolution in a Unique Environment*. Academic Press, Inc. San Diego. Pp. 322.

Fung, I., Doney, S.C., Lindsay, K. *et al.* 2005. Evolution of carbon sinks in a changing climate. *Proc. Nat. Acad. Sci. (USA)*, 102: 11201–11206.

Isabel, P.M., Tabor, N.J., Niemeier, D. *et al.* 2007. CO_2–Forced Climate and Vegetation Instability During Late Paleozoic Deglaciation. *Science*, 315 (5808): 87–91.

Jayasankar, P., Anoop, A., Krishnan, M. *et al.* 2007. A note on observations on cetaceans in the western Indian sector of the Southern Ocean (20–56°S and 45–57°30′E), January to March 2004. *Journal of Cetacean Research and Management*, 9(3): 263–267.

Moline, M.A., Claustre, H., Frazer, T.K. *et al.* 2004. Alteration of the food web along the Antarctic Peninsula in response to a regional warming trend, *Global Change Biol.*, 10: 1973–1980.

Root, T.L., Price, J.T., Hall, K.R. *et al.* 2003. Fingerprints of Global Warming on Wild Animals and Plants. *Nature*, 421: 57–60.

Sakshaug, E. and Skjoldal, H.R. 1989. Life at the ice edge. *Ambio.*, XVIII (1): 60–67.

Tasker, M.L., Jones, P.H., Dixon, T. *et al.* 1984. Counting Seabirds at Sea from Ships: A Review of Methods Employed and a Suggestion for a Standardized Approach, *The Auk*, 101(3): 567–577.

Watson, G.E. 1975. *Birds of the Antarctic and sub–Antarcitca*. American Geophysical Union, Washington DC.

Climate Change: Man and Environment (2012)   *Pages* 188–195
Editor: **Goutam Kumar Saha**
Published by: **DAYA PUBLISHING HOUSE, NEW DELHI**

13

Impact of Climate Change on Soil and Soil Biota

☆ *Subrata Roy**

Introduction

In the last two or three decades environmentalists throughout the globe are really worried about the trend of climatic change experienced by mother earth. The history of earth reveals that the climate on earth was not always uniform rather it exhibited some great changes in different geologic era. The climatic changes in pre historic earth were probably due to astronomical or geologic reasons. But with the advent of human civilization and advancement of technological know how the climatic scenario has shown a rapid change, even at a faster rate than the earlier ones. All the ecosystems with their biotic and abiotic components have either been affected by this climatic change or going to be affected very soon. Soil, as a part of terrestrial ecosystem, is not an exception to this. In this article attempts have been made to highlight the nature and properties of soil, its biotic component and impact of climate change on soil and its biota.

Soil–A Living and Dynamic System

Soil as a habitat, as a part of terrestrial ecosystem and as an abode of innumerable number of organisms is tremendously complex and has often been considered to be a living and dynamic system. It is the shallow upper layer of earth's crust and is derived from the massive rocks through a process of weathering. Both climate and biosphere are believed to play a significant role in the formation of soil and in the development of soil profile. The colour, texture, physicochemical nature and biotic

---

* Depatment of Zoology, University of Burdwan. E-mail: prof_roy_subrata@yahoo.co.in

components of soils exhibit wide range of variation in different climatic regions on earth surface. Besides being considered as a living and dynamic system, soil has also been regarded as a three phase system, the solid, liquid and gaseous phases. The solid phase is being represented by soil particles (of varying sizes) and organic matter. The particles, sand, silt and clay are derived from weathered rocks and are present in varying proportions in different types of soils. The soil organic matter, an important source of nutrients is a heterogeneous mixture of comminuted and partly decomposed fragments of litter, animal carcasses and dead microbial cell (Swift, 1995). The water present in small or large amount in any soil represents the liquid phase. Soil air found in soil pores represents the gaseous phase of soil. The oxygen and nitrogen ratio in soil air is more or less identical to atmospheric air. But the content of carbon dioxide is comparatively higher in soil air, it is about 0.03 per cent by volume in atmospheric air while it is more than 0.2 per cent in soil air. But water vapour content in both atmospheric air and soil air exhibits variation and depends on climatic factors like temperature, humidity, rainfall, etc.

The formation of soil depends on five factors *viz.*, parent material, topography, time, climate and biosphere. The first three are called passive factors since they represent soil forming mass and conditions affecting it. The last two *viz.* climate and biosphere are termed active factors since they supply energy to act upon the soil forming mass for the process of soil formation. Temperature and moisture are two important climatic agents that take active part in soil formation. The biosphere which includes both plant and animal world exerts significant influence on soil genesis.

The weathered materials or parent materials (developed through rock weathering) combine with organic matter and lead to formation of soil for which soil has often been referred to as 'organo–mineral complex'. But the soil formation is not complete or it is not considered to be mature until the materials of soil are arranged in distinct layers or horizons forming the soil profile. It has now been established that exact nature of mature soil profile depends upon the interactions between climatic factors, the biological agents, the topography of the land and nature of the parent rock.

Biotic Components of Soil–Number and Activities

Soil as an important habitat provide shelter to wide diversities of organisms including both flora and fauna. The floral components (often referred to as micro flora) include bacteria, actinomycetes, fungi and algae (Alexander, 1961). Viruses though present have not always been considered to be important for study. Soil microbiologists, have shown special interest to plant roots rather than macro flora or surface vegetation since the regions close to plant roots (rhizosphere) are considered to be very important site for floral and faunal concentrations. The higher plants or macrofloral component contribute significantly to the development of litter layer on soil surface. The faunal group present belong mainly to different invertebrate phyla like Protozoa, Helminths, Annelida, Arthropoda and Mollusca (Keven, 1965). Vertebrates include mammals like rodents which exist almost every where and make their burrows in soil. Other types of mammals may be associated with soils but their number and distribution is restricted. The soils also provide shelter to amphibians, reptiles and few species of birds which spend whole or part of their life in soil.

Microflora

Among the microbial organisms present in soil, bacteria outnumber all other groups in numerical abundance and are the most important participant of varied biological activities going on in soil. The bacteria in soil have numerous morphological and physiological adaptations that enable them to utilize their habitat effectively and successfully. The most important physiological adaptation is their capability to slow down the metabolic rate in order to obtain relatively high growth rate from meager resources available in soil (Killham, 1994). The decomposition of animal and plant residues in soil are effectively carried out by heterotrophic bacteria which are numerous in number and their substrate selectivity varies from species to species. In addition to their participation in decomposition process some heterotrophic bacteria are known to take part in non symbiotic nitrogen fixation.

The actinomycetes (intermediate forms between bacteria and fungi) are mostly free living saprophytic soil–form that are capable of decomposing various carbonaceous substances. In addition, they are known to degrade many polymers such as chitin, cellulose and hemicelluloses at a relatively higher soil pH. Many of the soil actinomycetes are known to exude antibiotics like streptomycin which is effective in controlling many human diseases.

The fungal flora in soil are major contributors of soil microbial biomass. They are obligate aerobes and heterotrophs and their main role in soil is the decomposition of organic matter from the simplest sugars and amino acids to the most resistant polymers such as lignin and complex soil humic acids. Since fungi exhibit greater tolerance to acidity than bacterial heterotrophs, the decomposition of organic matter in more acid soil is predominantly a fungal process (Killham, 1994). For which in acid forest soils, fungal population is more favoured than bacterial flora. The symbiotic as well as mycorrhizal association of fungi with plant root is of considerable importance in regulating nutrient uptake, disease resistance, water relations and ultimately growth of associated plant partner.

The algal population in soil is photoautotrophic in metabolism and are largely restricted to surface soils or large cracks where sunlight can reach. The algae are known to produce large amount of polysaccharides that act as aggregating agents and help in developing and maintaining stable soil structures. The microbial community in soil is considered to be the main driving force in soil ecosystem because it is this biomass that controls the rate of decomposition, mineralization, humification, nitrification and other biogeochemical processes in soil. But they are not evenly distributed and prefer certain niches or rhizospheric regions which may contain organic and inorganic substances as well as some amount of moisture. But the activity and performance of these microbial components in soil depend to a considerable extent on the presence of faunal community that are closely associated with the microbes.

Soil Fauna

The faunal component in soil is represented by animals of different phyla which according to body size are divided into three – microfauna, mesofauna and macrofauna (Wallwork, 1970). The protozoans belong to microfauna category and include flagellates, ciliates and amoebae. These forms secrete mucin or mucilaginous substances which help to bind the particles together leading to the formation of soil aggregates.

Soil nematodes have received much attention of the researchers for their damaging role as plant parasites in agricultural fields. But many of them are known to be free–living. They are abundant in soil and take active part in destruction of dead plant materials and also in the decomposition of animal matter (Chitwood and Chitwood, 1974). Earthworms and enchytracids (potworms), two important component of annelidan fauna, contribute a lot in decomposition of litter. They are known to possess enzymes in their gut that accelerates the degradation of organic matter and thus help in releasing nutrients (Dash *et al.*,1981). In addition, the burrows and channels created by them allow easy penetration of air and water and make soil relatively porous. The casts released by earthworms are rich in mineral contents and they are in a more easily available form. Earthworms help considerably in aggregate formation in soil which is evident from large number of aggregates in earthworm rich soil.

The arthropod fauna in soil is fairly rich and include varieties of forms such as acarines, collembolans, centipedes, millipedes, isopods, pseudoscorpions and different insects notable among them are ants, termites, beetles, gryllotalpids, earwigs etc. In addition, different types of insect larvae are also well represented. Some component of arthropod fauna by accelerating the fragmentation of litter (comminution) leads to the completion of decomposition of litter layer that has already been initiated by microorganisms. The burrowing forms like ants and termites not only help in aeration and movement of water but also help in mixing and translocation of materials in soil profile. Non-borrowers like mites and collembolans help in aeration and drainage by eating decayed plant roots and thus leaving channels containing organically rich faecal matter. The microphytic feeders and carnivorous forms of arthropods may help in keeping down the population of harmful fungi and nematodes.

The land snails and slugs constitute important component of molluscan fauna in tropics and subtropics. They are found either on soil surface or in crevices or may be encountered below the soil surface. Land snails are known to prefer decomposed plant material probably because of the growth of fungal hyphae (Raut and Ghosh, 1984). Thus feeding on surface vegetation and subsequent movement downwards indicate the probable role of molluscans in incorporating organic matter into the mineral structure of soil (Wallwork, 1970). But their role in decomposition of litter has not been established. They probably take part in breaking and altering the physicochemical nature of the litter which subsequently promotes fungal and microbial growth.

Vertebrates like snakes, lizards, toads, frogs and burrowing mammals take shelter in soil temporarily or make this habitat as their permanent abode. But their role in modification or change of the physicochemical or biological properties of soil rather appear to be insignificant and has not attracted the attention of soil biologists for extensive study.

Interaction between Climate, Soil and Soil Biota

It is well known that air currents, water currents, snow cover and land mass influence the global climate while the climate of a region is influenced by latitude, altitude and availability of large bodies of water. The regional climate is a dominant factor that controls the formation of all soil (Birkeland, 1984), since it circumscribes the forms and rates of local weathering, the translocation of weathered products and other pedogenetic processes. In case of global weathering climate is the major determinant of weathering process through its control of temperature and moisture, water flux and decomposition rates of organic material. In extremely cold climate weathering is practically a physical process while in cool and temperate climate a limited biochemical weathering may occur. But in hot and humid tropical climate geochemical weathering dominates that extend over very long period (Duchaufour, 1997).

Climate is known to have a significant bearing on the decomposition process participated by soil biota. Decomposition of organic residues and the maintenance of soil structure are largely complimentary process in most soils. Because organic matter exerts significant effect on soil structure by acting as important cementing agent which binds soil particles together to form larger aggregates. Decomposition, a very important biological process in soil may be defined as the sequence of organic matter transformation occurring after death of organism. These transformations involve two simultaneous but complimentary processes: mineralization and humification. The first one is a catabolic process through which the elements contained in organic form within the biological tissues are converted to inorganic forms such as nitrate, phosphate and sulphate ions while the second one is an anabolic process where organic molecules are condensed into degradation resistant polymers called 'humus' which may persist little altered for decades and centuries (Lavelle and Spain, 2007). In order to assess the degree of importance in the involvement of decomposition process, Di Castri (1988) made a hierarchical analysis of the factors and placed climatic factors (moisture and temperature) in the first position being followed by edaphic factors, physical and chemical properties of decomposition resources and biological interaction between macro and microorganisms respectively.

We know that under normal situation of the environment the biotic components of the soil through their activities release different types of nutrients in the soil that are being taken up by the plants for their growth and development. This transfer process becomes more effective if there is a synchronization between the release of nutrients and their uptake by plants. Thus the biotic components with the release of nutrients make the soil more fertile since the soil is able to satisfy plant demands for nutrients, water and an adequate aerated physical matrix for the growth of the plant roots (Swift, 1995). The activities of the biota become more significant when the two most important climatic agents like temperature and moisture are at the optimum level. With the seasonal variation of these two factors the activity also undergoes changes. But when the climate experiences irregular and completely different changes all the activities are jeopardized and the fertile soil either becomes infertile or less productive and may even be converted into a barren land.

Climate Change and its Impact on Soil Biota

Global climate change which occurred recently or changes that are expected in near future are the major topic of discussion by the present day environmentalists. The climate change may be natural or it may be brought by human interference. But man made changes have been found to be more intense. The changes that are happening rather appears to be rapid which include rise in average global temperature, melting of polar ice and rise in sea level, prevalence of heat waves, flash floods, etc. Overall effects are more horrifying, storms, cyclones, gales, hurricanes and typhoon will become more frequent and stronger as oceans heat up causing more water to evaporate (Khan, 2010). Kamboj and Khare (2009) while analyzing the impact of climate change suggested three primary causes for recent climate change; they are industrialization, deforestation and population explosion.

Soil biologists believe that some of the changes experienced by earth have either been reflected in the soil or its biotic components or going to be evident in near future. The impact of changes already visualized or expected may be as follows :

1. Extensive cultivation practices followed to meet up the food requirements of increased population will involve the use of huge amount of pesticides, fungicides and other agrochemicals which on reaching the soil with run-off water may destroy the non target soil biota.

2. Prevalence of an extreme drought situation for a longer period will either stop or slow down the decomposition process by soil biota and the release of nutrients will be hampered.

3. Heavy rainfall may destroy the soil aggregates or erode the surface layers to a considerable extent. The surface horizons being the site of maximum biological activity, the activity of fauna and flora will be significantly disturbed. In addition, heavy rainfall may lead to intense leaching and percolation resulting in the movement of nutrients from surface to deeper horizon.

4. Due to frequent floods, the surface horizons of arable lands are either washed away along with biota and nutrients or remain under the deposits of unwanted sand, silt or mud carried by floodwater. In addition, water logged situation may be created by floodwater in areas with heavy clay soil (with little percolation). Such situations are neither good for soil health nor conducive for the presence of numerically high soil biota.

5. Release of greenhouse gases from fossil fuel burning increases earth's temperature (global warming). The increased temperature may enhance the temperature of surface soils (particularly the sandy soils without proper vegetation cover) that induces the soil fauna to move downwards.

6. Rise in sea level due to melting of polar ice as a result of global warming, may inundate the coasts and adjoining areas affecting the soil biota. Intrusion of saline sea water in the riverine system of land masses may

occur which makes the river saline, soil fauna and microflora in the adjoining areas of such rivers may be exposed to some stress due to flooding of saline water. If the salinity is beyond the tolerance range of soil biota their survival chance will be diminished.

Remarks

Global climatic changes and their subsequent impact on soil and its biotic components reveal that most soils are somehow affected and their biological activities are disturbed leading to a probable decline in fertility and productivity. This in turn will affect human society since man depends solely on soil for food and shelter. Factors related to unwanted climate change should be controlled or checked since it is related to the degradation or deterioration of soil, which is considered to be the most important national wealth of any country. Hope human society will make an all out effort to stop their negative activities. The sooner it is, the better it will be.

Acknowledgement

The author expresses his deep sense of gratitude to his respected teacher late Prof. D. K. Choudhuri of Burdwan University who infused in him the special interest for soil zoology.

References

Alexander, M. 1961. *Introduction to Soil Microbiology*. John Wiley and Sons, Inc., New York. Pp. 472.

Birkeland, P.W. 1984. *Soils and Geomorphology*. Oxford University Press, New York.

Chitwood, B.G. and Chitwood, M.B. 1974. *Introduction to Nematology*. Baltimore Univ. Park Press, USA.

Dash, M.C., Nanda, M. and Misra, P.C. 1981. Digestive enzymes in three species of Enchytrasidae (Oligochaeta). *Oikos*, 36: 316–318.

Di Castri, F. 1988. Enhancing the credibility of Ecology; intersecting along and across hierarchieal scales. *Geo. Journal.*, 17(1): 5–35.

Duchaufaour, P. 1997. *Abrege de Pedologic. Sol. Vegetation. Environment*, (5th ed.). Mason, Paris.

Kamboj, M. and Khare, R. 2009. The impact of climate change. *Everyman's Science*, 4: 234–239.

Kevan Mc., D.K. 1965. The soil Fauna. In: *Ecology of soil borne plant pathogens. Prelude to Biological Control*. (Baker, K.F. and Snyder, W.C. eds.). Univ. of California Press, Berkley. Pp. 33–57.

Killham, K. 1994. *Soil Ecology*. Cambridge Univ. Press, Cambridge, UK.

Khan, Z.A. 2010. Global Environmental Challenge. *Science Reporter*, September: 19–22.

Lavelle, P. and Spain, A.V. 2007. *Soil Ecology (Indian edn.)*. Springer Science Publication. Pp. 654.

Raut, S. K. and Ghose, K.C. 1984. Pestiferous land snails of India. *Tech. Monograph. Zool, Surv. India*, 11: 151.

Swift, M.J. 1995. Soil biology and soil fertility in tropics. In: *Soil organisms and litter decomposition in tropic* (Reddy, M.V. ed.). Oxford and IBH Publishing House. Pp. 1–12.

Wallwork, J.A. 1970. *Ecology of soil animals.* McGraw Hill. London, N.Y., Panama.

Climate Change: Man and Environment (2012)   *Pages* 196–199
Editor: Goutam Kumar Saha
Published by: DAYA PUBLISHING HOUSE, NEW DELHI

14

Climate Change and its Impact on Sundarban Biosphere Reserve

☆ *Pranabes Sanyal**

Introduction

Climate change is one of the most important global environmental challenges facing humanity. Is the climate changing irrevocably or is it a naturally recurring phenomenon that has been part of the earth's deep past for centuries? What is the way ahead for us in a warming world? It has now been established by careful scientific observation that the earth's climate has warmed by 0.6°C in the last 100 years (Sanyal, 2009). Confirmation of global warming comes from warming of the oceans, rising sea levels, melting glaciers, retreating sea ice in the Arctic and diminishing snow cover in the northern hemisphere. Fragile and delicate ecosystems like coral reefs are extremely sensitive to even the slightest variation in climate.

In the awake of awareness about global climate change, Sundarban biosphere reserve with its 4.1 million strong human population features as one of the most vulnerable areas. Being the world's largest delta Sundarban receives second highest quantity (1670 million tons) of sediments (McDowell, 1995) and is thus, subjected to gradual subsidence. This is giving rise to additional acceleration to the eustatic rate of sea–level rise in Sundarban. As a result, sea facing islands are vanishing, for which famous Gangasagar Temple had to be shifted northward twice (Bandopadhyay, 1998). The salinity of surface and subsurface waters of the estuary is also on the rise (Milliman *et al.*, 1989). Several endangered mammals, reptiles became extinct during last 100 years. The accelerated rate of sea level rise has given rise to increased surface

* School of Oceanographic Studies, Jadavpur University. E-mail: pranabes@gmail.com

water salinity in Sundarban (CPCB, 2000; CICFRI, 2004) resulting in tiger migration northwards near villages leading to more man–animal confrontation (Sanyal, 2009). The freshwater loving gangetic dolphins of Sundarban are being replaced by salt water loving Irrawady dolphin (Sanyal, 2009). Likewise, the mangrove plant community migrates from outer to mid estuary, sundari trees are stressed (Sanyal *et al.*, 1984). Nearly half a million inhabitants of Sundarban are facing the hazards of losing land and house to become climate refugees. Fifteen per cent of the embankments of Sundarban are vulnerable to over toppling of tidal inundation within the agricultural areas (Hazra *et al.*, 2002). Bangladesh coast is subsiding and the sea level rise is at the rate of 10 mm/year at Khulna, which much exceeds the minimum estimate of global sea–level rise *i.e.*1.8 to 2 mm/year (Milliman *et al.*, 1989). In case of Indian Sundarban the rate of sea level rise has been at the rate of 3.14 mm/year at Sagar point (Hazra *et al.*, 2002) and 5 mm/year at Pakhiralay (Stanley and Hait, 1999).

Effects of Climate Change on Sundarban Ecosystem

Sundarban being a subsiding estuary (Sanyal, 2006) the rate of sea level rise is much faster as well as salinity of surface water is increasing. From a study of last two decades it is revealed that a 20 per cent increase in salinity has taken place along the coast. Coastal salt wedge has now extended upstream (CPCB, 2000). There is a tendency of northerly shift of present mangrove vegetation due to rapid development of shallow marine conditions created due to this accelerated rise of sea–level (Sanyal *et al.*, 1984). Salinity increased in creek waters. Climax species Sundari (*Heritiera fomes*) does not like shallow marine condition, succumbed to 'Tip dying' (Chaudhuri and Choudhury, 1994). Outer estuarine *Avicennia marina* migrated to mid estuary where velocity of water is much higher and developed abnormal stilt roots (Sanyal *et al.*, 1984; Sanyal and Bal, 1986). Besides, fishing cat (*Prionalirus viverrina*) and tiger (*Panthera tigris*) could not cope up with increased salinity and migrated northwards nearer to human habitation leading to more man–animal conflict (Sanyal, 2009). During last 100 years six important animals got extinct (Java rhino, great one horned rhino, swamp deer, barking deer, water buffalo, gharial) (Sanyal, 2006; 2009). From 2000 onwards, Gangetic dolphins (*Platinista gangetica*) have been largely replaced by salt tolerant Irrawadi dolphins (*Orcella bravirostris*) (Sanyal, 2006). Ghost crabs are replacing the useful Fiddler crabs (hatchlings are food for white pomphrets) due to warming effect (Sanyal, 2009).

Another effect is apparent on the subsurface waters salinity. There is salt water incursion into the shallow aquifer of Sundarban. At nearly 35 per cent islands of Sundarban, the shallow aquifer (100 m deep) was useful for the irrigation purposes two decades ago. But now they became saline as observed in Bali Island (Sanyal, 2006). The third effect is vanishing islands. The inhabited islands of Lohachara, Supuribhanga have vanished; 85 per cent area of Ghoramara island has vanished and as a result 7000 families became "climate refugees" (Hazra *et al.*, 2002). They have been rehabilitated at the Gangasagar colony, Jimbantala colony, Bankimnagar colony, Mahendragang colony, Hiradenpur colony and Ramghar colony under Sagar Island with only about 2.5 *bighas* of land for each family. The mangrove covered islands of south facing region also got 20 per cent reduced during last six decades.

Thus, the area of tiger habitat got reduced (Sanyal, 2006, 2009; Hazra *et al.*, 2002). The monsoon has been delayed by a month leading to a reduction in crop cultivation time (Hazra *et al.*, 2002). The intensity of cyclones has increased over last two decades causing more saline flooding of croplands (Hazra *et al.*, 2002).

Remarks

Following steps are to be taken to cope up with these adversities of climate change:

☆ Massive mangrove plantation on northern mudflats to make up the loss of tiger habitat.

☆ Large scale plantations of sundari (*Heritiera fomes*), dhundul (*Xylocarpus granatum*) and passur (*X. mekongensis*), golpata (*Nypa fruiticans*) along the fresh water canals of reclaimed Sundarban.

☆ Increasing the heights of the vulnerable embankments (about 15 per cent) mapped by School of Oceanographic Studies, Jadavpur University (Hazra *et al.*, 2002).

☆ Advance planning for resettlement of the *Envirorefugees*. There can be a bilateral arrangements to resettle them even in Andaman and Nicobar Islands where many of them have comfortably settled after several past disasters.

☆ Massive rain water harvesting to combat salinisation of ground water by:

1. Digging number of shallow ponds; pond depth in Sundarban cannot exceed 3 m otherwise saline oozing starts from below.

2. Re–excavation and fresh excavation of the age–old practice of shallow linear channels (for rain water accumulation) in Sundarban can be increased. During monsoon these channels serve as comfortable country boat transport routes.

3. Recharging the shallow water aquifer in order to dilute the salinity. This has been a very recent effort started by a local NGO with financial support from National Bank for Agricultural and Rural Development (NABARD).

4. Reintroduction of indigenous saline paddy varieties of Sundarban like *Nonabokhra, Talmogra, Matla, Hamilton, Sada Ketu, Kala Ketu*, etc. Although productivity of these saline paddy is less (7 md/*bigha*) but the rice fetch 1.5 times price and with less use of pesticide, the straw of indigenous paddy is 3 times costlier.

In view of above, coping up with the climate adversities in Sundarban became expedient. Massive rainwater harvesting can dilute the shallow aquifer and make it fit for farm irrigation. Large–scale mangrove plantation on new *char* lands can make up for the drowned mangrove covered south facing Islands. Height of vulnerable embankments can be raised to prevent inundation of farmlands by saline tidal waters. Advance rehabilitation planning for the possible climate refugees of inhabited vanishing Islands and reintroduction of indigenous saline–paddy varieties are the coping up steps. Thus, the above measures can help the distressed people of Sundarban. Biosphere Reserve sustainably coping up with the adversities of climate change in Sundarban.

References

Bandopadhyay, S. 1998. Coastal Erosion and its Management in Sagar Island, South 24 Parganas, West Bengal. *Indian journal of Earth Science*, 24 (3–4): 51–69.

Central Pollution Control Board (CPCB). 2000. *Pilot Study on Integrated Coastal Zone Management Plan for West Bengal*. Pp. 12.24–12.26.

Central Inland Captured Fisheries Research Institute (CIFCRI). 2004. *Annual Report*. Pp. 40.

Chaudhuri, A. B. and Choudhury, A. 1994. *Mangroves of the Sundarbans (Vol. 1)*. IUCN–The World Conservation Union. Pp. 165.

Hazra, S., Ghosh, T., Das Gupta, R. *et al.* 2002. Sea level and associated changes in the Sundarbans. *Science and Culture*, 68 (9–12): 309–321.

McDowell M.C. 1995. *The Development of the river Hooghly for navigation. Commemoration volume of Port of Calcutta, 125 years*. Pp. 69.

Milliman, J., Broadus, D.J.M. and Frank, G. 1989. Environmental and economic implications of rising sea level and subsiding deltas. The Nile and Bengal examples. *Ambio*, 18(6): 340–345.

Sanyal, P. 2006. Forest Conservation–Climate Change–Threat to Wildlife". In: *Environmental Awareness and Wildlife Conservation* (Basu, R.C. *et al.* ed.), Zoological Survey of India, Kolkata. Pp. 191–195.

Sanyal, P. 2009. Climate Change vs Wildlife with Special Reference to West Bengal. *Proc. Humbolt Kolleg on Global Warming in context to the Indian Subcontinent*, 54–57.

Sanyal, P., Banerjee, L.K. and Choudhury, M.K.1984. Dancing mangals of Indian Sundarbans. *J. Indian Soc. Coastal Agric. Res.*, 2(l): 10–16.

Sanyal, P. and Bal, A.K. 1986. Some observations on abnormal adaptation of Mangrove in Indian Sundarbans. *J. Ind. Soc. Agric. Res.*, 4(1): 9–15.

Stanley, D.J. and Hait, A.K. 1999. Holocene Dipositional Patterns, Neotectonics and Sundarban Mangroves in the Western Ganges–Bramhaputra Delta. *Journal of Coastal Research*, 16(1): 26–39.

Climate Change: Man and Environment (2012)
Editor: Goutam Kumar Saha
Published by: DAYA PUBLISHING HOUSE, NEW DELHI

Pages 200–205

15

Climate Change and its Impact on Wetland Ecosystem

☆ *Tapan Saha**

Introduction

Wetlands are among the most important ecosystems on Earth. Wetlands are a major feature of the landscape in almost all parts of the world and act as sources, sink transformers of a multitude of chemical, biological, and genetic materials. The several values of wetlands for the benefit of human society have been identified. Wetlands cover a heterogeneous spectrum of aquatic habitats, widely recognized as biodiversity hotspots and key components of the global carbon budget. According to Ramsar Convention on Wetland, definition of wetlands is "areas of marsh, fen, peat land or water, whether natural or artificial, permanent or temporary, with water that is static or flowing, fresh, brackish or salt, including areas of marine water, the depth of which at low tide does not exceed six meters". Wetlands include all inland aquatic habitats (permanent or temporary, whether fresh, brackish, or saline, and including lakes, streams, rivers and inland seas), coastal systems shallower than 6 m depth at low tide (lagoons, estuaries, marshes, mangroves, seagrass beds, mud flats and coral reefs) and human–made systems such as reservoirs, salt pans, etc.

Hydrologic conditions are extremely important for the maintenance of a wetland's structure and function. Hydrology is probably the single most important determinant and specific types of wetlands and wetland processes. The high degree of variability in the structure of wetland systems is mainly due to their individual hydrology, varying from peatland bogs in high–latitude boreal forests, through tropical monsoon

* Institute of Environmental Studies and Wetland Management, Kolkata. E-mail: tsaha1958@gmail.com

wetlands to high latitude wetlands in the Tibetan Andean Mountains (IPCC, 2007). The hydrology of a wetland creates the unique physicochemical conditions that make such ecosystems. Hydrologic pathways such as precipitations, surface run-off, groundwater, tides, and flooding rivers transport energy and nutrients to and from wetlands. Water depth, flow patterns and duration and frequency of flooding, which are the result of all of the hydrologic inputs and outputs, influence the biochemistry of the soils and are major factors in the ultimate selection of the biota of wetlands. Biota ranging from microbial communities to vegetation to water fowl are all constrained and enhanced by hydrologic conditions (Mitsch and Gosslink, 2007).

The transport and transformation of chemicals in ecosystems, known as *biogeochemical cycling,* involve a great number of interrelated physical, chemical and biological processes. The unique and diverse hydrologic conditions in wetlands markedly influence biogeochemical processes. These processes result not only in changes in the chemical forms of materials but also in the spatial movement of materials within wetlands, as in water–sediment exchange and plant uptake and with the surrounding ecosystems, as in organic exports. These in turn, determine the overall wetland productivity. Wetlands act as a net "carbon sink" and liberate oxygen through photosynthesis of its vegetation.

Wetlands perform physical, chemical and biological functions depending on hydrology and biogeochemical processes, which provides ecosystem goods and services. Wetlands provide many services and commodities to humanity. Wetland values can be considered from the perspective of three levels – population, ecosystem and global. As the *population level*, wetland–dependent fish, shellfish, fur animals, waterfowls and timber provide important and valuable harvests and millions of days of recreational fishing and hunting. At the *ecosystem level*, wetlands moderate the effects of floods, improve water quality and have aesthetic and heritage value. They also contribute to the stability of *global level* of available nitrogen, atmospheric sulfur, carbon dioxide and methane (Mitsch and Gosslink, 2007). Wetlands are highly dependent on water levels and so changes in climatic conditions that affect water availability will highly influence the structure and function of wetlands.

Impact of Climate Change on Wetlands

Climate change will have its pronounced effects on freshwater wetlands as well as coastal wetlands (IPCC, 2007). The decrease of wetland areas is a worldwide phenomenon and seems to progress faster than in other ecosystems. Non-climatic direct anthropogenic impacts have been more noteworthy than impacts directly attributed to climate change (Scavia *et al.,* 2002). The major non-climatic impacts include drainage of wetlands, water withdrawal, deforestation and land reclamation, habitat fragmentation, discharge of sewage, eutrophication and pollution, overharvesting and overexploitation and the introduction of invasive alien species. Climate change is expected to intensify the loss and degradation of many wetlands not only because of the complex relationships and feedbacks between climate and wetlands, but also because of the likely synergistic effects of the non-climatic and climatic drivers, especially in coastal and low lying areas. It is widely accepted that climate change will increase the likelihood of potentially abrupt changes in wetlands, which can be large in magnitude and difficult, expensive, or impossible to reverse (Millennium Ecosystem Assessment, 2005).

Climate change will have its most pronounced effects on inland freshwater wetlands through altered precipitation and more frequent or intense disturbance events (droughts, storms, floods). Relatively small increase in precipitation variability can significantly affect wetlands plants and animals at different stages of their life cycle (IPCC, 2007). Monsoon fed areas are more likely to be affected by more intense rain events over shorter rainy seasons, exacerbating flooding and erosion in catchments and the wetlands themselves. An ecosystem is a dynamic complex of biotic communities and the abiotic environment, interacting as a functional unit. Climate change is likely to affect both the abiotic and the biotic components (from individual organisms, populations, to communities) of the ecosystems, and in turn, their whole structure and functioning. The major consequences of the climate change will depend on the temporal and spatial scales at which effects are assessed. Transitional scales pertaining to landscape–level shifts in location, morphometry and persistence of wetlands and their biota over decades to centuries (Carpenter *et al.*, 1992).

All the biological processes are generally either enhanced or constrained by temperature. Therefore, a change in the temperature can directly regulate growth and behavioural pattern, *e.g.* phenology, and influence habitat preference. Thus increased temperatures may exceed the thermal tolerance for some coldwater species, mainly in summer. As because species have varying tolerance ranges for temperature, changes in temperature can produce shifts in species composition that, in turn, can affect the overall metabolism and productivity of wetlands (IPCC, 2002). The feasibility of species to migrate into the new expanded ranges will depend on both the habitat availability and the capability to move along dispersal corridors. However, both circumstances will largely depend on wetland type and on the habitat fragmentation. In fact, the degree of fragmentation of the landscape in which wetlands occur is historically unparalleled (Dale, 1997). Wetlands are increasingly isolated and disconnected. Hence, making adjustment to rapid climate change through animal and plant dispersal becomes very difficult. Thus, climate change clearly represents an additional, significant threat to wetlands, one that will interact in complex ways with existing human–caused stresses (Poff *et al.*, 2002).

Hydrology directly influences wetland ecosystem structure and functions by determining the extent of suitable habitat for species and many aspects of water quality. Precipitation and run–off regimes of a region determine the seasonal dynamics of the water volume of certain wetland. By the middle of the 21st century, annual average run–off and water availability are projected to increase by 10–40 per cent at high latitudes and in some wet tropical areas and decrease by 10–30 per cent over some dry regions at mid–latitudes and in the dry tropics, with an increase in heavy precipitation events (IPCC, 2007). Such seasonality strongly influences the species composition that thrives in a wetland. Therefore, a change in climate that alters the existing hydrologic regime has the potential to greatly modify habitat suitability for many species and cause significant ecological changes (IPCC, 2007). Climate changes driven by global warming has an immense impact on the hydrological cycle as well as a rise in sea level and increase in extreme events like storm surges, which will result in the enhanced erosion of shores and coastal habitats, salinization of groundwater and estuaries, altered tidal ranges in rivers and creeks, changes in sediment inputs and nutrient loads and increased coastal flooding and consequently in a decrease of freshwater availability for human and ecosystems in coastal areas. Wetlands, such as mangroves, saltmarshes and floodplains, can play a critical role in the physical buffering of climate change impacts (Scavia *et al.*, 2002). Effects of climate change, on precipitation, wind patterns and the frequency and intensity of storms, have a key influence on wetlands response to climate change (IPCC, 2007).

Climate interacts with lake morphometrics and wind (which affects the depth of the epilimnion) to determine the strength and extent of thermal stratification. Due to wind–driven water mixing and photosynthetic activity the oxygen levels of the epilimnion are high, whereas the hypolimnion may undergo oxygen depletion due to decomposition of dead organic material settling from the productive epilimnion, influencing habitat suitability for many species as stated below. Lakes will be affected by climate change through warmer temperatures, longer ice–free seasons and modifications in the hydrologic cycle. The response to climate change will largely depend on lake morphometrics (surface area and depth), exposure to wind and location (altitude, latitude and elevation). Such factors will determine the occurrence, strength and duration of thermal stratification, and thus, the seasonal extent of cool and warm water habitats available for species. Since the 1960s, epilimnion of many lakes around the world has warmed by 0.2°C to 2.0°C and the hypolimnion (which reflect long-term trends) has increased by 0.2°C to 0.7°C (IPCC, 2001). In warmer years, epilimnion temperature is higher, evaporation increases, summer stratification occurs earlier in the season and thermocline becomes shallower. In several lakes in Europe and North America, the stratified period has advanced by up to 20 days and lengthened by 2 to 3 weeks, with increased thermal stability. The extent of ice cap for northern hemisphere lakes during the past 150 years is one of the strongest evidences of climate change. On average, from 1846 to 1995 the freeze data was 8.7 days later and the ice breakup 9.8 days earlier (Magnuson *et al.*, 2000). The supply of water to lakes may be differently affected by climate change depending on whether the water originates from glacier, snowmelt, rain or aquifers. Water levels will probably increase in high latitude lakes, where climate models forecast increased precipitation, whereas water levels at mid and low latitudes are projected to decline. Fauna and flora constitute an important component of the freshwater biota. Acknowledging that, there may be complex and unpredictable changes in species composition (Stefan *et al.*, 2001), an overall increase in system productivity would likely be a common response to climate warming if other factors do not limit the potential for enhanced productivity. However, if warming and eutrophication occurs together, the respiration of the whole community can easily counterbalance the primary production and the whole system becomes heterotrophic, *i.e.* a potential source of carbon. Global warming will modify the extent of habitat availability for cool and warm–water species.

Coastal swamps and marshes absorb wave energy, reduce erosion on estuarine shoreline, buffer the land from storms. Coastal wetlands are among the most productive ecosystems on earth. The high productivity of coastal wetlands is largely supported by nutrients delivered by freshwater run–off and the tidal mixing. Changes in the timing and volume of freshwater run–off will affect salinity, sediment and nutrient availability and moisture regimes in coastal ecosystem. Freshwater inflows into estuaries also influence water residence time, vertical stratification and control of phytoplankton growth rates. Climate change can affect each of these variables by altering precipitation and locally driven run–off or more importantly, run–off from watersheds that drain into the coastal zone. Hydrology has a strong influence on the distribution of coastal wetland plant communities, which typically grade inland from salt, to brackish, to freshwater species (IPCC, 2007). Coastal wetlands are also among the most altered and threatened natural systems due to multiple human pressures. Many coastal wetlands are likely to change as a consequence of projected sea level rise, increased storm and tidal surges, changes in storm intensity and frequency and subsequent changes in river flow regimes and sediment transport (Poff *et al.*, 2002). Over the next 100 years, the sea level is estimated to increase at a rate up to 2 to 9 mm per year, deposition of sediment will also increase at a much higher rate than that observed over the last century (average values, 1.3 to 2.1 mm y^{-1}). Submergence and erosion of the substrate are causing a substantial loss of coastal wetlands (Kesel, 1989). Such loss will produce negative effects in the ecosystem. For example, coastal fisheries will be negatively affected, since wetlands are suitable habitats (nursery, feeding and living) for many fish and shellfish species. In addition, sea–level rise will lead to increasing salinity in coastal wetlands (reinforced if river discharge decreases), thereby tending to displace existing coastal plant and animal communities inland if migration is not blocked and if the rate of change does not exceed the capacity of communities to adapt or migrate. Climate change impacts on one or more cornerstone species, however, can result in abrupt community changes. Saltwater intrusion into freshwater aquifers is also potentially a major problem (Llorens, 2008). Changes in river discharges into shallow near–shore marine environments due to climate drivers will lead to changes in turbidity, salinity, stratification and nutrient availability (Justic *et al.*, 2005). Of all ecosystems, however, freshwater wetland ecosystems and coastal ecosystems appear to have the highest proportion of species threatened with extinction by climate change (Millennium Ecosystem Assessment, 2005).

References

Carpenter, S.R., Fisher, S.G., Grimm, N.B. *et al*. 1992. Global Change and Freshwater Ecosystems. *Annual Review of Ecology and Systematics*, 23: 119–139.

Dale, V.H. 1997. The Relationship between Land–Use Change and Climate Change. *Ecological Processes*, 7: 753–769.

Intergovernmental Panel on Climate Change (IPCC). 2001. *Climate Change 2001: The Scientific Basis*. (Houghton, J.T., Ding, Y., Griggs, D.J. *et al*. eds.) Cambridge University Press, Cambridge, UK. Pp. 881.

Intergovernmental Panel on Climate Change (IPCC). 2002. *Climate change and biodiversity*. IPCC Technical Paper V, Geneva, Switzerland. Pp. 77.

Intergovernmental Panel on Climate Change (IPCC). 2007. *Climate change 2007: The Physical Science Basis.* (Solomon, S., D. Qin, M. Manning, Z. *et al.* eds.). Cambridge University Press, Cambridge, UK. Pp. 996.

Justic, D., Rabalais, N.N. and Turner, R.E. 2005. Coupling between climate variability and coastal eutrophication: evidence and outlook for the northern Gulf of Mexico. *Journal of Sea Research*, 54(1): 25–35.

Kesel, R.H. 1989. The role of the Mississippi River in wetland loss in Southeastern Louisiana, USA. *Environmental Geology and Water Sciences*, 13:183–193.

Lloréns, J.L.P. 2008. Impacts of Climate Change on Wetland Ecosystem. In: Proc. Climate Change and Water Extremes: Impact of Climate Change on Wetland Ecosystems. Expo Zara Goza, University of Cadiz (Spain), 21–24 July, 2008.

Magnuson, J.J., Robertson, D.M., Benson, B.J. *et al.* 2000. Historical trends in lake and river ice cover in the Northern Hemisphere. *Science*, 289: 1743–1746.

Millennium Ecosystem Assessment. 2005. *Ecosystems and human well–being: wetlands and water.* Synthesis. World Resources Institute, Washington DC. Pp.155.

Mitsch, W.J. and Gosselink, J.J. 2007. *Wetland* (4th ed.), John Wiley and Sons Inc., New Jersey, USA.

Poff, L.N., Brinson, M.M. and Day, J.W.J. 2002. *Aquatic Ecosystems and Global Climate Change: Potential Impacts on Inland Freshwater and Coastal Wetland Ecosystems in the United States.* Pew Center on Global Climate Change, Arlington, VA. Pp. 44.

Scavia, D., Field, J.C, Boesch, D.F. *et al.* 2002. Climate change impacts on U.S. coastal and marine ecosystems. *Estuaries*, 25: 149–164.

Stefan, H.G., Fang, X. and Eaton, J.G. 2001. Simulated fish habitat changes in North American lakes in response to projected climate warming. *Transactions of the American Fisheries Society*, 130: 459–477.

Climate Change: Man and Environment (2012)   *Pages* **206–221**
Editor: **Goutam Kumar Saha**
Published by: **DAYA PUBLISHING HOUSE, NEW DELHI**

16

Climate Change and its Impact on Aquatic Biodiversity

☆ *Ramakrishna and C. Raghunathan**

Introduction

Biological diversity is identified as the number of taxa within an area, as the number of different life–history stages, the different species interactions within an area, or simply the number of individuals of a selected species or group of species and the geographic distribution of those individuals. In other words biodiversity is the variety and variability of life and its many processes. It includes all life forms from bacteria, fungi and protozoa to higher plants, insects, fishes, birds and mammals. Biodiversity also includes countless millions of races, subspecies and local variants of species and the ecological processes and cycles that link organisms into populations, communities, ecosystems and ultimately the entire biosphere. Biological diversity exhibits at three levels *viz.*, Genetic, Species and Ecosystem. Genetic diversity means that species contain characteristic levels of genetic variation within and among their population, including that of extreme population. The foundation of biodiversity is genetic variation. Genetic variation affects a species' physical characteristics, productivity, resilience to stress and long-term evolutionary potential.

Changing Climate

Climate varies for reasons that are natural, such as volcanic eruptions, changes in solar output and natural changes in the orbital characteristics of the earth, which are known to have profound impacts on global change (Hulme and Barrow, 1997). However, there is now convincing evidence for a growing human influence on global

* Zoological Survey of India, Kolkata. E-mail: ramakrishna.zsi@gmail.com, kscraghu@gmail.com

climate. The industrial economy, through its reliance on carbon–based fuels has continued to alter the properties of the earth's atmosphere. In less than two hundred years, human have managed to increase the atmospheric concentrations of greenhouse gases by *ca*. 50 per cent relative to pre–industrial levels. Given the inertia in the energy systems and the long memory exhibited by the climate system, this human–induced climate change will become increasingly important relative to natural climate variability during the century to come (Hulme *et al.*, 2002). Certain naturally occurring gases, such as carbon dioxide (CO_2) and water vapour (H_2O), trap heat in the atmosphere causing greenhouse effect. Burning of fossil fuels, like oil, coal and natural gas is adding CO_2 to the atmosphere. The current level is the highest in the past 650,000 years. The Fourth Assessment Report of the Intergovernmental Panel on Climate Change (IPCC) concludes, "that most of the observed increase in the globally averaged temperature since the mid–20th century is very likely due to the observed increase in anthropogenic greenhouse gas concentrations." The findings of the IPCC demonstrate that even if by 2050 emissions would be reduced to below half of 1990 levels, a temperature rise of up to 2°C above pre–industrial levels will be difficult to avoid. Unmitigated climate change beyond 2°C will lead to unprecedented security scenarios as it is likely to trigger a number of tipping points that would lead to further accelerated, irreversible and largely unpredictable climate changes.

Impact of Climate Change

Historically, earth's climate has undergone significant changes, by way of warming or cooling. This is evident with the fossil records depicting mass extinctions of species over the past 500 million years, thus affecting a series of ecological changes. However, the present scenario is different from the previous one, as they are induced anthropogenically by emission of greenhouse gases, mainly in the form of carbon dioxide, methane, nitrous oxide, water vapour etc. According to CBD document (2008), climate change is projected to increase species extinction rates, with approximately 10 per cent of the species assessed so far at an increasingly high risk of extinction for every 1°C rise in global mean surface temperature within the range of future scenarios typically modelled in impacts assessments (usually <5°C global temperature rise). The impact will be mainly on wetlands, mangroves and coral reefs.

Impacts of Climate Change on Aquatic Biodiversity

Climate change is considered to be one of the principal threats to biodiversity and to the structure and functioning of ecosystems (McCarthy *et al.*, 2001; Vitousek, 1994). Although the causes and likely impacts are subject to debate (Sharp, 2003; Veizer, 2005), the scientific consensus is that climate change is a hard reality (Houghton *et al.*, 2001; Walther *et al.*, 2005). Average global temperatures have increased by ~0.6°C over the past century. During this period, both marine (Levitus *et al.*, 2000) and freshwater systems (Winder and Schindler, 2004) have warmed. Over geological time, climate has varied (Crowley, 1983), influencing the distribution and suitability of habitats, which in turn, have influenced the distribution and dispersal of species. It is therefore, realistic to expect that further climate change will have a strong controlling effect on habitats, communities, species and individual organisms in the future (Parmesan and Yohe, 2003; Root *et al.*, 2003).

Lakes

Lake primary productivity is closely linked to mean air temperature and the length of the growing season (Brylinsky and Mann, 1973) and fish production is positively correlated with mean annual air temperature (Schlesinger and Regier, 1982). In the North American Great Lakes, Meisner *et al.* (1987) suggested that an increase in mean air temperature by two per cent could lead to an increase in fisheries yield of ca. 25 per cent. An increase in potential fish production following climate change may be limited by a greater probability of hypolimnetic oxygen depletion in productive lakes (Carpenter *et al.*, 1992). The timing and intensity of lake stratification is likely to change (De Stasio *et al.*, 1996), with implications for lake fishes (Lehtonen, 1996), their parasites (Marcogliese, 2001) and their prey (Winder and Schindler, 2004). Recent modelling studies suggest that the negative effects of nutrient enrichment on lake algal dynamics may become increasingly problematic as temperatures increase (Elliott *et al.*, 2006). Fishes found in shallow habitats or habitats with restricted water exchange, *e.g.* shallow lakes and ponds, will be affected by increased water temperatures following climate warming and in extreme cases, loss of habitat, or death if these systems dry out. Some systems may become ephemeral following future climate change and become fish–free or only partly utilised by fish. Increased lake levels following winter precipitation will improve access to additional spawning or feeding habitats for some species *e.g.* pike (*Esox lucius*) (Billard, 1996).

Rivers

Apart from increases in water temperature, climate change is likely to impact riverine systems following shifts in precipitation patterns (*e.g.* increases in winter, reductions in summer (Arnell, 1998; Hulme *et al.*, 2002), including an increased probability of extreme events (Christensen and Christensen, 2003). Residence times, import and export of organic matter, dilution of pollutants, primary production and dissolved oxygen concentrations are all likely to be altered (Conlan *et al.*, 2005; Mohseni *et al.*, 2003). Riverine fishes display a complex array of environmental requirements (Crisp, 1996; Mann, 1996) and major changes in seasonal flow patterns are likely to have significant consequences (Hulme *et al.*, 2002). Migratory species have evolved to utilise predictable floods for migrations (Crisp, 1996) and changes in the frequency or intensity of floods may impact the ability of adult fishes to successfully reach spawning areas. The character of running waters depend fundamentally on changing patterns of rainfall or melting snow, while floods or droughts can have major effects on the ecology of these lotic ecosystems. Their small size means limited capacity to absorb atmospheric heat, so that stream temperatures often correspond closely to changing air temperature. Warming effects could be of major significance in such habitats where most animal species are cold–blooded invertebrates, fishes or amphibians.

Climate change scenarios predict significant increases in extreme precipitation events (Hulme *et al.*, 2002), where flood intolerant species or sensitive life stages, *e.g.*

eggs or larvae, could become displaced or killed (Jager *et al.*, 1999; Poff, 2002). However, in some river systems fishes have proved to be remarkably resilient to flooding (Heggenes, 1988; Lobón–Cerviá, 1996) and increased winter flooding may prove beneficial to certain species, providing additional feeding or spawning opportunities (Masters *et al.*, 2002). If hydrological regimes shift, *e.g.* reduced surface or groundwater flows during periods of drought, hydrological marginal habitats such as floodplains or wetlands may become disconnected from the main river channel, with subsequent impacts on habitat availability for fish and their production and diversity (Robinson *et al.*, 2002). Reductions in stream flow during warm periods may lead to increased stream temperatures, decreased concentrations of dissolved oxygen and reduced dilution of pollutants (Mohseni *et al.*, 2003).

Estuaries

Estuaries represent the interface between marine, freshwater and terrestrial environments and are extremely complex ecosystems where salinity, temperature and oxygen fluctuate according to tidal amplitude and season (Ketchum, 1983). Estuarine communities (Haedrich, 1983) are often structured according to salinity resistance and are well adapted to fluctuations in salinity, temperature and oxygen (Vernberg, 1983). Estuarine habitats are likely to experience very different hydrological regimes under future climate conditions (Struyf *et al.*, 2004) and the effects of climate change will potentially be complex (Scavia *et al.*, 2002). Decreased summer precipitation will affect freshwater inputs, which will increase residence times and the time taken to flush nutrients and pollutants from the system (Struyf *et al.*, 2004) and lead to increased intrusion by saline waters (Kennedy, 1990; Roessig *et al.*, 2004). Predicted increase in sea levels (Hulme *et al.*, 2002) will also lead to shifts in salinity profiles, which are likely to result in changes in estuarine fish community structure towards salinity–tolerant estuarine fishes or those typical of fully marine habitats. Although there is considerable variation in nutrient load between regions (Nedwell *et al.*, 2002), some estuaries in Britain and Ireland have undergone eutrophication (Mathieson and Atkins, 1995). The risk and frequency of estuarine algal blooms may increase in nutrient–rich estuaries following climate change. Reduced freshwater inputs during hot dry summer months could increase residence times and reduce the dilution of dissolved nutrients. This, combined with increased summer temperatures might lead to increased phytoplankton production and the risk of low oxygen conditions. Predicted increase in winter precipitation will result in greater run–off to surface waters, including estuarine waters (Arnell, 1998; Hulme *et al.*, 2002). Increased freshwater discharge during winter will result in an overall decrease in salinity and shifts in estuarine salinity gradients. Residence time will fall and nutrients, contaminants, and organic material will be transferred to coastal waters more rapidly (Struyf *et al.*, 2004), potentially reducing the productivity of estuarine habitats.

Marine and Coastal Habitats

The impacts of climate change on marine ecosystems extend beyond increased water temperature (Harley *et al.*, 2006) and include changes in oceanic circulation (Scavia *et al.*, 2002), sea level rise (Hulme *et al.*, 2002), increasing frequency of storm surges (Hulme *et al.*, 2002), changes in chemistry including acidification (Royal Society,

2005) and nutrient availability (Scavia *et al.*, 2002). The likely ecological consequences of these changes to marine ecosystems are understandably diverse, but include changes in the phenology of species that form the base of marine food webs (Heath, 2005), *e.g.* phytoplankton and zooplankton (Hays *et al.*, 2005; Steingrund and Gaard, 2005), with clear implications for fishes and other taxa (Edwards and Richardson, 2004). If changes in the biotic (*e.g.* seasonal availability of food) and abiotic (*e.g.* water temperature, salinity, circulation) environments of marine fishes are significant, it is likely that interactions between individuals and species will be modified, impacting population and community dynamics and leading to shifts in the structure of marine fish assemblages (Genner *et al.*, 2004; Perry *et al.*, 2005). Harley *et al.* (2006) suggested that changes in the chemistry of marine waters may be more important than changes in temperature. For instance, the oceans have absorbed large volumes of CO_2 which has led to significant acidification of seawaters (Royal Society, 2005). If global emissions of CO_2 continue, it is feared that the average pH of the oceans could fall by 0.5 pH units (equivalent to a threefold increase in H^+ ions) by 2100 (Royal Society, 2005). Although the impact of such acidification is likely to be less extreme in the temperate seas than in tropical or southern seas, it has clear potential to impact ecologically important calciferous organisms, such as molluscs, cold-water corals, echinoderms, foraminifera and coccolithophores (Royal Society, 2005). Increased concentrations of dissolved CO_2 also have the potential to impact the physiology and reproductive success of water breathing organisms including larger invertebrates and fishes (Ishimatsu *et al.*, 2004; Pörtner *et al.*, 2004). Increases in sea level due to thermal expansion of seawater (Hulme *et al.*, 2002) and the melting of polar ice (Overpeck *et al.*, 2006) may reduce the area of inter–tidal habitats as coastal waters encroach especially if coastal defences are present (Galbraith *et al.*, 2002).

Sea surface warming has been associated with increased and decreased phytoplankton abundance in cooler and warmer regions, respectively (Richardson and Schoeman, 2004). Through bottom–up effects, this impact can propagate up food webs through herbivorous and carnivorous zooplankton. That this will have an effect on higher trophic levels seems inevitable and it is likely that fish and other top predators will have to adapt to changing spatial distribution of primary and secondary productivity within marine pelagic ecosystems following climate change (Richardson and Schoeman, 2004). Results from continuous plankton recorder surveys have shown that south of 59° N in the northeast Atlantic, (*e.g.* in the seas around Britain and Ireland) phytoplankton has shown a significant response to climate change, with increased abundance and a marked extension of the growing season (Reid *et al.*, 1998). The timing (phenology) of major oceanic trophic events such as spring blooms, seasonal peaks in zooplankton abundance and the timing of hatching of fish eggs can be of central importance to fish stocks. Variation in pelagic food webs can be driven by fluctuations in plankton production and effects of climate change on plankton dynamics are transmitted to upper trophic levels (*e.g.* fishes). Temperate marine environments may be particularly vulnerable to changes in phenology because the level of response to climate change may vary across functional groups and trophic levels. This is important because recruitment success of higher trophic levels is highly dependent on synchronisation with pulsed planktonic production.

The copepod *Calanus finmarchicus* is of key trophic importance in the northeast Atlantic, but is in pronounced decline. It is being gradually replaced by its warm–temperate congener *C. helgolandicus* with some negative impacts on fish recruitment in species including cod (Beare *et al.,* 2002; Beaugrand *et al.,* 2003). In the North Sea, *C. finmarchicus* has shown a rapid and almost complete collapse (Hays *et al.,* 2005), with a twenty–fold decrease recorded in the northern North Sea between 1958 and 1998 (Beare *et al.,* 1998) and an increasing overall prevalence of temperate Atlantic and neritic (shallow–water) taxa (Beare *et al.,* 2002; Edwards and Richardson, 2004). Atlantic inflow into the North Sea is increasingly thought to be the main regulator of long-term abundance of *C. finmarchicus* in the North Sea (Heath *et al.,* 1999; Planque and Taylor, 1998). Unlike temperate Atlantic taxa such as *C. helgolandicus, C. finmarchicus* cannot overwinter in large numbers in the North Sea because it is too shallow and cold and must therefore, migrate to deeper over–wintering areas (*e.g.* the Faroe–Shetland Channel) (Heath and Jonasdottir, 1999). Rising temperatures would result in increased winter survival of temperate and neritic species in the North Sea. However, two ecological features clearly differentiate these two species of calanoid copepod: their temperature preferences and over wintering strategy (Planque and Taylor, 1998).

In the north Atlantic phytoplankton and zooplankton species plus communities have been associated with Northern Hemisphere Temperature (NHT) trends and variations in the North Atlantic Oscillation (NAO) index. These have included changes in species distribution and abundance, the occurrence of sub–tropical species in temperate waters, changes in overall phytoplankton biomass and season length, changes in the North Sea ecosystem, community shifts, phenological changes and changes in species interaction. Over the last decade numerous other investigations have established links between the NAO and the biology of the north Atlantic including the benthos, fish, seabirds and whales. While the NAO index integrates variability in many hydro–climatic parameters, it is assumed that the increase in temperature over the last decade has had a primary role in influencing the ecology of the north Atlantic. Indirectly the progressive freshening of the Labrador sea region, attributed to climate warming and the increase in freshwater input to the ocean from melting ice, has resulted in increasing abundance, blooms and shifts in seasonal cycles of dinoflagellates due to the increased stability of the water–column. Sex determination in an atherinid fish, the Atlantic silverside (*Menidia menidia*), is under the control of both genotype and temperature during a specific period of larval development. The sex ratios of the progeny of different females are variable and differ in their responsiveness to temperature. This demonstrates that sex ratio in fishes that normally have separate sexes can be influenced by the environment.

Like the North Atlantic, many long-term biological investigations in the Pacific have established links between changes in the biology and regional climate oscillations such as the El Nino Southern Oscillation (ENSO) and the Pacific Decadal Oscillation (PDO). In the case of the Pacific these biological changes are most strongly associated with El Nino events which can cause rapid and sometimes dramatic responses to the short–term Sea Surface Temperature (SST) changes accompanying El Nino events. Changes in mesozooplankton abundance have also been related to large–scale climate influences in the Californian upwelling system. In the Southern Ocean the long-term decline in krill stock has been linked to changes in the extent of winter ice, which has been related to warming temperatures.

Effect on Seagrass Habitat

There are about 60 species of sea grass or eel–grass, the only known flowering plant to live permanently submerged in coastal waters in the world, that grow from the mid–intertidal to 50–60m depth. They inhabit all types of substrate but the most extensive sea–grass beds are on soft sediments like sand and mud. The roots and rhizomes of sea grass stabilise large areas of shallow water sediment and these meadows are highly productive and important nursery grounds for many species including commercially exploited fish and crustaceans. These grasses are the main food for many threatened species like manatees, dugongs and green sea turtles, all threatened species as well as of great public interest. The extent of meadows has been greatly reduced due to such human activities. Climate change is likely to have significant consequences, not only due to changing temperature but also to changing sea levels with effects on coastal hydrography and increasing frequency of damaging storms. While the changes may make new areas available for colonisation by sea grasses and improve conditions for them in others, in the short–term, the overall effects of climate change are likely to be negative.

Kelp forests are marine ecosystems dominated by large brown algae (*Phaeophyta*), which are among the most biologically productive habitats in the marine environment are thickest in summer, declining during winter storms and high surf, though some plants may survive for up to three years. These forests provide a habitat utilized by many species, including marine mammals, fish, other algae and a vast numbers of invertebrates. Many of the species characteristic of the forests is highly adapted to the conditions provided by the kelps, the keystone species of these ecosystems. Climate change may influence kelp forests through the direct effects of temperature, moving the boundaries of the areas where they have the potential to survive. In addition, nutrient inputs necessary for the kelp to thrive may be altered due to changes in the geographical or seasonal distributions of rainfall and up welling. In kelp forests of southern California, the proportion of cold water fish such as the green spotted rockfish, have fallen and warm water species like the garibaldi have increased since the 1970s. An increase in the numbers of sea urchins has also contributed to the decline. These are normally passive grazers of algae but graze the kelp when the waters become warmer, as in El Nino events. In the Northern Hemisphere former kelp forests have become 'sea urchin barrens' due to over fishing, removing the predators that would normally control the urchins. Disruption of ecosystems due to climate change may reinforce the imbalances that have resulted in the overgrazing of the kelps.

Effect on Intertidal Organisms

The main impact of fluctuating climate on intertidal ecosystems would be through changes in sea level and temperature. There is considerable evidence over past epochs of changes in sea level; deposits of shells of intertidal molluscs and barnacles are found above and below present sea levels, including the well–known submerged forests and raised beaches. Owing to the high heat capacity of the ocean, sea temperature is a more conservative parameter, so that at low tide intertidal organisms are exposed to the greater fluctuations of air temperature. Hence, observed changes in intertidal biodiversity can show east–west trends in species range and abundance as well as north–south trends. For example, during rising temperatures in the last two decades, recognised southern species of barnacles and molluscs have moved eastward along the English Channel and from north to south down the east coast of Scotland (Hiscock *et al.*, 2004). Barnacles, which have planktonic larvae, are known as good indicators of climatic variation as once they have settled they are immobile. Rapid sea level change resulting from the melting of ice caps as well as thermal expansion of seawater may significantly alter habitat distributions due to topographical factors, the changed shapes of the coastline and depth distributions near to the shore, changing the hydrography.

Shifts in the Distribution of Marine Fishes

There have been numerous historical accounts of apparent distribution shifts of fish in response to climate change, of which the herring/pilchard shift is a particularly clear illustration. It appears that the recent warming of the north Atlantic is responsible for a shift in distribution of some species. Counts of novel immigrant or vagrant species have been positively correlated to the increased water temperatures in the north Atlantic over the last 40 years (Stebbing *et al.*, 2002). Perry *et al.* (2005) demonstrated that many exploited and unexploited North Sea fishes have apparently demonstrated a marked response to recent increases in sea temperature: nearly two–thirds of species (21/36) shifted mean latitude and/or depth over a 25 year period. Approximately half of the species with a latitudinal boundary of distribution (northerly or southerly) in the North Sea showed a northerly shift in their boundary. The most significant shift was demonstrated by the blue whiting, whose southern limit moved northwards ca. 820 km in only 25 years (Perry *et al.*, 2005). They speculated that if temperatures continue to increase in the North Sea, blue whiting and redfish will probably be lost from the North Sea and bib will extend their range to encompass the whole region. They also highlighted that species with "faster" life histories, *e.g.* those with significantly smaller body sizes, faster maturation and small sizes at maturity, tended to shift their distribution and that it was those species that responded most strongly to climate change.

Effect on Coral Reefs

Corals most often exist as colonial organisms composed of thousands of individuals, called polyps. All species of coral secrete calcium carbonate ($CaCO_3$), and the majority of coral species form reef structures over time. Coral reefs harbour more than 25 per cent of all known fish and provide our oceans with the highest biodiversity of any marine ecosystem. Coral reefs are an ecosystem that have the second biggest biodiversity in the world and have been described as the rainforests of the ocean. For years they have been threatened by man, physical destruction from boats, fishing and pollution. Now their biggest threat may be from climate change. 15 per cent of the world's reefs have already been lost and 30 per cent may die in the next

30 years. Surface warming and acidification of the oceans adversely affect the health of coral reefs. Surface warming increases the likelihood of coral bleaching (stress–induced expulsion of unicellular algae resulting in the loss of coral color) and, if conditions are warm for long, can cause reef mortality.

Ocean warming is directly reducing coral cover through coral bleaching. Reef–building corals harbours zooxanthellae that live symbiotically within their tissue. Zooxanthellae provide their coral host with food and oxygen and in return, the zooxanthellae receive nutrients, carbon dioxide, and an enemy–free shelter. This symbiotic relationship evolved tens of millions of years ago and has been critical to the success and evolutionary radiation of corals and to the development of reef ecosystems. When summer time water temperatures are just a degree or two warmer than usual for a few weeks, this critical yet delicate symbiotic relationship breaks down and the zooxanthellae are expelled, often leading to the coral's death. The greater the magnitude or duration of the warming leads the greater the mortality and effect on coral populations. The phenomenon is called "coral bleaching" because the coral animal appears to turn white after the zooxanthellae loss. This is because without their zooxanthellae symbionts, which contain various photosynthetic pigments, corals are nearly transparent and the white, external calcium carbonate skeleton that the coral polyps live on becomes plainly visible. It is also stated that corals could become rare on tropical and sub–tropical reefs by 2050 due to the combined effects of increasing dissolved carbon dioxide (CO_2) and increasing frequency of bleaching events. Since the impacts of increased CO_2 are greater at higher latitudes, cold–water corals are likely to show large reductions in geographical range this century.

The increase in ocean temperature is variable and quite subtle: on the order of 1°C over the last several decades. But even such modest changes have caused mass coral mortality events around the world during some of the especially warm summers we have all experienced over the last ten years. In 1998, when an intense El Nino greatly warmed much of the western Pacific and Indian Oceans, coral bleaching was widespread, causing mass coral mortality in many countries. For example, in Palau, more than 90 per cent of the corals on some reefs bleached and at least 50 per cent perished. Even some isolated reefs were impacted. In the *Maldives*, in the east Indian Ocean, bleaching caused coral cover to plummet to only about 5 per cent.

Undersea surveys made by the scientists of National Coral Reef Research Institute of ZSI, Port Blair during 8th to 15th May 2010 indicated that most of the coral colonies in several islands of Andaman and Nicobar Islands were bleached. The main causative factor for the bleaching event is, the temperature in the Andaman sea stood at 31–32°C for a long period this year, making the sea warmer than the previous two years. It is believed that the warmer–than–usual sea temperature is a consequence of the late onset of the monsoon over the Bay of Bengal and Andaman sea. Such bleaching phenomenon is also reported in Gulf of Thailand, Malaysia and Burma during May 2010.

Ocean warming can also indirectly kill corals by magnifying the effects of infectious diseases, which are one of the primary causes of coral loss, particularly in the Caribbean. The number, prevalence and impacts of diseases of corals and many other types of marine animals have been increasing over the last 20–30 years. The severity of marine diseases could increase with temperature for several reasons. Because elevated water temperature causes corals physiological stress, it can also compromise their immune system, potentially making them more susceptible to infections. Additionally, increased temperature could also benefit bacterial and fungal pathogens, making them more fit and/or virulent. A recent study found that anomalously high ocean temperatures greatly increased the severity of the coral disease white syndrome on the Great Barrier Reef. Disease outbreaks only occurred on reefs with high coral cover after especially warm years. The disease was largely absent on cooler reefs. The temperature increases required to trigger a white syndrome outbreaks were relatively modest as most disease outbreaks occurred on reefs where the temperature was only 1–2°C warmer than usual. Other evidence also points to temperature as an important driver of coral epizootics. For example, some coral diseases such as black band disease become more prevalent or spread faster in the summer. However, not all coral epizootics are caused by anomalously high temperature. Some major outbreaks have occurred during relatively cool periods or years, such as white band disease, which decimated the then–dominant branching corals *Acropora palmata* and *Acropora cervicornis* in the Caribbean in the 1980s.

Effect on Mangroves

Mangroves are intertidal communities of primarily tree species of the tropics and sub–tropics. These forests protect coasts against storms, tides and cyclones. Mangroves are important in relation to animal and plant productivity, as nutrient sinks, for substrate stabilisation and as a source of wood products. Warming may promote expansion of mangroves to higher latitudes but they are susceptible to frost and drought. Many mangrove forests are excessively exploited, thereby reducing resilience in the face of sea–level rises. With a 1m sea level rise in Cuba, more than 300,000 hectares of mangroves would be threatened. The most luxuriant growth of mangroves with the greatest diversity is in the Indo–Pacific region. The Sundarbans, a "World Heritage Site" in India and Bangladesh, account as the largest single mangrove unit globally. The area under Sundarbans was large in the past, but has been reduced to the present level due to degradation that took place during last two hundred years. Logging operation, aquaculture, reclamation of swamps, paddy cultivation on the east coast of India and salt production on the west coast are the main reasons for degradation, resulting into shrinking of tidal forests throughout the Indian coast. Climatic factors like temperature fluctuation, humidity, precipitation, number of rainy days, regular wind flow, radiation and fresh water flow in the region act as the most significant factors for development and succession of mangroves. Mangroves in tropical region are extremely sensitive to global warming and depend upon the rate of sea level rise relative to growth rates and sediments supply, space for and obstacle to horizontal migration, changes in climate–ocean environment. Sea level rise will affect mangroves by eliminating or modifying their present habitats and creating new tidally inundated areas to which some mangrove species may shift. Extent of high tidal mudflats constitutes major share of the tidal mudflats, especially in Gujrat. This will provide great potential to the mangroves of the region for adjustment and adaptation against sea level rise. Climate unsuitability is another factor responsible for mangrove's change and disappearance. Hardy species like *Avicennia* sp. recolonised the area fast in the area like the Gulf of Kutch, but this is not true in case of the species of genus *Rhizophora, Ceriops, Sonneratia* and *Aegiceras* in the Gulf of Kutch and other regions and *Heretia* sp. in Sundarbans as they were unable to recover from the effect within short period of time. Global warming and sea level rise would bring changes in most of the region, resulting in alternation in mangrove setting. In deltaic mangroves, there is consistent seaward movement of the mangroves but this may take a reverse change due to rise in sea level. Topography of deltaic mangroves, especially in Sundarbans is such that landward migration of mangroves, coupled with accretion of sediment might more or less keeps pace with rising sea level.

It is also expected that average global rainfall will increase with marked regional variations. If this happens, climate change is likely to lead to an increase in species migration pole wards. This may result into better environment for mangroves in semi–arid region like Gulf of Kutch. Many species are sensitive to fast changes, especially to anthropogenic disturbance and sea level rise. If pace of sea level is high, these species may not be successful to compete and may loose in favour of hardy and great coloniser, especially *Avicennia marina, A. alba, Acanthus ilicifolius* and *Suaeda* sp. in semi–arid in Gujarat and *A. officinalis* and other species in the moist region. It is expected that species diversity may suffer in some areas, especially in Andaman and Nicobar Islands.

Most predictions suggest that future rises in relative sea level will be of the order of 100–200cm/100 years. If this projection becomes reality, mangroves of the world may suffer serious loss and majority of the species may fail to adapt new environment. The extensive mangrove systems of the Sundarbans in the Bay of Bengal are examples of river–dominated systems where relative sea level may raise less owing to the influx of large amount of silt. There is little information on this aspect of mangrove ecology, but evidence suggests that mangroves can cope with accretion rates of the order of 10cm/100years, albeit with some change in community structure and species composition.

Remarks

There is a growing scientific consensus that human activities have modified the composition of the atmosphere and that these changes have and will continue to, cause significant shifts in the climate including changes in air and water temperatures, precipitation, solar radiation and wind speed (Hulme *et al.*, 2002; Sweeney *et al.*, 2003). As might be expected, human–forced climate change has and will continue to affect as humans rely heavily on aquatic systems for many goods and services, *e.g* food production, recreation, nutrient recycling and gas regulation. Impacts of climate change on aquatic systems and their inhabitants are therefore likely to have widespread implications for future human populations around the globe. Existing international agreements and legislation, *e.g.* the Ramsar convention (UNESCO, 1971) and the International Convention on Biological Diversity (UNCED, 1992) provide a clear obligation for Government and managers to respond to this challenge. In order for Government to react, they rely on scientists to provide them and other interest groups with reliable information regarding the responses of natural systems to climate change. Harley *et al.* (2006) noted that ecologists face immense challenges if they are to predict how natural systems will respond to environmental conditions that have no parallel in recent time. The studies on long-term monitoring will provide baseline data against which climate change can be quantified. These data are essential for modelling responses to climate change at the individual, population or community level. However, to date, little work has examined how interspecific interactions might influence the response of aquatic taxa to climate change and there is a pressing need for such studies.

India being largest developing country with nearly two–thirds of the population depending directly on the climate sensitive sectors such as aquatic, agriculture and forest resources. The projected climate change under various scenario is likely to have implications on food production, water supply, biodiversity and livelihoods. Thus, India has a significant stake in scientific advancement as well as an international understanding to promote mitigation and adaptation. This requires improved scientific understanding, capacity building, networking and broad consultation processes.

Acknowledgement

The authors wish to express sincere gratitude to the Secretary, Ministry of Environment and Forests for their support.

References

Arnell, N.W. 1998. Climate change and water resources in Britain. *Clim. Change*, 39: 83–110.

Beare, D.J., McKenzie, E. and Soeirs. D.C. 1998. The unstable seasonality of *Calanus finmarchicus* in the Fair Isle current. *J. Mar. Biol. Assoc. UK*, 78: 1377–1380.

Beare, D.J., Batten, S., Edwards, M. *et al.* 2002. Prevalence of boreal Atlantic, temperate Atlantic and neritic zooplankton in the North Sea between 1958 and 1998 in relation to temperature, salinity, stratification intensity and Atlantic inflow. *J. Sea. Res.*, 48: 29–49.

Beaugrand, G., Brander, K.M., Lindley, J.A. *et al.* 2003. Plankton effect on cod recruitment in the North Sea. *Nature*, 426: 661–664.

Billard, R. 1996. Reproduction of pike: gametogenesis, gamete biology and early development. In. *Pike, Biology and Exploitation* (Craig, J.F. ed.), Chapman and Hall, London. Pp. 13–43.

Brylinsky, M. and Mann, K.H. 1973. An analysis of factors governing productivity in lakes and reservoirs. *Limnol. Oceanogr.*, 18: 1–14.

Carpenter, S.R., Fisher, S.G., Grimm, N.B. *et al.* 1992. Global change and freshwater ecosystems. *Annu. Rev. Ecol. Syst.*, 23: 119–140.

Christensen, J.H. and Christensen, O.B. 2003. Severe summertime flooding in Europe. *Nature*, 421: 805–806.

Conlan, K., Lane, S.N., Ormerod, S. *et al.* 2005. Preparing for climate change impacts on freshwater ecosystems (PRINCE): literature review and proposal methodology. *Science Report: SC030300/PR* Environment Agency, Bristol. In. *Biogeography: an Ecological and Evolutionary Approach* (Cox, C.B. and Moore, P.D. eds.). Blackwell Science, Oxford.

Crisp, D.T. 1996. Environmental requirements of common riverine European salmonid fish species in fresh water with particular reference to physical and chemical aspects. *Hydrobiologia*, 323: 201–221.

Crowley, T.J. 1983. The geologic record of climate change. *Rev. Geophys. Space. Phys.*, 24: 828–877.

De Stasio, B.T., Hill, D.K. Jr. and Kleinhans, J.M. 1996. Potential effects of global climate change on small north–temperate lakes: physics, fish, and plankton. *Limnol. Oceanogr.*, 41: 1136–1149.

Edwards, M. and Richardson, A.J. 2004. Impact of climate change on marine pelagic phenology and trophic mismatch. *Nature*, 430: 881–884.

Elliott, J., Jones, I. and Thackeray, S. 2006. Testing the sensitivity of phytoplankton communities to changes in water temperature and nutrient load, in a temperate lake. *Hydrobiologia*, 559: 401–411.

Galbraith, H., Jones, R., Park, R. *et al.* 2002. Global climate change and sea level rise: potential losses of intertidal habitat for shorebirds. *Waterbirds*, 25: 173–183.

Genner, M.J., Sims, D.W., Wearmouth, V.J. *et al.* 2004. Regional climatic warming drives long-term community changes of British marine fish. *Proc. R. Soc. Lond,(B)*, 271: 655–661.

Haedrich, R.L. 1983. Estuarine fishes. In. *Ecosystems of the World 26: Estuaries and Enclosed Seas* (Ketchum, B.H. ed.), Elsevier, Amsterdam. Pp. 183–207.

Harley, C.D.G., Randall H.A., Hultgren, K.M. *et al.* 2006. The impacts of climate change in coastal marine systems. *Ecol. Lett.*, 9: 228–241.

Hays, G.C., Richardson, A.J. and Robinson, C. 2005. Climate change and marine plankton. *Tree*, 20: 337–344.

Heath, M. and Jonasdottir, S.H. 1999. Distribution and abundance of overwintering *Calanus finmarchicus* in the Faroe–Shetland Channel. *Fisheries Oceanogr.*, 8: 40–60.

Heath, M., Backhaus, J.O., Richardson, K. *et al.* 1999. Climate fluctuations and the spring invasion of the North Sea by *Calanus finmarchicus*. *Fisheries Oceanogr.*, 8: 163–176.

Heath, M. 2005. Changes in the structure and function of the North Sea fish foodwebs, 1973–2000. *ICES J. Mar. Sci.*, 62: 847–868.

Heggenes, J. 1988. Effects of short–term flow fluctuations on displacement of and habitat use by, brown trout in a small stream. *Trans. Am. Fish Soc.*, 117: 336–344.

Hiscock, K., Southward, A., Tittley, I., *et al.* 2004. Effects of changing temperature on benthic marine life in Britain and Ireland. *Aquat. Conserv. Mar. Freshwat. Ecosyst.*, 14: 333–362.

Houghton, J.T., Ding, Y., Griggs, D.J. *et al.* (eds.). 2001. *Climate Change 2001: the Scientific Basis*. Cambridge University Press, Cambridge, UK. Pp. 881.

Hulme, M. and Barrow, E. (eds.). 1997. *Climates of the British Isles: present, past and future*. Routledge, London. Pp. 454.

Hulme, M., Jenkins, G.J., Xianfu, L. *et al.* 2002. *Climate Change Scenarios for the United Kingdom: The UKCIP02 Scientific Report*, Tyndall Centre for Climate Change Research, School of Environmental Sciences, University of East Anglia, Norwich, UK.

Ishimatsu, A., Kikkawa, T., Hayashi, M. *et al.* 2004. Effects of CO_2 on marine fish: larvae and adults. *J. Oceanogr.*, 60: 731–741.

Jager, H.I., Van Winkle, W. and Holcomb, B.D. 1999. Would hydrologic climate changes in Sierra Nevada streams influence trout persistence? *Trans. Am. Fish Soc.*, 128: 222–240.

Kennedy, V.S. 1990. Anticipated effects of climate change on estuarine and coastal fisheries. *Fisheries*, 15: 16–25.

Ketchum, B.H. 1983. Estuarine characteristics. In: *Ecosystems of the World 26: Estuaries and Enclosed Seas* (Ketchum, B.H. ed.), Elsevier, Amsterdam. Pp. 1–14.

Lehtonen, H. 1996. Potential effects of global warming on northern European freshwater fish and fisheries. *Fish Mgmt. Ecol.*, 3: 59–71.

Levitus, S., Antonov, J.I., Boyer, T.P. *et al.* 2000. Warming of the World Ocean. *Science*, 287: 2225–2229.

Lobón–Cerviá, J. 1996. Response of a stream fish assemblage to a severe spate in northern Spain. *Trans. Am. Fish Soc.*, 125: 913–919.

Mann, R.H.K. 1996. Environmental requirements of European non-salmonid fish in rivers. *Hydrobiologia*, 323: 223–235.

Marcogliese, D.J. 2001. Implications of climate change for parasitism of animals in the aquatic environment. *Can. J. Zool.*, 79: 1331–1352.

Masters, J.*E.G.*, Welton, J.S., Beaumont, W.R.C. *et al.* 2002. Habitat utilisation by pike *Esox lucius* L. during winter floods in a southern English chalk river. *Hydrobiologia*, 483: 185–191.

Mathieson, S. and Atkins, S.M. 1995. A review of nutrient enrichment in the estuaries of Scotland: Implications for the natural heritage. *Aquat. Ecol.*, 29: 437–448.

McCarthy, J.J., Canziani, O.F., Leary, N.A. *et al.* (eds.). 2001. *Climate Change 2001: Impacts, Adaptation, and Vulnerability*. Cambridge University Press, Cambridge, UK. Pp. 1031.

Meisner, J.D., Goodier, J.L., Regier, H.A. *et al.* 1987. An assessment of the effects of climate warming on Great Lakes basin fishes. *J. Great Lakes Res.*, 13: 340–352.

Mohseni, O., Stefan, H.G. and Eaton, J.G. 2003. Global warming and potential changes in fish habitat in US streams. *Climate Change,* 59: 389–409.

Nedwell, D.B., Dong, L.F., Sage, A. *et al.* 2002. Variations of the nutrients loads to the mainland U.K. estuaries: correlation with catchment areas, urbanization and coastal eutrophication. *Estuar. Coast Shelf Sci.,* 54: 951–970.

Overpeck, J.T., Otto–Bliesner, B.L., Miller, G.H. *et al.* 2006. Paleoclimatic evidence for future ice–sheet instability and rapid sea–level rise. *Science,* 311: 1747–1750.

Parmesan, C. and Yohe, G. 2003. A globally coherent fingerprint of climate change impacts across natural systems. *Nature,* 421: 37–42.

Perry, A.L., Low, P.J., Ellis, J.R. *et al.* 2005. Climate change and distribution shifts in marine fishes. *Science,* 308: 1912–1915.

Planque, B. and Taylor, A.H. 1998. Long-term changes in zooplankton and the climate of the North Atlantic. *ICES J. Mar. Sci.,* 55: 644–654.

Poff, N.L. 2002. Ecological response to and management of increased flooding caused by climate change. *Phil. Trans. Roy. Soc.(A).,* 360: 1497–1510.

Pörtner, H.O., Langenbuch, M. and Reipschläger, A. 2004. Biological impact of elevated ocean CO_2 concentrations: lessons from animal physiology and earth history. *J. Oceanogr.,* 60: 705–718.

Reid, P.C., Edwards, M., Hunt, H.G. *et al.* 1998. Phytoplankton change in the North Sea. *Nature,* 391: 546–546.

Richardson, A.J. and Schoeman, D.S. 2004. Climate impact on plankton ecosystems in the Northeast Atlantic. *Science,* 305: 1609–1612.

Robinson, C.T., Tockner, K. and Ward, J.V. 2002. The fauna of dynamic riverine landscapes. *Freshwater Biology,* 47: 661–677.

Roessig, J.M., Woodley, C.M., Cech, J.J. *et al.* 2004. Effects of global climate change on marine and estuarine fishes and fisheries. *Rev. Fish Biol.,* 14: 251–275.

Root, T.L., Price, J.T., Hall, K.R. *et al.* 2003. Fingerprints of global warming on animals and plants. *Nature,* 421: 57–60.

Royal Society. 2005. *Ocean acidification due to increasing atmospheric carbon dioxide.* Royal Society, London.

Scavia, D., Field, J.C., Boesch, D.F. *et al.* 2002. Climate change impacts on U.S. coastal and marine ecosystems. *Estuaries,* 25: 149–164.

Schlesinger, D.A. and Regier, H.A. 1982. Climatic and morphoedaphic indices of fish yields from natural lakes. *Trans. Am. Fish Soc.,* 111: 141–150.

Sharp, G.D. 2003. Future climate change and regional fisheries: a collaborative analysis. *FAO Fish Tech. Pap.,* 452: 1–75.

Stebbing, A.R.D., Turk S.T.M., Wheeler, A. *et al.* 2002. Immigration of southern fish species to south–west England linked to warming of the North Atlantic (1960–2001). *J. Mar. Biol. Assoc. UK.,* 82: 177–180.

Steingrund, P.S. and Gaard, E. 2005. Relationship between phytoplankton production and cod production on the Faroe Shelf. *ICES J. Mar. Sci.*, 62: 163–176.

Struyf, E., Van Damme, S. and Meire, P. 2004. Possible effects of climate change on estuarine nutrient fluxes: a case study in the highly nutrified Schelde estuary (Belgium, The Netherlands). *Estuar. Coast Shelf Sci.*, 60: 649–661.

Sweeney, J., Brereton, T., Byrne, C. *et al.* 2003. *Climate Change: Scenarios and Impacts for Ireland (2000–LS–5.2.1–M1). Final Report* Environmental Protection Agency, Johnstown Castle, Wexford.

UNCED. 1992. *International Convention on Biological Diversity, June 5, 1992* UNCED, Rio.

UNESCO. 1971. *Convention on Wetlands of International Importance especially as Waterfowl Habitat.* United Nations Educational, Scientific and Cultural Organization, Ramsar, Iran.

Veizer, J. 2005. Celestial climate driver: a perspective from four billion years of the carbon cycle. *Geosci. Can.*, 32: 13–28.

Vernberg, W.B. 1983. Responses to estuarine stress. In: *Ecosystems of the World 26: Estuaries and Enclosed Seas* (Ketchum, B.H., ed.), Elsevier, Amsterdam. Pp. 43–63.

Vitousek, P.M. 1994. Beyond global warming: ecology and global change. *Ecology*, 75: 1861–1876.

Walther, G. R. 2005. Ecological responses to recent climate change. *Nature,* 416 (6879): 389–395.

Walther, G.R., Hughes, L. and Vitousek, P. 2005. Consensus on climate change. *Tree*, 20: 648–649.

Winder, M. and Schindler, D.E. 2004. Climate change uncouples trophic interactions in an aquatic ecosystem. *Ecology*, 85: 2100–2106.

Climate Change: Man and Environment (2012)   *Pages* 222–226
Editor: Goutam Kumar Saha
Published by: DAYA PUBLISHING HOUSE, NEW DELHI

17

Climate Change: Impacts on Species Conservation

☆ *Dipankar Ghose**

Introduction

Perhaps the most significant current conservation challenge is represented by climate change. There are increasing concerns on negative impacts of climate change on vulnerable ecosystems in the world and so is in India. Modern global temperature and land cover and projected future temperatures suggest that tropical forest species will be particularly sensitive to global warming (Wright *et al.*, 2009). There are visible impacts of climate change in different ecosystems. Phenology of plants in the Himalayas has changed, for example rhododendron flowering in Eastern Himalaya has been erratic over the past few years. Climate change is not only impacting forests and wildlife, but also human populations which is apparent from a statement issued by WWF in 2008 that reads 'global warming is melting glaciers in every region of the world thereby putting millions of people at risk from floods, droughts and lack of drinking water'. Glacier melting will not only endanger the survival of those species inhabiting icy habitats, but also result in changes in water flows and rising sea level causing inundation of islands as well as low productivity of underwater coral reefs. Recent studies have shown that an entire ecosystem on the crumbling ice shelves of the Arctic has been lost due to climate change.

The Eastern Himalaya

The ecologically fragile Eastern Himalaya in India is part of the Himalaya Biodiversity Hotspot (Conservation International, 2008). This is a resource rich

* WWF–India, New Delhi. E-mail: dipankarghose@rediffmail.com

ecosystem providing a multitude of ecological services apart from harbouring a wide range of biological diversity including several endemic species. The forests of the Eastern Himalaya in India are either managed by the State Governments' Forest Department in the protected areas and reserve forests or are owned by the local indigenous communities who have their traditional customs and regulations to manage them. The Eastern Himalaya region is highly vulnerable to climate change (*e.g.* melting of glaciers), with significant impact for entire northeast India as a consequence. Due to the mountainous terrain forests provide indispensable ecological functions and services. In this region local communities still heavily rely on local produce for fuel, food, timber, and other forest products and services. In many instances this has led to overuse and forest degradation.

The need of the hour is enhancement of the protected area management, thereby improving biodiversity conservation; restoration of forest cover and unique forest ecosystems; absorbing additional amounts of atmospheric carbon through restored forest stands; improved and sustained supply of forest products (wood and non-wood) to local communities; regulation of watersheds and enhancement of biodiversity and erosion control, especially in hilly areas; development of replicable pilots for adaptation to climate change and development of adequate adaptive resource management strategies for vulnerable ecosystems in India.

Are Tropical Species more Vulnerable?

Wright *et al.* (2009) suggested that wild species' response to global warming would be manifested through acclimation, evolutionary adaptation and range shifts to cold refuges. There is a trend of temperature increase of an average 0.26°C per decade since 1970s in the tropical areas (Malhi and Wright, 2004). To escape global warming, the temperature sensitive species may disperse to cool refuges (Bush, 2002; Colwell *et al.*, 2008). A major deciding factor for this dispersal is the distance to cool refuges which is largest near the equator, shortest for mid–temperate latitudes and large for north temperate and arctic latitudes. Distances to refuges are more than 2000 km in some tropical mountain areas like the Western Ghats in India. As species' abundances change in response to climate change, the influences of negative density dependent factors on their vital rates will shift accordingly, stabilizing the new communities (Wright *et al.*, 2009).

Latitude more Determining than Longitude

Species' ability to respond to global warming is likely to vary with latitude (Wright *et al.*, 2009) because of four factors due to latitudinal differences; these are expected increase in temperature, inherent sensitivity of species to temperature change, proximity to cooler refuges and the potential for interaction with species adapted to warmer climates. Wright *et al.* (2009) suggested that species in the low latitude areas are likely to be more sensitive to temperature change. According to Terborgh (1973), lowland mean annual temperatures range from 24–27°C in a 31 million sq km area and 47° of latitude between the Tropic of Cancer and Tropic of Capricorn. Plant communities in the tropical regions are not likely to adapt well to climate change and as a result the animals depending on them will be affected. Deutsch *et al.* (2008) found that warming in the tropical areas, despite being small in magnitude is likely to have severe consequences as the insects inhabiting this region are relatively sensitive to temperature change and the present temperature here is close to their optimal range. Inadequate capability of tropical species to acclimate to global warming is going take its toll in the tropics over a longer period. Since the tropical areas have a greater concentration of species (Terborgh, 1973; Gaston and Blackburn, 2000; Colwell and Lees, 2000), any shrinkage in their habitat resulting from climate change would have detrimental effect on the population. Deutsch *et al.* (2008) tested frogs, lizards and turtles using published documents where critical thermal limits were experimentally determined for at least 12 population of closely related taxa across large climate gradients. They found that in each taxonomic group, warming tolerance of an organism increases strongly with the seasonal temperature variability of its habitat, which indicates that warming will cause tropical vertebrate ectotherms to approach their critical maximum temperatures proportionately faster than similar species inhabiting the higher latitude areas. As a consequence, these taxonomic groups will likely experience severe changes in thermal performance during global warming. Bush (2002) suggested that as global temperature increase, immigrants from warmer areas are likely to be better adapted to new and warmer climates than are the original residents of the former cooler areas. Species in the mountains in the tropical region are to be mostly affected because the surrounding lowlands provide a nearby source of species adapted to warmer climates; however, lowland tropics would not be threatened by the arrival of species adapted to warmer climates as this area already supports the highest mean annual temperature and species adapted to warmer climates are not here originally (Wright *et al.*, 2009). Considering 5,257 terrestrial mammalian species after Schipper *et al.* (2008) and their habitat associations as per the IUCN Red List (IUCN, 2008), Wright *et al.* (2009) came up with a total of 4,351 terrestrial mammalian species that are fully or partly distributed in the tropical latitudes; out of which a total of 2,814 species had ranges restricted to tropical latitudes and among these 1,224 species were restricted to the narrow belt of low–latitude tropics (10"N to 10"S). Wrights *et al.* (2009) described these later 1,224 species to be sensitive to global warming.

The Case of Flagship Species

The most accepted hypothesis is Bengal tigers (*Panthera tigris tigris*) dwelling the mangrove islands of Sundarbans in Bangaladesh and India face severe threats from climate change. Agarwala *et al.* (2003) and Mohal *et al.* (2005) predicted that permanent inundation, drainage congestion, salinity intrusion and more frequent storm induced inundation would be probable climate change impacts in the Sundarbans. Since these mangrove islands, which are also habitats for tigers and their prey, are surrounded by high density human habitations, except the southern side which has Bay of Bengal, the chance of these animals' dispersal into safer and higher grounds is very slim and is also going to increase the thriving human – tiger conflict. Tigers are no safer from effects of climate change outside Sundarbans. Beniston (2003) and Cruz *et al.* (2007) have noted impacts of climate change in the greater Himalayas and it is highly likely that the two major rivers, Ganga and Bramhaputra, arising out of this region will also bear the brunt of these impacts. Cruz *et al.* (2007) predicted a decline of water flow in these rivers due to diminishing glacial ice source over long term resulting in water shortage for downstream wildlife. They also suggested that the two above–mentioned rivers could have seasonal flows in near future. As vast extent of grasslands and riparian habitats are supported by these two rivers' downstream flows which constitute important habitats for tigers and their prey in India and Nepal, climate change is surely going to impact the population thriving there.

As mentioned above, impacts of climate change on the grasslands and riparian habitats of tigers and their prey are also going to have adverse impacts on populations of Asian elephant (*Elephas maximus*). This is reiterated by Seidensticker (2008) who noted that changes in river hydrology with reduced runoff will impact the forests and grasslands found along the outer range of the Himalayas, in the Shivalik hills and the associated narrow strip of lowland forests in the bhabar and the terai. Most affected areas will be the Terai Arc Landscape from the river Yamuna in India in the west to river Bagmati in Nepal in the east that is spread in the Indian states of Uttarakhand, Uttar Pradesh and Bihar and adjoining low–land areas of Nepal; the dooars in northern part of West Bengal and the Bramhaputra floodplains in Assam starting from Manas National Park in the west through Pobitora Wildlife Sanctuary, Kaziranga National Park, Laokhowa–Burachapori Wildlife Sanctuary, Nameri National Park to Dibru–Saikhowa Wildlife Sanctuary.

The Indian rhinoceros (*Rhinoceros unicornis*) are also affected by the stresses in grassland habitats as mentioned in case of elephants. In addition, rhinos are also affected due to flood in their prime habitats, more so in Kaziranga National Park and Pobitora Wildlife Sanctuary in Assam, India.

Remarks

Tigers, elephants and rhinos which are already threatened by poaching because of a thriving demand for trade in their body parts, high level of human–animal conflict and resulting retaliatory killing, habitat fragmentation and loss of corridors due to agricultural expansion, rapid human population growth and an unplanned development regime are going to be less resilient to impacts of climate change. Floods add another stress to the already threatened species. Animals from flood inundated areas migrate to higher grounds in search of dry land and in doing so, sometimes disperse out of the protected areas where there safety is compromised; they either come in conflict with human beings, get killed or hit by trains and speeding vehicles.

Climate change might not be reversed, but global warming can be kept within tolerable limits if proactive actions are taken right away. The first path towards this is to reduce emission of greenhouse gases including CO_2. It is a fact that CO_2 is released during burning and combustion of fossil fuels such as coal, oil and natural gas required for transportation or production of electricity. Most damaging is coal as it produces 70 per cent more CO_2 emissions than natural gas for the same energy output. Electricity generation is the single largest source of manmade CO_2 amounting to 37 per cent of worldwide emissions (WWF, 2008). To reduce these emissions, WWF is calling for actions that include improving the energy efficiency of power plants; increasing use of renewable energy sources; stopping investment in new coal plants and coal mining; making consumers aware to use green power and energy efficient appliances; and requesting policy makers to ease the transition to a carbon–free energy industry by passing suitable legislation that creates favourable market conditions and ensure that the 'Kyoto Protocol' enters into force as soon as possible.

References

Agarwala, S., Ota, T., Ahmend, J.S. *et al.* 2003. *Development and climate change in Bangladesh: focus on coastal flooding and the Sundarbans, working party on global and structural policies, working party on development co–operation and environment.* Organization for Economic Co–operation and Development, Paris.

Beniston, M. 2003. Climatic change in mountain regions: a review of possible impacts. *Climatic change*, 59: 5–31.

Bush, M.B. 2002. Distributional change and conservation on the Andean flank: a paleoecological perspective. *Global Ecology and Biogeography*, 11: 463–473.

Colwell, R.K. and Lees, D.C. 2000. The mid–domain effect: geometric constraints on the geography of species richness. *Trends in Ecology and Evolution*, 15: 70–76.

Colwell, R.K., Brehm, G., Cardelus, C.L. *et al.* 2008. Gobal warming, elevational range shifts and lowland biotic attrition in the wet tropics. *Science*, 322: 258–261.

Conservation International. 2008.

http://www.biodiversityhotspots.org/xp/hotspots/himalaya/Pages/default.aspx

Cruz, R., Harasawa, L., Lal, M., *et al.* 2007. Asia. In: *Climate Change 2007: Impacts, Adaptation and Vulnerability* (Parry, M.L., Canziani, O.F., Palutikof, J.P. *et al.* eds.). Cambridge University Press, Cambridge, UK. Pp. 469–506.

Deutsch, C.A., Tewksbury, J.J., Huey, R.B. *et al.* 2008. Impacts of climate warming on terrestrial ectotherms across latitude. *Proc. Nat. Acad. Science, USA*, 105: 6668–6672.

Gaston, K.J. and Blackburn, T.M. 2000. *Pattern and processes in macroecology*. Blackwell Scientific, Oxford, UK.

Malhi, Y. and Wright, J. 2004. Spatial patterns and recent trends in the tropical rainforest regions. *Philosophical Transations of the Royal Society, B*, 359: 311–329.

Mohal, N., Khan, Z.H., Khan, N. *et al.* 2005. *Impact of sea level rise on coastal rivers of Bangladesh.* Institute of Water Modelling, Dhaka, Bangladesh.

Schipper, J., Chanson, J.S., Chiozza, F. *et al.* 2008. The status of the world's land and marine mammals: diversity, threat and knowledge. *Science*, 322: 225–230.

Seidensticker, J. 2008. Wild cats and climate change. *Wild Felid Monitor*, 1(1): 1–6.

Terborgh, J. 1973. On the notion of favourable in plant ecology. *The American Naturalist*, 107: 481–501.

Wright, S.J., Muller–Landau, H.C. and Schipper, J. 2009. The future of tropical species on a warmer planet. *Conservation Biology*, 23(6): 1418–1426.

WWF. 2008. http://www.panda.org/climate/glaciers

Climate Change: Man and Environment (2012)   *Pages* **227–235**
Editor: **Goutam Kumar Saha**
Published by: **DAYA PUBLISHING HOUSE, NEW DELHI**

18

The Changing Global Climate and Future of Marine and Inland Fisheries

☆ *Anilava Kaviraj* *

Introduction

Recent changes in global climate are apprehended to cause massive alterations in the environment, unprecedented in the history of human civilization. Millions of people around the globe have already been affected and hundreds of millions more are threatened to be affected in the next few decades due to climate–related impacts. Productivity of the terrestrial and aquatic ecosystem are threatened to be disturbed seriously, which in turn may jeopardize food security of man in the next century. The extent to which fisheries sector has been affected by such climate changes has not been evaluated thoroughly. But there is strong evidence that aquatic ecosystem and the fisheries resources are most vulnerable to climate changes. To understand the ability of the aquatic ecosystems to adapt to these changes and predict what might occur in the future careful observations of the present and past conditions of the climate are essential.

Global Climate System and the Changing Scenario

Rise in temperatures and changes in precipitation are the two most important factors of climate change that affect the biodiversity, distribution of species and productivity of the terrestrial and aquatic ecosystem. Associated with this are rise in

* Department of Zoology, University of Kalyani, Kalyani. E-mail: akaviraj@gmail.com

sea level and increase in frequency and intensity of extreme events such as storms, droughts and flood. The ability of the terrestrial and aquatic ecosystems to adapt to such climate changes is dependent on the rate and extent of these changes. Climate is a result of dynamic interrelationship among atmosphere, oceans, the ice cover, living organisms and the geosphere consisting of soil, sediment and rock. Of these five components, the atmosphere plays the most crucial role and any change in atmosphere influences the other components and the climate as a whole. Global temperature is determined by the balance of solar and terrestrial radiation budget of the earth. The atmosphere, which has a role in maintaining this balance, is practically a mixture of gases and suspended liquid and solid particles. It envelops the earth in four distinct layers: the troposphere, stratosphere, mesosphere and thermosphere. The troposphere and stratosphere play the most important role in controlling the energy budget through thermodynamic process. The troposphere contains more than 75 per cent of the atmospheric gases. But density of the gases and the suspended particles decrease with the altitude resulting in decrease of temperature from an average of 15°C at the sea level to –56°C at the end of troposphere (an average decrease of 6.5°C per km height). Temperature begins to increases in the stratosphere from –56°C to –2°C due to absorption of UV radiation by the ozone layer, the characteristic constituent of stratosphere. The entire atmosphere is otherwise almost transparent to visible spectrum of light (little absorption). As a result, visible solar radiation enters the earth atmosphere of the troposphere unhindered and heat the surface of the earth. When the earth surface radiates the infrared radiation, a significant part of it is absorbed by water vapour, CO_2, methane and other trace gases. Absorption of terrestrial infrared radiation is important to the energy budget of earth. Some of this is released into space while most is radiated back to earth. The neat effect is storage of energy near the surface of earth. The process is known as greenhouse effect and is an essential phenomenon to maintain optimum temperature and diurnal fluctuation required for the survival of life on earth. The gases which absorb the outgoing infrared radiation are known as the greenhouse gases (GHG), which include the water vapour, CO_2, methane and others. Despite the absorption, there is an atmospheric window through which terrestrial infrared radiation can pass. It is gradually closing as a result of anthropogenic emissions of greenhouse gases (Figure 18.1).

The changing composition of the atmosphere, including its greenhouse gas and aerosol (suspended liquid and solid particles) content is a major internal forcing mechanism of climate change. The anthropogenic factors, like burning of fossil fuels, forest cleaning and other industrial processes have increased the amount of carbon dioxide and other greenhouse gases since the eighteenth century. CO_2 alone has increased from 280 ppm during the pre–industrial age to 379 ppm in 2005, while the total GHG emissions have been increased by 70 per cent between 1970 and 2004 (IPCC, 2007). The developed nations have contributed the major share of the global concentration of GHGs in contrast to meager contribution by India and other developing or underdeveloped nations. Per capita consumption of CO_2 in India is only 1.2 ton as against 20.6 ton in USA, 20 ton in Canada and around 10 ton each in UK, Germany and Japan. The global share of CO_2 emissions by India is only 4.6 per

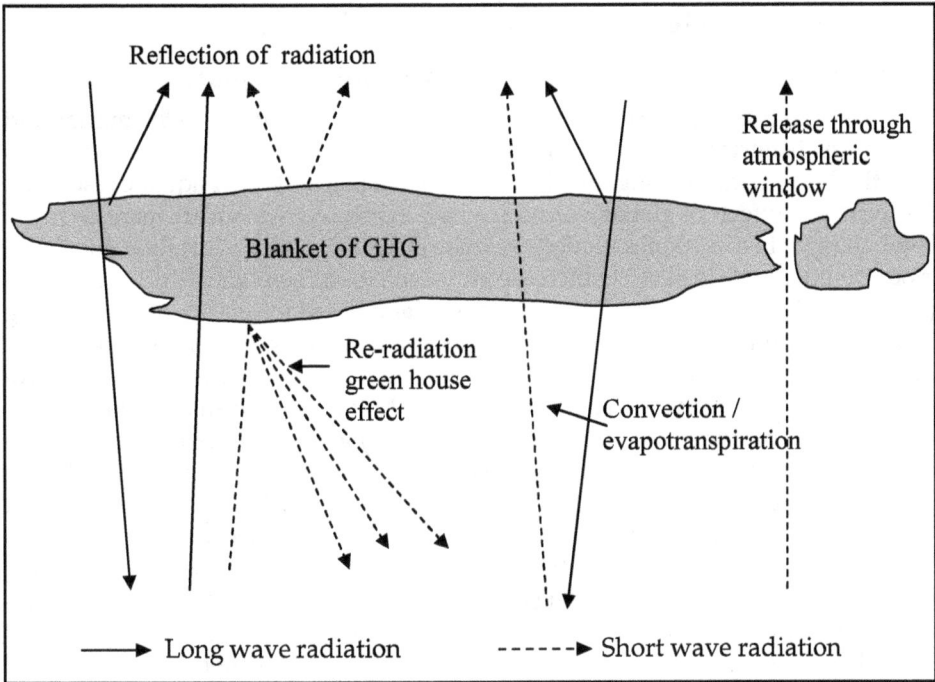

Figure 18.1. Global energy transfer.

cent as against 20.9 per cent by USA and 17.4 per cent by China. However, the most dangerous are the chloroflurocarbons (CFCs), which are entirely anthropogenically produced. CFCs were absent in the atmosphere before 1930s, but their concentrations have steadily increased over the last few decades. They not only possess longer life time in atmosphere, but also are thousand of time stronger than CO_2 as greenhouse gas. Since CFCs enter into stratosphere and destroy ozone layer, these are more dangerous and being phased out as part of Montreal protocol and replaced by hydroflurocarbons (HCFCs).

The neat effect of increase in GHGs and CFCs is the rise of global mean temperature, which has been recorded as 0.75 °C during 1906 to 2005. However, model based projections made in the Fourth Assessment Report of the International Panel on Climate Change (IPCC) predict a rise of average global temperature between 1.1 and 6.4°C by the year 2100. Warming in the mid troposphere and cooling in the much of the stratosphere are useful detection fingerprint of greenhouse warming. The cooler stratospheric temperature would be an expected consequence of the increased trapping of terrestrial radiation in the troposphere. However, stratospheric cooling may not solely be attributed to greenhouse forcing and may also result from ozone depletion.

Impact of Climate Change on Aquatic Resources

Melting of Ice, Sea Level Rise and Oceanic Circulation

The changing scenario of climate has the potential to influence the marine and freshwater fisheries indirectly through alterations in the aquatic ecosystem and directly through influencing the physiology of the fish and other aquatic resources. The principal effects of global warming on the marine ecosystem are increase in sea level, changes in oceanic circulation, increase in salinity and acidification. Increased global temperature is likely to increase global mean sea level partly due to thermal expansion of sea and partly due to melting of land based ice masks. There is strong evidence of melting of ice core that has survived warming for long period. These include breakdown of Larsen ice shelf on the Antarctic continent, disintegration of sea ice blocking the Gustav channel between the Antarctic Peninsula and James Ross Island and substantial recession of mountain glaciers since the later half of the nineteenth century. Global sea level rise has been recorded at 1.8 mm/year during 1961–2003. The average rise during the recent past (1993 to 2003) was even faster (at the rate of 3.1 mm/year). The current global climate mean projects a 4 cm rise of sea level per decade. If it is reasonable, it may cause serious threats to the low lying islands and coastal zones increasing the risk of coastal flooding, damage of fisheries and salination of fresh ground water supplies. India having a long coast line of about 8000 plus km is most vulnerable to such rise in sea level. One meter rise in sea level can put as many as 7.1 million people, including all coastal fishing communities whose livelihood is directly linked to the ocean, of the country at risk of displacement. Current records indicate an average of 1.30 mm sea level rise per year along the coasts in India (URL–1).

Changes in oceanic circulation, a consequence of changing global climate, influence the marine ecosystem and its fisheries. Increasing frequency of El Nino episode is considered as evidence of such changes. El Nino is defined as sustained sea surface temperature anomalies of magnitude greater than 0.5°C across the central tropical Pacific Ocean. El Nino's warm current of nutrient–poor tropical water, heated by its eastward passage in the equatorial current during Christmas, replaces the cold, nutrient–rich surface water of the Humbolt current, also known as the Peru current, which support great populations of food fish. In most years the warming lasts only a few weeks or a month, after which the weather patterns return to normal and fishing improves. During this period walker circulation is seen at the surface as easterly trade winds which move water and air warmed by the sun towards the west. This also creates ocean upwelling off the coasts of Peru and Ecuador and brings nutrient rich cold water to the surface, increasing fishing stocks. However, when El Nino conditions last for many months, more extensive ocean warming occurs and their impacts on local fishing become serious. Some fish population increase while others decrease during El Nino years. The horse mackerel and scallop population increased, while the jack mackerel and anchoveta population decreased during the conspicuous El Nino episodes in 1971, 1982, 1991 and in recent years. The shrimps and sardines have been found to migrate southwards during the episode. Based on paleoclimatic studies (records of major episodes of climate change during the course of earth history) and pattern of circulation change (Boyle and Keigwin,1987; Boyle,1988) it has been predicted that El Nino episode will possibly increase in their intensity or/and frequency in the coming decades.

Sea Surface Temperature, Salinity and Acidification

Records of past 25 years indicate that sea surface temperature has positively increased and this has been ascribed to greenhouse warming. Dealing with recent Atlantic hurricane activities, Saunders and Lee (2008) predicted that an increase in 0.5°C sea surface temperature could contribute a more or less 40 per cent increase in hurricane frequency and activity of seas. Several hurricanes of the recent past such as Katrina, Rita, Aila etc. are assumed to be effects of increasing sea surface temperature. But, based on measured refractive index of sea water it has been revealed that infrared radiation can not penetrate into sea water and contribute to ocean warming. Therefore, effect of greenhouse warming on increase of sea surface temperature is questionable. However, there is no doubt that oceans are warming and the near surface water in the more evaporative regions is increasing in salinity in almost all ocean basins. But high latitudes are showing decreasing salinity due to greater precipitation, higher run-off, melting ice and advection.

Increasing concentration of CO_2 in the atmosphere has resulted in increasing concentration of carbonic acid, a product of reaction between CO_2 and water, in the water and decrease of pH. About 1/3 of the CO_2 from fossil-fuel burning is absorbed by the world's oceans. Ocean acidification poses a great threat to marine life and ecosystems, especially organisms that use calcium carbonate to form shells and skeletons, like phytoplankton and corals. In a more acidic marine environment, the sea urchin's ability to multiply goes down by 25 per cent, as its sperm swim more slowly and move less effectively in acidic water. An estimated decrease of 0.075 pH has occurred in ocean water during the last 250 years. Atmospheric CO_2 concentrations need to remain at less than 500 ppm for the ocean pH decrease to stay within the 0.2 limit set forth by the U.S. Environmental Protection Agency. Atmospheric CO_2 levels presently stand at 380 ppm, but are expected to reach 500 ppm by mid-century (Caliera, 2007).

Precipitation

Precipitation is likely to increase with the increase of global temperature due to greater rates of evaporation of sea surface water. But this is only a prediction at present. Neither there is any reliable estimate of evaporation nor is there adequate instrumental data of precipitation (Buchdahl, 1999). However, some studies indicated that precipitation tended to increase in the mid-latitudes, decrease in the northern hemisphere subtropics and increase in general through out the southern hemisphere (Buchdahl, 1999).

Impact on Fisheries

Individual and Population Effects

The fish and invertebrates are cold blooded animals and their body temperature varies with the ambient temperature. Thus any change in habitat temperature will significantly influence metabolism of these animals. The consequence of global warming on individual fish and aquatic invertebrates is exhibited in the alterations of physiological functions such as thermal tolerance, growth, metabolism, food consumption, reproductive success (Roessig *et al.*, 2004) and susceptibility to diseases and toxins. It influences spatial distribution of fishing and aquaculture activities and their productivity and yields. In general, the cold water tolerant species are more susceptible to warming than their warm water tolerant counterparts.

One classical example of cold water species being victim of warming is the outbreak of disease called "excretory calcinosis" in American lobster (*Homaras americanas*) inhabiting northern end of the Long Island in USA. This is a gill tissue blood disorder resulting directly from warm temperatures (Dove *et al.*, 2004) and causing massive catastrophic summer fall mortalities since August 1999. The disease appears to be moving northward. Another example of climate induced effects on fisheries involves the northward expansion of a disease known as "Dermo" that affects the oyster and is caused by a parasite which kills 50 per cent of the oyster in the Gulf of Mexico. From 1980 to 1990 the disease moved 500 km northward from the south of lower Cheaspeake Bay to the Gulf of Maine (Ford and Smolowitz, 2007). In recent years the Dermo has been recorded very high from Delaware Bay to Cape Cod with no signs of abating.

However, global warming may also prove useful for certain species of fish as well as to some localities. Increase in temperature may cause increase in rate of consumption and growth of certain species up to a certain point. Once the temperature increases past the point where the consumption and growth are maximized, cost of respiration is increased and consumption rate is decreased resulting in rapid decline of growth. In sockeye salmon (*Oncorhynchus nerka*) food consumption triples between 2.5 and 17.5°C, but decreases above 17.5°C which is thermal optimum of this species (Brett, 1971). Similar effects of temperature are found in many species of fish. If the temperature is high enough to cause thermal stress the fish may face problems in osmoregulation (Boyd and Tucker, 1998), possibly due to increased permeability of gills (Somero and Hoffman, 1997). The fish populations may have to achieve new equilibrium dictated largely by the energy costs of coping with a new thermal environment. Achievement of new equilibrium depends on ecological status of the fish (whether stenothermal or eurythermal), the magnitude and rate of thermal change of the ecosystem and opportunities to migrate.

Impact on Marine Fisheries

Shifting in locality has been detected as a tendency in many species in recent years. As a result many areas are being benefited with catch of fisheries that were not present in the area earlier. But this is at the cost of some other areas from where the fish population is shifting. During the last 40 years, many familiar species have been shifting to the north where ocean waters are cooler, or staying in the same general area, but moving into deeper waters than where they traditionally have been found. It is apprehended that by 2050 large number of marine fish will migrate from tropical seas to cooler waters specifically the Arctic and Southern oceans.

Investigations by Central Marine Fisheries Research Institute on Indian marine fisheries resources indicate that climate changes have altered the production and distribution of some commercially important pelagic fishes from Indian waters. Historically, the distribution of sardines and mackerels were restricted to the Malabar upwelling system along the southwest coast of India. However, a clear cut distribution shifts in these two species have been observed since 1989. Oil sardine (*Sardinella longiceps*) has emerged as a major species along southeast coast of India, while mackerel (*Rastrelliger kanagurta*) fishery has emerged along the northwest coast. Like many other tropical pelagic fishes, the Indian mackerel and the Indian oil sardine have shown population crashes and sudden recoveries and very strong inverse relationship. Shifts in the vertical distribution of the Indian mackerel have also been observed. This pelagic species is now caught by bottom trawlers. Demersal species such as threadfin bream (*Nemipterus japonicus*) appear to shift the month of peak spawning toward colder months off Chennai. Shifts in the relative abundance of finfish bear signature of ocean warming. Some species are able to adjust to the immediate challenge of rise in temperature for a shorter or longer duration. But, the species which are not able to adjust the climatic changes or has hurdles in migration may face severe consequences in the coming decades. The cods are fast disappearing from New England. The effect of ocean warming on spatial distribution of Bombay duck, whose northward distribution is blocked by the terrestrial land, is not known. The oceanic tunas are strongly influenced by thermocline. Effects of ocean warming on their distribution are also not clearly known.

Impact on Inland Fisheries

Concrete evidence of global warming on inland fisheries is lacking. But many lakes across the globe are showing tendency of warming since last fifty years. African lakes are of great concern, because average temperature of this continent is higher than the global average. A minute increase in average temperature of this continent may result in drying of many lakes. However, warm water fishes are rather tolerant to warming in contrast to the cold water counterparts. Most of the cold water fishes are stenothermal and are most sensitive to even minute fluctuations in temperature. The trout and salmon thrive in streams with temperatures ranging between 10–18°C. In many areas, these species are already living at the upper end of their thermal range, meaning even modest warming could render streams uninhabitable to them. In India, the snow trouts (*Schizothorax* sp.), loaches (*Nemacheilus* sp., *Botia* sp. etc), lesser barils (*Barilius* sp.), mahaseers (*Tor* sp.) and the exotic trouts (*Salmo* sp.), which are distributed in the rivers and lakes of the Himalayas, are vulnerable to even minute warming of the environment.

The wetlands and shallow rivers are also susceptible to changes in temperature and precipitation and water levels may drop to the point of completely drying out in dry summers. Recent surveys on two floodplain lakes of Ichamati river basin of West Bengal in India have indicated an alarming trend in decline of depth of the lakes (Mondal *et al.*, 2009). Similar decreases in depths have been recorded in many other Indian floodplain lakes. But in most cases the reasons behind such decrease have been ascribed mainly to eutrophication, macrophyte infestation and discharge of organic debris from human settlements (Goswami *et al.*, 1999). Impact of climate change on these water bodies has hardly been evaluated. However, it has been established that climate change produces most pronounced effect on wetlands through alterations in hydrological regimes. Increased temperature may lead to stronger, earlier and longer stratification of lakes and reservoirs and with limited or seasonal turnover, greater deoxygenation (*i.e.* hypoxia) of the bottom layer, which in turn may influence the community structure of the floodplain lakes. Therefore, it is very important to maintain hydrology, reduce pollution, control exotic vegetation and protect wetland biological diversity and integrity to maintain and improve the resiliency of wetland ecosystems so that they continue to provide important services under changed climatic conditions.

References

Boyd, C. and Tucker, C. 1998. *Pond aquaculture and water quality management*. Kluwar Academic publishers, Norwell, Mass.

Boyle, E.A. and Keigwin, L.D. 1987. North Atlantic thermocline circulation during the past 20,000 years linked to high–latitude surface temperatures. *Nature*, 330: 35–40.

Boyle, E.A. 1988. Vertical oceanic fractionation and glacial/interglacial CO_2 cycles. *Nature*, 331: 55–58.

Brett, J. R. 1971. Energetic responses of salmon to temperature. A study of some thermal relations in the physiology and freshwater ecology of sockeye salmon (*Oncorhynchus nerka*). *American Zoologist*, 11: 99–113.

Caliera, K. 2007. CO_2 emissions cause ocean acidification, threaten sea life. *Geophysical Research Letter*, September : 21.

Buchdahl, J. 1999. Global climate change student guide – a review of contemporary and pre–historic global climate change. Atmosphere, climate and environment information programme, Manchester. Available at: http://www.doc.mmu.ac.uk/aric/

Dove, A.D.M., LoBue, C., Bowser, P. *et al.* 2004. Excretory calcinosis: a new fatal disease of wild American lobster *Homarus americanus*. *Disease of Aquatic Organism*, 58 (2–3): 215–223.

Ford, S.E. and Smolowitz, R. 2007. Infection dynamics of an oyster parasite in its newly expanded angle. *Mar. Biol.*, 151: 95–97.

Goswami, M.M., Deka, T.K., Shing, P.K. *et al.* 1999. Studies of some wetlands of Assam with reference to the eutrophication stresses. *Journal of Inland Fisheries Societies of India*, 31(2):39–43.

Intergovernmental Panel on Climate Change (IPCC). 2007. Summary for Policymakers. In: *Climate Change 2007: The Physical Science Basis* (Solomon, S., Qin, D., Manning, M. *et al.* eds.). Cambridge University Press, Cambridge, UK. Pp. 1–18.

Mondal, D.K., Kaviraj, A. and Saha, S. 2009. Water quality parameters and fish biodiversity indices as measures of ecological degradation: A case study in two floodplain lakes of India. *Journal of Water Resource and Protection*, 2: 85–92.

Roessig, J.M., Woodley, C.M., Cech, J.J. Jr., *et al.* 2004. Effects of global climate change on marine and estuarine fisheries. *Reviews in Fish Biology and Fisheries*, 14: 251–275.

Saunders, M.A. and Lee, A.S. 2008. Large contribution of sea surface warming to recent increase in Atlantic hurricane activity. *Nature*, 451: 557–560.

Somero, G. N and Hoffmann, G.E. 1997. Temperature thresholds for protein adaptation: when does temperature starts to hurt? In. *Global warming: Implications for Freshwater and Marine Fish* (Wood C.M. and McDonald, D.G. eds.), Cambridge University Press, Cambridge, UK. Pp. 1–24.

URL–1. Sea–level changes along the Indian coast – Impacts and Vulnerability. Avaialable at: http://moef.nic.in/downloads/others/Sealevel_Unnikrishnan_Manimurali percent20and percent20Kumar.pdf

Climate Change: Man and Environment (2012)   *Pages* **236–245**
Editor: **Goutam Kumar Saha**
Published by: **DAYA PUBLISHING HOUSE, NEW DELHI**

19

Impact of Climate Change on Human Physiology vis-à-vis Human Health

☆ *Somenath Roy**

Introduction

Climate change is likely to have a greater impact in India compared to other countries. This is due to the unique combination of its geography, varied ecological conditions, diverse population and extremely high carbon related energy dependence. Climate change is bound to affect the basic requirements for maintaining health–clear air, water, food and freedom from diseases. Over the last fifty years, human activities particularly the burning of fossil fuel have released sufficient quantities of carbon dioxide and other greenhouse gases to trap additional heat in the lower atmosphere and affect the global climate. In the last hundred years, the world has warmed by approximately, 0.75°C. It is predicted that over the last twenty–five years, the rate of global warming has accelerated, at over 0.18°C per decade (Hadley Research Centre, 2008). Heat waves directly contribute to deaths from cardiovascular and respiratory diseases, rapidly among the elderly people. In the heat wave of 2003 summer in Europe, for example, more than 70,000 excess deaths were recorded (Robine *et al.*, 2008). High temperature also raises the levels of ozone and other pollutants in the air that exacerbate cardiovascular and respiratory diseases. Pollens and other aero–allergens levels are also higher in extreme temperature regime, which can trigger asthma and other allergic complains. Changes in the climate are likely to lengthen

* Department of Human Physiology with Community Health, Vidyasagar University, Midnapore.
  E-mail: sroy.vu@hotmail.com

the transmission seasons of important vector-borne diseases and thus help to alter their geographic ranges. Malaria is strongly influenced by climate. Transmitted by *Anopheles* mosquitoes, malaria kills almost one million people every year. The *Aedes* mosquito, vector of dengue, is also highly sensitive to climate conditions. Studies suggest that climate change could expose an additional two billion people to dengue transmission by the 2080s (Hales *et al.*, 2002). Higher ambient temperature accelerates the growth of pathogens that thrive in or on food, such as, *Salmonella* spp. A cholera epidemic struck Peru and rapidly spread to other southern countries in 1991. This is due to the warm ocean waters in the eastern Pacific, which is termed as El Nino. A meningitis epidemic struck western Africa after a spell of hot and dry weather.

Impact on Human Health

Aila and Climate Change

Global climate change put a serious impact on natural system and thus intensity of high magnitude storms have increased worldwide. Rainfall becomes erratic and more concentric. Concentration of rain to few spells causes floods during peak rain and invites long duration drought and water scarcity for the rest of the seasons. Since 1891, there were 1148 depressions and cyclones occurred in Bay of Bengal, of which 216 are severe cyclonic storms. This indicates a high frequency of cyclones at costal area, which leads to storm surge, generating high energy waves and consequent costal erosion and flood. The recent cyclone, Aila, of 25th May, 2009 at Sundarban took a toll of 75 people from Sandeshkkhali I and II and Hingalganj block only. All the property and belonging of the local habitants are damaged. The social and economic system are so altered that many people have lost their livelihood and became 'climate refugee'.

Brown Cloud in Asia

In 1995, Paul Crutzen, a Nobel Laureate working on ozone holes, reported that two million people in India were dying each year from respiratory disease caused by air pollution. According to the report of UN, the 'brown cloud' covers over south Asia spreading from Afghanistan to Sri Lanka and was three kilometers thick. It was a cocktail of fly ash, nitrates, sulphates, black carbon, acid aerosols, soot and other pollutant spread over an area of ten million square kilometers. It is the product of forest fires, burning of agricultural wastes, fossil fuels, vehicular emissions and *chullahs* using wood, coal or cow dung. The cloud is cutting down direct sunlight by 10–15 per cent, reducing rain fall, increasing risks of drought and also of acid rain.

Acid Rain and Human Health

Acid rain is produced when sulfur dioxide (SO_2) and/or the oxides of nitrogen (NO_x) and their oxidation products react with the moisture in the atmosphere and come down to earth as rain or any other form of precipitation. Sulfur oxides are usually the main pollutants responsible for the formation of acid rain. They are produced by burning of coal, especially coal with high sulfur content, mainly in power station. Oxides of nitrogen usually originate from power station and from vehicular emissions. Acid rain affects surface water (such as rivers and lakes) and ground water, as well as soil and vegetation resulting in changes in the chemistry and ecology of lakes, rivers and soil chemistry, productivity of plants and ultimately affecting human health. Smoke produced by domestic fires, power stations and coal burning industries was the important pollutants but the most dangerous was sulfuric acid, flowing free in aerosol form or attached to the smoke particles. The infamous London smog (1952) developed as a result of meteorological condition which allowed the build up of pollutants within the urban atmosphere. The sulfuric acid present in the smog aggravated breathing problems leading to 4000 deaths.

Impact of Air Pollution on Human Health

It has been found that a man can live for five weeks without food and five days without water, but only for five minutes without air. The effect of air pollution on human health is worst during winter season. The pollution level reach maximum in winter as more pollutants are trapped in the near–surface atmosphere due to 'temperature inversion'. The impact of air pollution on human health generally occurs as a result of contact between the air pollutants and physiological system. Usually, body contact occurs at the surface of the skin and exposed membranes. Exposed membranous surfaces are more susceptible due to their high absorptive capacity compared to that of the skin. Vapours, fumes, air–borne gases and dust may cause irritation of the membranes of eye, nose, throat, larynx, tracheo–bronchial tree and lung. Some of the pollutants even reach the mucosa of the digestive tract. Probable health damages include:

1. Irritation of the eye, nose, throat and respiratory tract.
2. Danders, pollen in the air initiate asthmatic attacks.
3. Chronic pulmonary disease, like bronchitis and asthma are aggravated by high concentration of SO_2, NO_2 particulate matters and photochemical smog.
4. Carbon monoxide combines with haemoglobin in the blood and consequently increase stress on those suffering from cardio–vascular and pulmonary diseases.
5. Hydrogen fluoride causes disease of the bone (fluorosis), melting of teeth.
6. Cancer causes due to carcinogenic chemicals like benzo[a]pyrene, tobacco-specific nitrosamines, etc.
7. Dust particles causes respiratory disease *e.g.* silicosis, asbestosis etc.
8. Heavy metal like copper, lead and zinc enter through lungs and cause poisoning.
9. Pesticides like DDT, PCB etc. can cause the reproductive damages.

Effect of Pollutants on Physiological Systems

The experimental exposure of different types of animals under controlled condition to various concentrations and dosages of air pollutants can provide valuable information regarding the mode of infection of pollutants. Types of effects noticed and measured in human include detection of odour, irritation of eyes, nose and throat, change in pulse rate, breathing frequency, reduction in physical activity and many other physiological responses. Studies of truck drivers, traffic policemen, factory–workers and other occupational groups may give valuable information. The following pollutants may have some impact on human health.

Sulphur Dioxide

This chemical is produced by making paper and metals. It is an irritant gas which affects the mucous membrane. Some of the air borne sulphur dioxide gas is oxidized to sulphur trioxide. It is also very strong and more irritant than sulphur dioxide, causing severe broncho–spasms at relatively low concentrations.

Carbon Monoxide

Carbon monoxide is produced from the partial oxidation of carbon–containing compounds, mainly from improper burning of various fuels. It causes blood vascular problems and respiratory failure. This gas has a strong affinity for haemoglobin (about two hundred times more affinity than the affinity of oxygen for haemoglobin) and readily binds together to form carboxy–hemoglobin (COHB), reducing the oxygen carrying capacity of the haemoglobin. Carbon monoxide also affects central nervous system. It is also responsible for cardiac arrest.

Oxides of Nitrogen

This chemical is also formed by burning of fuels, like gas, coal, oil and leads to the formation of smog. It is one of the major causal agent of 'acid rain' and can damage trees, buildings and monuments. Nitric oxide and nitrogen dioxide are thought to affect human health. Manufacture of nitric acid, high nitrate fertilizer and electric arch welding are the occupations with NO_2 hazards. This compound causes problem in the respiratory system. It is estimated that after exposure to about 15 ppm of nitrogen dioxide eye and nasal irritation will be experienced and pulmonary discomfort will start to occur after a brief exposure of 25 ppm.

Ozone

It is the basic element of smog and causes many different kinds of health issues, related to lung and skin. It can damage plants and limit sight. In contrast to the polluting role of tropospheric ozone, the most obvious cause for concern is the depletion of the stratospheric ozone, as it plays a vital role of filtering the Sun's ultraviolet radiation. Correlations between the increase in biologically active ultraviolet ray (UV–β) and its estimated ozone loss may affect rates of skin cancer. It is suggested that every one per cent decrease of ozone column (O_3) will result in a three per cent rise in the incidence of non-melanoma skin cancers, which translates into some 12,000–15,000 extra cases a year in the USA, together with a possible one per cent increase in mortality from melanoma. Exposure to enhanced levels of UV–β can also have other direct harmful effects on the human body, the two most serious being a tendency to suppress the body's immune response and to cause damage to the eyes, especially in the development of cataracts. Although even more difficult to quantify, it was noticed that these effects would touch all populations, with some consequences–possible increases in the incidence of severity of infectious diseases.

The following effects have been found due to enhanced exposure of UV–β radiation:

Immuno-suppression

Ultraviolet–β radiation can initiate a selective down regulation of cell mediated immunity in mammals, including human. Immuno–suppressive effect of UV–β radiation play an important role in UV–β induced skin cancer by preventing the destruction of highly antigenic skin cancer by immune system. A recent report indicated that binding receptor for cis–UCA (transurocanic acid) has been identified as the neurotransmitter 5–hydroxytryptamine (5HT) (Nghiem *et al.*, 2002).

Skin Cancer

There are three main types of skin cancer. Two tend to neutralize and are known as basal and squamous cell carcinoma and are often referred to collectively as non-melanoma skin cancer. The third type, which shows a higher mortality and which can metastasize aggressively is malignant melanoma of which several sub types exist (Fears *et al.*, 1976; McGovern *et al.*, 1973). There is such experimental evidence of a clear connection between sunlight exposure and non-melanoma skin cancer which implicates UV– radiation as a carcinogen (Armstrong *et al.*, 1997).

Non-Hodgkin's Lymphoma

Epidemiological evidence suggests that there is a link between non-Hodgkin's lymphoma and sunlight exposure. This is suggested to be in the immunosuppressive effects of UV– β radiation. A correlation between the occurrence of skin cancer and occurrence of non-Hodgkin's lymphoma has also been described (Cliff and Mortimer, 1999). Lymphatic malignancies, autoimmune diseases (like Type–I diabetes, multiple sclerosis) may also have been an immunosuppressive connection with UV– β radiation (Staples *et al.*, 2003).

Cataracts

Cataract is a major cause of blindness. There were estimated 180 million people worldwide who were visually disabled in 2002 and 40–45 million people are blind. Cataract formation is a complicated process with many risk factors. The precise mechanism of action is not known, although UV–β radiation is very strongly implicated and associated with latitude and climate of different countries have been reported. Severe types of cataract exist with a varying degree of association with sunlight (Hockwin *et al.*, 1999). The importance of dietary factors and cataract also require further research.

Vitamin D

Vitamin D or Calciferol is a fat soluble vitamin. It is present in food, but can also be made in human skin after exposure to UV–β radiation from the sun (Holick, 2001). Many factors can affect vitamin D production such as season, latitude, age, skin colour, sun angle and time spent. The major function of vitamin D is to maintain normal blood levels of calcium and phosphorus. Vitamin D contributes calcium absorption, helping to form and maintain a strong skeletal structure. Without vitamin

D, bones may become thin, brittle, soft or deformed. Vitamin D prevents rickets in children and osteomalacia in adults.

Lead

This chemical is used in paint, leaded gasoline, smelters and in lead storage batteries. The main source of lead in atmosphere is the automobile exhausts. It creates urban concentration of inorganic lead of about 1–3 µg/m^3, with higher values in area of heavy traffic. Exposure to lead may result in several neural and cerebral disorders and digestive problems. Inorganic lead may cause liver and kidney damage, abnormalities in fertility and pregnancy. In addition, lead damages the membranes of red blood cell and interferes with the metabolism in a way that shortens the survival of each individual cell, often resulting in anaemia. Research provided evidence that levels of lead exposure associated with central nervous system effects, particularly manifested in behavioral changes.

Insecticides

Insecticides are not only harmful for insects but also have ill– effect on man and its environment. In fact, DDT has been found in mother's milk in western countries and in our country. According to a study at the Industrial Toxicology Research Centre, Lucknow, the accumulation of pesticides in the environment due to their growing use for agricultural purpose can also cause premature labour and abortion. It can also affect the central nervous system and may attack other vital organs.

Radioactive Isotopes

The important radioactive isotopes that may reach ambient air are–iodine, sulphur 35, calcium 45 and uranium. The major source of radioactive air pollutants are nuclear reactors, testing of nuclear bombs, as well as scientific and medical use of radioactive isotopes as tracers. The serious health effects are anemia, leukemia and cancer. Radioactive isotopes also cause genetic defects, sterility, embryo defects and congenital malformation.

Environment–induced Diseases

Some of the diseases have long been recognized as due to occupational/ environment related diseases. Dr. Percivall Pott was the first to point out that the direct connection between an occupational exposure and the risk of a specific cancer (scrotum cancer) (Cole and Goldman, 1975). Miners, stone cutters and lens grinders developed respiratory disease from inhaling large quantities of dust. Many people suffered from brain damage as a result of absorbing highly toxic vapors from mercurial (mercury containing) compounds. Coal, tar, pitch workers, textile–dyers, shoe and leather workers are all having an increased risk of developing bladder cancer because of their exposure with coal products and aromatic amines (Maltoni and Selikoff, 1988). The most serious diseases are cancer, respiratory diseases, birth defects, heavy metal poisoning and defects in the reproductive system.

Heart and Respiratory Diseases

Airborne pollutants are formed in the urban environment in high concentration and ultimately affect the lungs (Fennelly, 1976). Chronic bronchitis, emphysema and lung cancer have became major public health problem in developed and major industrialized countries. In the United States of America, heart diseases have been known as the number one killer for several decades. Between the late 1950 and 1960, a large number of patients in Japan suffered from chronic obstructive pulmonary disease (COPD) like chronic bronchitis, bronchial asthma and emphysema particularly in highly industrialized cities near Tokyo and the areas, which were heavily polluted with SO_2 pollution causing acute respiratory diseases (Murakami, 1996).

Heavy Metal Induced Diseases

The production of these heavy metals like copper (Cu), lead (Pb) and zink (Zn) has increased tremendously (nearly ten fold). Another group of toxic elements like cadmium (Cd), mercury (Hg), nickel (Ni) and arsenic (As) accumulate within the limited geographic areas to excessive levels, because of industrial pollution (Nriagu, 1996). They have produced major outbreaks of chronic illness in human.

In ancient Rome, lead in pipes carrying drinking water and cooking vessels was a major source of excessive intake. Even today lead contamination in water supplies occurs in some area. Smoke containing lead from burning trash, coal and recently in automobile exhausts is probably even more hazardous as it is inhaled by human (Waldron,1974). The most prominent adverse effects are on nervous system, blood–vascular system and kidneys.

Public concerns over mercury pollution reached its peak in the 1970s, when the Minamata disaster in Japan sparked off widespread concerns about the pollution of the environment by mercury and other toxic metals. The first indication of mercury poisoning was the development of a nervous disease in many local cats and dogs with a number of fatalities. Later local people started to show a variety of symptoms like numbness of limbs, deafness and mental derangement. The consumption of sea food contaminated with mercury was found to be the causal agent of this disease. Inorganic mercury itself is not very toxic. However, once released into the environment, mercury undergo methylation by bacterial action.

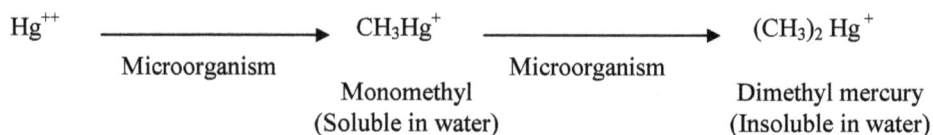

$$Hg^{++} \xrightarrow{\text{Microorganism}} CH_3Hg^{+} \xrightarrow{\text{Microorganism}} (CH_3)_2 Hg^{+}$$

	Monomethyl		Dimethyl mercury
	(Soluble in water)		(Insoluble in water)

Monomethyl mercury is soluble in water and very toxic. It is readily absorbed by fish. Human consuming this fish are at risk of mercury poisoning. It acts as a neurotoxin and causes several behavioral disturbances in affected individuals. It can induce permanent damage to the brain and kidneys, as well as loss of vision and disturbed cerebral function and in severe cases leading to coma and death.

Reproductive Damage

An increasing number of chemicals are now known to have ill effects on animal as well as human health. These are DDT, PCBs, dioxins, pesticides (such as carbamates, triazines, pyrethroids), cadmium, lead, mercury and organobrominate compounds, which can induce detrimental effects on reproductive system in animal and human. Chinese scientists have reported that shortening in gestation periods are due to the benzene exposure. They have claimed that pregnancies shortened in 542 births in women working at petrochemical plants in Beijing due to the exposure of benzene in the plants (Hileman, 2001). There are reports that in 2001, industrial chemicals, pesticides and air pollutants contributed to 23 per cent rise in premature birth in the USA since the early 1980s. One of the strongest causes was found in a study that measured the level of DDE (a metabolite of DDT) in the stored sera of mothers. The greater the level of DDE in the mother's blood, the higher was the risk for the infant. The toxicity of the pesticide DBCP (2,2–dibromo–3–chloropropane) in the reproductive system became clear in the late 1970s to early 1980s, when male farm workers in the banana growing region of Costa Rica were found to be sterile.

Birth Defects

Many chemicals like various organic solvents, pesticides, dioxins, heavy metals (lead, cadmium and mercury) are responsible for birth defects. It has been estimated that about 5–10 per cent of all birth defects are due to an uterine exposure to such known teratogenic agents. Retardation of the uterine growth is due to the number of agents including hypoxia, drugs, x–ray irradiation, maternal endocrine and nutritional factors and environmental chemicals. It was found 100,000 congenital abnormalities occur in a total three million live birth annually in USA (Kalter and Warkany, 1983). Congenital malformation is the leading cause of infant mortality rate in the United States. Obviously, enormous financial costs and emotional suffering are associated with these malformations.

Cancer Death/Risk

Scientists have found that industrial activities and environment pollution are responsible for the increase in the cancer incidences and cancer deaths. A New York Department of Health study has shown that women living within one kilometer of a chemical, petroleum, rubber or plastic factory were 60 per cent more likely to develop post menopausal breast cancer than those were living in other parts of country (American Chemical Society, 1994). According to the American Cancer Society, smoking, obesity and physical inactivity have a greater effect on individual cancer risk than do exposure to trace amounts of pollutants in air, food or drinking water (American Cancer Society, 2003). Obviously, the degree of risk from pollutants depends on the concentration, intensity and duration of exposure.

Remarks

Air pollution is a major cause of concern throughout the globe. Several air pollutants, especially the greenhouse gases (GHGs), not only cause global warming and subsequent climate change, they directly affect the physiology of several organs and often threaten the survival of mankind. It is widely documented that the global warming will raise sea levels, melt glaciers, shrink crop yields as well as show some direct impacts on human health through altered patterns of infection and vector-borne disease transmissions. But in a changed climate scenario and temperature regime the reaction and efficacy of several pollutants will change inevitably. Some pollutants might show more fatal effects to human organs, which would already be under severe stress in a warmer and changed climatic condition. Few other chemicals might react with each other and/or show synergistic effects giving rise to more deadly chemical pollutants. Warm and light air, as a consequence of global warming, will raise the pollutants to much higher height and violent weather events and storms may spread these harmful chemicals and gases to wider regions exposing more population to those pollutants. Forecasting the effects and efficacy of the pollutants in a changed climate scenario can help us to adopt appropriate mitigation and/or adaptation strategies.

Ill effects of various environmental pollutants are widely documented. Besides affecting individual health condition, several air pollutants (especially the GHGs) today pose a larger threat in the form of global warming and climate change and challenge the existence of human civilization in this planet earth. Therefore, prevention of air pollution is the need of the hour. Scientists claimed that even if all countries miraculously achieve the impossible feet of stopping all emission of carbon dioxide and other greenhouse gases, that trap the heat from the sun, some climate consequences are inevitable. Not only a wide consensus among the people, but a serious activism to stop pollution is essential as with every minute we are inhaling more poisonous air and the earth is becoming warmer. This mass activism throughout the globe can only force the country heads to seriously think about mitigating pollution vis-à-vis global warming and climate change.

References

American Cancer Society. 2003. *Cancer facts and figures 2003*. Retrieved from www.cancer.org

American Chemical Society. 1994. *Cancer and Environment*. Pp. 43.

Armstrong, B.K., Kricker, A. and English, D.R. 1997. Sun Exposure and skin cancer. *Australasian Journal of Dermatology*, 38 (Supplement):S1–S6.

Cliff, S. and Mortimer, P.S. 1999. Skin cancer and non-Hodgkins lymphoproliferative diseases: is sunlight to blame? *Clin. Expt. Dermatology.*, 24(1): 40–41.

Cole, P. and Goldman, M.B. 1975. Occupation. In: *Persons at High Risk of Cancer: An Approach to Cancer Etiology and Control* (Fraumeni, J.F., Jr. ed.). Academic Press, New York. Pp. 167–183.

Fears, T.J. Scotto, J. and Schneiderman, M. 1976. Skin cancer, melanoma and sunlight. *American Journal of Public Health*, 66: 461–464.

Fennelly, P.F. 1976. The origin and influence of airborne particulates, *Am. Scient.*, 64: 46–56.

Hadley Research Centre. 2008. Based on data from the United Kingdom Government Met office.HadCRTUT3 annual time series. Retrieved from: (http://hadobs.metoffice.com/hadcrut3/diagnostics/global/nhpercent2Bsh/index.html).

Hales, S., Wet, N., Maindonald, J. *et al*. 2002. Potential effect of population and climate changes on global distribution of dengue fever: an empirical model. *Lancet*, 360 (9336): 830–834.

Hileman, B. 2001. Causes of premature births probe. *Chem. Eng. News.*, 79(48): 21–22.

Hockwin, O., Kojima, M., Sakamoto, Y., *et al*. 1999. UV damage to the eye lens: further results from animal model studies: a review. *Journal of Epidemiology*, 9: S39–S47.

Holick, M.F. 2001. Sunlight "D"ilemma: risk of skin cancer or bone disease and muscle weakness. *Lancet*, 357: 4–6.

Kalter, H. and Warkany, J. 1983. Congenital malformations: etiologic factors and their role in prevention. *N. Engl. J. Med.*, 308: 424–431.

Maltoni, C. and Selikoff, I.J. (eds) 1988. Living in a Chemical World: Occupational and Environmental Significance. *Ann. NY Acad. Sci.*, 534: 1000–1020.

McGovern, V.J., Mihm, M.C. Jr, Bailly, C. *et al*. 1973. The classification of malignant melanoma and its histologic reporting. *Cancer*, 32(6): 1446–1457.

Murakami, M. 1996. Environmental health surveillance system for monitoring air pollution. *Environ. Sci.*, 4: 1.

Nghiem, D.X., Walterscheid, J.P., Kazimi, N. *et al*. 2002. Ultraviolet radiation induced immunosuppression of delayed–type hypersensitivity in mice. *Methods*, 28(1): 25–33.

Nriagu, J.O. 1996. History of global metal pollution. *Science*, 272: 223–224.

Robine, J.M. *et al*. 2008. Death toll exceeded 70,000 in Europe during the summer of 2003. *Les Comptes Rendus/Serie Biologies*, 331:171–178.

Staples, J.A., Ponsonby, A.L., Lim L.L. *et al*. 2003. Ecologic Analysis of Some Immune–Related Disorders, Including Type 1 Diabetes, in Australia: Latitude, Regional Ultraviolet Radiation, and Disease Prevalence. *Environmental Health Perspective*, 111: 518–523.

Waldron, H.A.1974. The blood lead level threshold. *Arch. Environ. Health.*, 29: 271–273.

Climate Change: Man and Environment (2012)   *Pages* **246–252**
Editor: **Goutam Kumar Saha**
Published by: **DAYA PUBLISHING HOUSE, NEW DELHI**

20

Climate Change and
Human Health

☆ *Krishnangshu Ray**

Introduction

In 2003, heat waves killed 35,000 people in Europe out of which 14,300 from France only. Majority of them were geriatric population more than 75 years of age. The nature of illness was heat exhaustion or heat stroke, cardiovascular and kidney diseases. City dwellers are at particular risk because of elevated temperature in cities known as urban heat island effect. Global warming is acknowledged by the scientists around the world to be a reality and to have been caused primarily by the human activities associated with civilization (Pachauri and Reisinger, 2007). As the earth warms, the delicate balance of climate, weather events of climate and life is disrupted. Consequences threaten human health and, ultimately, survival. WHO reports that long-term climate changes profoundly affect the pillars of public health including water, sanitation, air quality, food and level of allergens.

Human life is dependent on the dynamics of the earth's climate system. The interactions of the atmosphere, oceans, terrestrial and marine biospheres, cryospheres and land surfaces determine the earth's surface climate. The temporal and spatial changes in temperature and moisture are likely to change the ecology and biology of vectors as well as intermediate hosts (human beings, animals), which aggravates the incidences of disease transmission. Such reemergence of diseases are categorized as vector borne parasitic diseases (malaria, leishmaniasis, kala–azar, Chaga's disease, filaria, etc.), viral diseases (dengue fever, yellow fever, West Nile fever, Rift Valley

* Calcutta School of Tropical Medicine, Kolkata. E-mail : stm.director@gmail.com

fever, tick borne encephalitis, etc.) and bacterial rickettsial diseases like Lyme borreliosis, tularemia and plague. India is afflicted with six major vector borne diseases namely malaria, kala–azar, dengue, chikungunya, filariasis and Japanese encephalitis. Malaria ranks number one and has the largest disease burden (illnesses, deaths, financial costs and other indicators) of any vector-borne diseases worldwide (St. Louis and Hess, 2008).

Vectors and Vector Borne Diseases

Some 350–500 million people of the world are suffering from malaria, resulting in the deaths of one to three million people. If global temperature increases by 2–3°C, as expected, it is estimated that the population at risk for malaria will increase by 3–5 per cent, which means that millions of additional people would probably become infected with malaria each year (Emily, 2010). It is estimated that annually there are over 100 million cases of dengue infection worldwide. The case–fatality rate of its more serious form, dengue haemorrhagic fever (DHF), is about 5 per cent, mostly among children and young adults (CDC, 2010). As per National Vector Borne Disease Control Programme (NVBDCP), in 2007, cases and deaths due to Japanese encephalitis, dengue and visceral leishmaniasis were 4022 (963), 5534 (69) and 44001 (189) respectively. In India, around 60,000 suspected cases of chikungunya were also reported in 2007. In addition to mortality, vector borne diseases cause morbidity of millions of persons resulting in loss of man–days causing economic loss.

Climate change has the potential to alter the average exposure of human populations to vectors by changing the geographical distribution of conditions that are suitable for the vectors and disease pathogens. An increase in global temperatures will result in an expansion of warm temperature regimens into higher altitudes and latitudes. Any associated changes like rainfall in tropical and subtropical zones will also render habitats more or less suitable for vectors (Bouma *et al.*, 1994,1996; Burgos *et al.*, 1994; Sutherst, 2001). Climate change will affect both the invertebrate vectors and the development of pathogens in those vectors. Basic biological considerations indicate that with global warming, the duration of the growth season will increase, allowing more generations of vectors each year in cooler areas. Mosquitoes and other vectors are highly sensitive to the changes of temperature. The higher temperature causes:

☆ Boosting of reproductive rates, lengthening of the breeding seasons and increased susceptibility to bites.

☆ Shortening of the time to mature the pathogens to an infectious state.

☆ Expanding the mosquitoes' range to higher elevations and more northern latitudes, potentially putting previously unexposed population at risk.

Ticks maintain multiple and diverse disease agents (including bacteria, viruses, and parasites) and serve as bridging vectors between animal reservoirs of the disease and humans. Changing weather patterns associated with climate change have induced shifts in the distribution of vector ticks carrying Lyme disease. Modelling indicates that in the future, the tick that carries Lyme disease will expand to Canada (Brownstein *et al.*, 2005).

Fleas are best known as vectors of the causative agent of plague. Plague has caused millions of human deaths, especially during the black death of the middle ages. Plague still exists, notably in New Mexico, where its prevalence seems to correlate with the timing and amount of precipitation (Parmenter *et al.*, 1999; Enscore et al., 2002). Recently, studies of plague transmission in the United States found that the pattern of human plague cases has shifted northward as temperatures have warmed. It is predicted that some northward expansion of the disease could occur from New Mexico into Wyoming and Idaho (Nakazawa, *et al.*, 2007).

Reemergence of Diseases

Reemergence of diseases refers to disease outbreaks, which are previously controlled with ease by chemotherapy but now developed resistance and reappearing in the form of epidemics. Changes in the local ecological patterns not only increase vector densities but also facilitates the reintroduction of newer pathogens. Such changes in the habitats are further compounded with breakdown of the vector control measures. Alterations in the hydrological cycles (rainfall–evaporation cycle), melting of glaciers and failure in 'flushing mechanisms' of warm sea water from Gulf region to the Atlantic ocean are the added attributes.

The ecology and epidemiology of vector borne diseases can be better described by the concepts of disease triangle of host–pathogen–environment (Figure 20.1). This problems are further complicated with recently documented reports of pesticides menace. The latter includes physiological resistance, decrease in agricultural growth (especially rice), alteration in 'bird life cycle' and pyrethroids induced health hazards like hormonal changes and carcinomatous conditions due to reduced cholinesterase levels in the exposed population.

Successful control over the vector borne diseases could be summarized under following heads:

☆ Effective surveillance to provide feedbacks assuming that the benchmarks have been established.

☆ Viable public health infrastructure to deliver the services.

☆ Detection of actual cases and their reporting to the competent authority.

☆ Community ownership of the newly developed control measures.

☆ Climate forecasting and early warning system by developing suitable outbreak indicators.

☆ Research and development in the field of malaria genome project and developing newer antimalarial molecules or recombinant vaccines.

☆ Use of biopesticides (entomopathogens) like *Gambusia affinis* (mosquito fish) or *Mesocyclops asperiornis* (copepod).

Waterborne Infectious Diseases

Like vector borne diseases, waterborne infectious diseases are also strongly affected by changing climate. During drought, water scarcity results in poor sanitation and much of the population can be exposed to potentially contaminated water. For

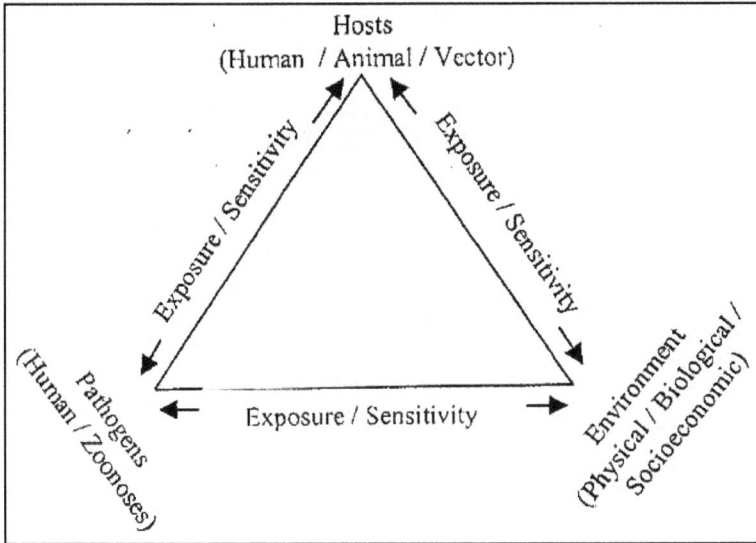

Figure 20.1: A host-pathogen-vector-environment framework for the assessment of risk to human from vector-borne diseases under global change.

example, there is currently an epidemic of cholera in northern Kenya during severe drought. Like drought, excess rainfall and flooding can also contribute to epidemics of waterborne infectious diseases, because of poor sanitation resulting from run–off from overflowing sewage lines or the contamination of water by livestock. An example is the 1993 epidemic of diarrhoeal disease due to cryptosporidium in Milwakee after heavy spring rains. Global warming also rises water borne infectious diseases due to change in water cycle. Infectious diarrhoea is caused by bacteria, virus and parasite due to contamination from food and drinking water. It is predicted that by 2030, there will be 10 per cent more diarrhoeal disease and is expected to be increased due to climate change and that will primarily affect the health of young children. Indeed, the impact on children might well be amplified by the effects of such diseases on malnutrition, development and cognition.

Cholera outbreaks usually occur from bacterial contamination from river, estuaries and coastal water. Scientists could observe that there exists a strong correlation between increase in sea surface temperature and onset of cholera epidemic. *Vibrio cholerae* is a natural inhabitant of the coastal, estuarine and riverine areas of tropical and temperate regions. The microorganism is known to have been traveled by land, water and air via international trade and migration of population. It has been observed that warm water along the coast, coupled with plankton blooms fostered by El Nino rains helped to multiply the cholera bacteria in the ocean. Sea surface temperature and plankton blooms could be considered as effective indicators in predictive markers of cholera outbreak. Higher water levels carry infected water further inland in one hand and reduction in salinity (due to influx of water) on the other, which worsen the situation. In this regard the declaration of Rita Colwel, the Director of the National Science Foundation (USA) may be quoted: "we can learn from the oceans as a vector of life and death, health and disease; a purveyor of environmental process; and a model of environmental and human health".

Remarks

Global change drivers are identified to have been left severe impacts on human health. These drivers are intensification of agriculture, variation of vegetation density, trade and travel, industrialization and population migration (Figure 20.2).

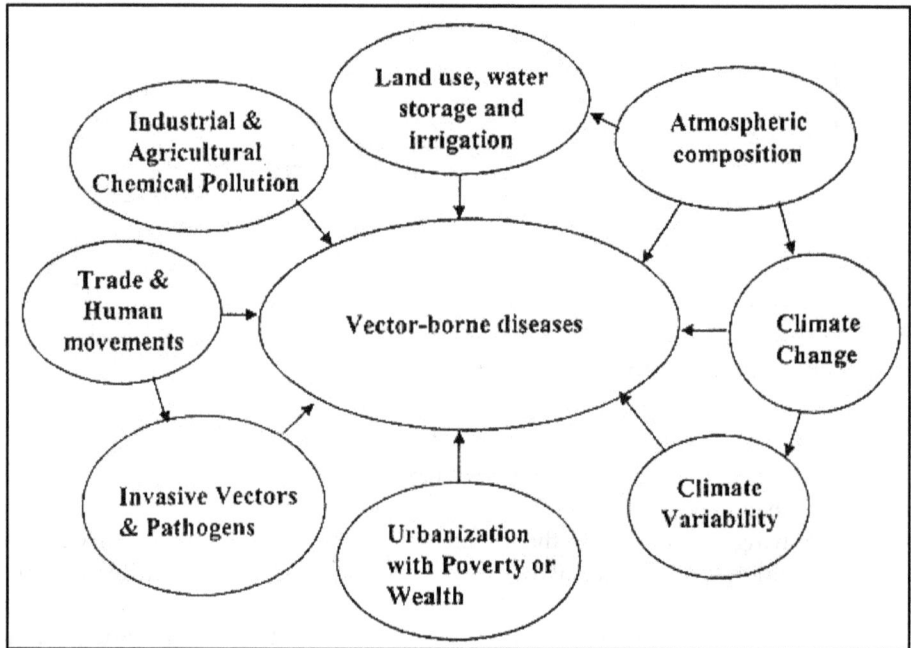

Figure 20.2: Drivers of global change considered in relation to potential changes in the status of vector-borne diseases.

Climate change is a new problem in the field of human health but little research effort has so far been directed to measure the effects of long-term climate change on vector borne and other reemerging diseases. Many of the principles and problems (variability in monitoring and surveillance, necessity for long data sets to overcome statistical noise cofounding variables, publication bias, etc.) applied to detect the changes in human diseases. However, there are significant additional challenges mainly involved with differentiating climate influences from the other multiple determinants of vector borne disease transmission. Therefore, field based epidemiological research on climate influences on disease causation requires even greater emphasis on differentiation between the effects of coexistent climate and non-climate factors. Many environmental and socioeconomic factors have affected the distribution and seasonality of vector borne diseases in recent years (Gubler, 1997; Morse, 1995).

The need of the hour is to develop holistic approach to risk assessment (outbreak indicators) of vector borne diseases under climate changes. Appropriate tools for the management response must be developed on the basis of adaptation technologies, cultural parameters, public health infrastructure and resources. Newer statistical models are to be innovated to forecast the resurgence of diseases in terms of meteorological data, local ecological indicators including vector density, immunological status of human hosts. To combat the consequences of climate change, we need to act in time by developing tools for early warning for seasonal forecasts, plan public health infrastructures, allocate resources and strengthen surveillance programs. Adaptation strategies must be considered to reduce disease burden, injuries, disabilities and death. It is concluded that culturally, economically, socially and environmentally sustainable programs may improve human health and personal development. Situation may also demand appropriate legislation in this regard.

References

Bouma, M.J., Dye, C. and Van der Kaay, H.J. 1996. Falciparum malaria and climate change in the Northwest Frontier Province of Pakistan. *Am. J. Trop. Med. Hyg.*, 55: 131–137.

Bouma, M.J., Sondorp, H.E. and van der Kaay, H.J. 1994. Health and climate change. *Lancet*, 343:302.

Brownstein, J.S., Holford, T.R. and Fish, D. 2005. Effect of climate change on Lyme disease risk in North America. *EcoHealth*, 2: 38–46.

Burgos, J.J., Curto de Casas, S.I., Carcavallo, R.U. *et al.* 1994. Global climate change influence in the distribution of some pathogenic complexes malaria and Chagas' disease in Argentina. *Entomol. Vect.*, 1: 69–78.

CDC. 2010. "Dengue Fever Fact Sheet." Centers for Disease Control and Prevention. http://www.cdc.gov/ncidod/dvbid/dengue/#current.

Emily K.S. 2010. Global Climate Change and Infectious Diseases. *The New England Journal of Medicine*, 362: 1061–1062.

Enscore, R.E., Biggerstaff, B.J., Brown, T.L. *et al.* 2002. Modelling relationships between climate and the frequency of human plague cases in the southwestern United States, 1960–1997. *Am. J. Trop. Med. Hyg.*, 66: 186–196.

Gubler, D.J. 1997. Dengue and Dengue hemorrhagic fever; its history and resurgence as a global health problem. In: *Dengue and Dengue Hemorrhagic Fever* (9th ed.) (Gubler, D.J. and Kuno, G. eds.). CABI, New York. Pp. 1–22.

Morse, S.S. 1995. Factors in the emergence of infectious diseases. *Emerg. Infect. Dis.*, 1: 7–15.

Nakazawa, Y., Williams, R., Peterson, A.T. *et al.* 2007. Climate change effects on plague and tularemia in the United States. *Vector Borne Zoonotic Dis.*, 7: 529–540.

Pachauri, R.K. and Reisinger, A. 2007. *Climate change 2007: synthesis report.* IPCC, Geneva.

Parmenter, R.R., Yadav, E.P., Parmenter, C.A. *et al.* 1999. Incidence of plague associated with increased winter–spring precipitation in New Mexico. *Am. J. Trop. Med. Hyg.*, 61: 814–821.

St. Louis, M. and Hess, J. 2008. Climate Change: Impacts on and Implications for Global Health. *Am. J. Prev. Med.*, 35(5): 527–538.

Sutherst, R.W. 2001. The vulnerability of animal and human health to parasites under global change. *Int. J. Parasitol.*, 31: 933–948.

Climate Change: Man and Environment (2012)   *Pages* 253–257
Editor: **Goutam Kumar Saha**
Published by: **DAYA PUBLISHING HOUSE, NEW DELHI**

21

Impact of Climate Change on Microbial Community with Special Emphasis on Cholera

☆ *Banwarilal Sarkar**

Introduction

The heat generate from sun and enters into the earth system. Sunlight travels through space and our atmosphere, heating up the land surface and the oceans. The warmed earth then releases heat back into the atmosphere. Heat exits the earth system as the earth's surface warmed by solar energy, radiates heat away. However, greenhouse gases in our atmosphere allow the lower atmosphere to absorb the heat radiated from the earth's surface, trapping heat within the earth system. Greenhouse gases, such as water vapour, carbon dioxide, methane and nitrous oxide, are an important part of our atmosphere because they keep earth from becoming an icy sphere with surface temperatures of about 0°F. Over the past century or so, the amount of greenhouse gases within our atmosphere has been increasing rapidly, mainly due to the burning of fossil fuel, which releases carbon dioxide into the atmosphere. Consequently, in the past one hundred years global temperatures have been increasing more rapidly than the historic record shows. Scientists believe this accelerated heating of the atmosphere is because increasing amounts of these greenhouse gases trap more and more heat and temperatures are expected to increase by another 1.8 to 5.8°C by the end of this century. The hydrologic cycle will be altered, since warmer air can retain more moisture than cooler air. Some geographic areas will have more rainfall

* National Institute of Cholera and Enteric Diseases, Kolkata. E-mail: bl_sarkar@hotmail.com

and some more drought and severe weather events–including heat waves and storms are expected to become more common. For these reasons, the term "climate change" is now preferred over the term "global warming."

Alteration of the global environment by human has triggered the sixth major extinction event in the history of life and caused widespread changes in the global distribution of organisms. These changes in biodiversity deter ecosystem processes and change the resilience of ecosystems to environmental change. This has profound consequences for services that humans derive from ecosystems. The large ecological and societal consequences of changing biodiversity should be minimized to preserve options for future solutions to global environmental problems. Strong evidence for changes in biodiversity and its ecosystem and societal consequences calls for similar international actions. The most important causes of altered biodiversity are factors that can be regulated by changes in policy like emissions of greenhouse gases, land–use change and species introductions etc. The climate has a potential impact on human health from vector borne and water borne infections, *e.g.* malaria and cholera. For example, there is currently an epidemic of cholera in northern Kenya in the wake of a severe drought. It concentrates on the impact of two possible changes to climate; increased frequency of heavy rainfall events with associated flooding and increased temperature. There have been numerous reports of outbreaks that followed flooding that led to contamination of underground sources of drinking water. Heavy rainfall also leads to deterioration in the quality of surface waters that could adversely affect the health associated with recreational water contact. Because of rising temperatures and changing rainfall patterns, climate change is expected to have a substantial effect on the burden of infectious diseases that are transmitted by insect vectors and through contaminated water.

Climate Change and Human Health

Researchers have found that there is a close link between local climate and the occurrence or severity of some diseases and other threats to human health (WHO, 2005). It is estimated that climate change contributes to 150,000 deaths and 5 million illnesses each year, and the World Health Organization estimates that a quarter of the world's disease burden is due to the contamination of air, water, soil and food. In the last quarter of the 20th century, the average atmospheric temperature rose by about 1°F. By 2000, that increase was responsible for the annual loss of about 160,000 lives and the loss of 5.5 million years of healthy life, according to estimates by the World Health Organization. The toll is expected to double to about 300,000 lives and 11 million years of healthy life by 2020. The biggest tolls were in Africa, on the Indian subcontinent and in southeast Asia. Most of the increased burden of death and disease were from malnutrition, diarrhoea, malaria, heat waves and floods. But those diseases will play a minor role, at best, in many regions that nevertheless will feel the effects of global warming. What is readily apparent from these data is that developing regions of the world have been disproportionately affected by climate change relative to developed regions. Climate–sensitive diseases are among the largest global killers. Diarrhoea, malaria and protein–energy malnutrition alone caused more than 3.3 million deaths globally in 2002, with 29 per cent of these deaths occurring in the region of Africa. Deadly diseases like the West Nile virus, cholera and Lyme disease, often associated with hot weather, are spreading rapidly throughout North America and Europe because increased temperatures in these areas allow disease carriers like mosquitoes, ticks and mice to thrive.

Therefore, we must focus our efforts on mitigating the effects of climate change, including its potential impact on the global burden of infectious diseases. Additional research is needed on the ecology and epidemiology of infectious diseases that will probably be affected by climate change. The best means for accomplishing this aim would be to incorporate research on the effect of climate change into existing infrastructures, such as the ambitious malaria–eradication program recently launched by the Bill and Melinda Gates Foundation. It is imperative that organizations such as the WHO continue their missions of treating and preventing otherwise neglected infectious diseases, as part of a multifaceted approach in improving global health. Effective treatments and vaccines will go a long way in preventing human suffering that could otherwise occur as a result of climate change.

The impact of climate change on water availability is likely to be one of the most significant for the health of populations. Higher temperatures are hastening rates of evaporation of surface water thereby reducing the availability of fresh water. Lack of fresh water compromises hygiene and hence increasing incidence of diarrhoeal disease. On the other hand, too much water, in the form of floods, causes contamination of freshwater supplies. Extreme events like sea level rise coupled with stronger storm surges and coastal flooding can be followed by outbreak of diseases such as cholera. Other water–borne and food–borne infectious diseases pose a great threat to public health in the tropical monsoon region. In 1995, in eight countries of the region (Bangladesh, Bhutan, India, Indonesia, Myanmar, Nepal, Sri Lanka and Thailand), the total death toll from diarrhoea was estimated to be 1.03 million; 71 per cent and 12 per cent of the deaths occurred in India and Bangladesh, respectively. Children under five years of age are accounted for about 25 per cent of these deaths (WHO, 1996).

The cyclone Aila that lashed the city on May 25th, 2009 is an indication. Kolkata is among the world's most vulnerable metropolises. Scientists studying the vulnerability of Kolkata from events triggered by climate change have warned that vital services, like water and electricity, are extremely susceptible to cyclonic storms and sea level rise due to its proximity to the Bay of Bengal, 80 km away from the city limit. Due to climate change, deltas particularly Sundarbans will be threatened by flooding, erosion and salt intrusion. Loss of coastal mangroves will have an impact on fisheries. The major delta area of the Ganga, Brahmaputra and Indus rivers, which have large populations reliant on riverine resources will be affected by changes in water regimes, salt water intrusions and land loss. It is now well documented that the rise in temperature and change in humidity will adversely affect human health in India. Heat stress could result in heat cramps, heat exhaustion, heat stroke and damage physiological functions, metabolic processes and immune systems. Increased temperatures can increase the range of vector borne diseases such as malaria, particularly in regions where minimum temperatures currently limited pathogen and vector development.

Several reports have highlighted other effects of climate change on infections. For example, El Nino since 1975, and more recently its frequency, has been shown to strengthen the relationship between the cycle and cholera outbreaks in Indian subcontinent. According to Colwell (1996), the major rivers of the Indian subcontinent that discharge into the Bay of Bengal carry huge amounts of agricultural and industrial waste, providing nutrients sufficient to convert coastal waters to eutrophic conditions. Brackish water extends some distance upriver for all rivers. Salinities between 5 ppm and 30 ppm, which were detected in inland coastal areas and in sea water are favorable for the growth of *Vibrio cholerae.*

In the Bay of Bengal, evidence has been found by synthesizing satellite remotesensing, *in–situ* hydrographic and meteorological data sets, that cholera cases occur with a rise in ocean temperature (Colwell, 1996). Outbreaks during the monsoon have been substantially smaller. Colwell (1996) also indicated a possible link between outbreaks of cholera in Peru and neighboring countries and a warming El Nino/ Southern Oscillation (ENSO) event. Although the ENSO phenomenon has a substantial effect on the Indian summer monsoon, its linkage with cholera has not been investigated. In addition to enteric pathogens, climate also influences the abundance and ecology of nonenteric and other pathogens which are naturally present in the environment. The best examples of this include the pathogenic *Vibrio* spp., which are autochthonous in estuarine ecosystems. Along with *V. cholerae*, the bacteria thrive in warm waters of moderate salinity and are closely associated with aquatic invertebrates. An estimated 94 per cent and 40 per cent of deaths were due to diarrhoeal diseases and malaria, respectively that has relation with environmental factors. In Kolkata, a survey of 13 years observation on cholera revealed that gradual changes in shifting of cholera incidence with single peak (April) followed by two peaks (September–October and continued till December) in a year. The above cholera infection periodicity has been related to climatic factors like relative humidity, temperature, rainfall, sea surface temperature, sun spot numbers along with influence of El Nino and La Nino. Therefore, with a changing climate, the geographic range of these pathogens may also change, potentially resulting in increased exposure and risk of infection for humans. Furthermore, changes in plankton populations and other hosts for which vibrios are commensals or symbionts, would similarly alter the ecology of these pathogens that are autochthonous to the aquatic environment. Furthermore, the incidence of cholera may occasionally be surprisingly low at times when climate seems to favour cholera transmission.

We, the Indians, are facing more and more heat during summer and the temperature is increasing day by day. At the same time, the monsoon and other seasons are changing due to global warming. The changing climates affect the human health as well as the environs. As a result, emerging and reemerging diseases are coming up with a great concern. Inspite of extensive research studies, several programme launched by the government and non-governmental agencies for the protection of our earth, still, the nature is progressing in its own way. In reality, we are totally under the influence of nature. It is the nature, does harm to the mankind and also may protect us from the disaster.

References

Colwell, R.R. 1996. Global climate and infectious disease: the cholera paradigm. *Science*, 274 (5295): 2025–2031.

World Health Organization (WHO). 1996. *The World Health Report 1996*. WHO, Geneva.

World Health Organization (WHO). 2005. *Climate and health: fact sheet http://www.who.int/globalchange/news/fsclimandhealth/en/*

Climate Change: Man and Environment (2012)
Editor: **Goutam Kumar Saha**
Published by: **DAYA PUBLISHING HOUSE, NEW DELHI**

Pages **258–262**

22

Some Thoughts on Environment and Arthropod Borne Diseases in India

☆ *Amiya Kumar Hati**

Introduction

Environment is always intimately related to the diseases that are transmitted by arthropods. The spread of vector borne diseases depends on density of the vectors. When environment becomes favourable for breeding of the vector above its threshold level, there will be effective transmission of the disease, but when, environment is not suitable for efficient breeding of the vector and optimum density of the vector is not reached, the disease will not be transmitted, even if the vector is present. When *Aedes index* (number of premises on which it is breeding) is below 5 per cent, epidemic of yellow fever subsides (Hati, 2010). Through source reduction only, yellow fever has been eliminated from Panama region (Fox, 2008). Singapore has previously succeeded in controlling DF/DHF following this procedure (Yoshikawa *et al.*, 2009).

In India, important vector borne diseases, such as malaria, filaria, Japanese encephalitis, dengue and dengue haemorrhagic fever (DHF), chikungunya, kala–azar and other cutaneuous leishmaniases, Kyasanur Forest Disease (KFD), etc. are major public health problems, even after sixtythree years of achieving independence. The problems have become complicated mainly due to the changing environment that favours for unabated growth of vector populations. Pollution of environment favouring growth of vector population may occur mainly due to:

1. **Unplanned urbanization**, which is rapid and alarming in India. One of the reasons of spread of *Culex quinquefasciatus* mosquitoes, transmitter of

* Calcutta School of Tropical Medicine, Kolkata. E-mail : amiya_hati@rediffmail.com

brugian filariasis, is related to this haphazard urbanization (Hati, 2010). This is also true for malaria, dengue, DHF and chikungunya.

2. **Faulty irregular interrupted water supply** is another important aspect for polluting environment inviting insect breeding. Storage of water or collection of water in several types of small containers helps prolific breeding of *Anopheles stephensi* (important vector of urban malaria in India) and *Aedes aegypti* (vector of dengue), specially in the monsoon and post monsoon seasons, tallying with transmission seasons of malaria and dengue.

3. **Rapid Industrialisation** also poses problem, specially in our country, where standard of living of industrial workers and labourers is far below normal. Moreover, aggregation of people from different areas surrounding an industry is often responsible for initiating or spreading insect borne diseases.

4. **Migration** of infected person(s) in a new virgin area, when effective vector exists, can help to introduce the disease. This is a normal mode of transmission so far as kala–azar is concerned. From Bihar, kala–azar entered Malda and from Malda to about 10–13 other districts of West Bengal, migration is responsible for opening up new foci of kala–azar (Hati *et al.,* 1985).

5. **False reliance** on some so–called insect–abating appliances that pollute environment, examples are mats, coils, vehicle mounted or hand operated fogging instruments etc. These appliances mostly help the insect populations directly and indirectly to be more environment friendly, so to say. On the one hand, they are becoming resistant to those poisonous chemicals and on the other hand, their spread is also becoming facilitated as from one place they are forcefully driven to the other place. A study in central Kolkata reveals that of 700 persons, using mats, coils, etc. 80 per cent suffered from malaria and of 300 persons using mosquito nets 3.29 per cent only suffered from malaria (Hati, 1999).

6. **Agriculture** sometimes adversely affects environment to spread insect borne diseases. The classical example is Japanese encephalitis. The vector mosquitoes belonging to *Culex vishnui* group are rice field breeders. In this regard, it may be said that due to indiscriminate use of pesticides and chemical manure natural enemies of mosquitoes such as frogs, larvivorous fishes, bugs, odonates, etc. have almost disappeared (Hati, 1986).

7. **Deforestation** is severely destabilizing environment. Change in the ecology is taking place rapidly. Some forest mosquitoes for example, *Anopheles dirus* (Nagpal and Sharma,1995) have become very fond of taking human blood and transmitting *P. falciparum* parasites. People residing near the forest are facing this danger. As the ecology will be grossly changed, the foci of campestral plague may be shifted more towards human habitations. Kyasanur forest disease may become sylvatic or campestral.

8. **Zoonosis**: Zoonotic diseases may suddenly appear due to environmental lapses. Japanese encephalitis was not present in India in epidemic form

before 1973, in which year it devastatingly spread throughout West Bengal (Hati, 1986). Plague has disappeared from India, though a hue and cry was raised in 1994 throughout the country about invasion of plague, which has not been adequately substantiated. However, as stated earlier, it is not impossible that plague may appear in certain niche as a result of a link between campestral and urban situation, where interaction between agent, host and vector can take place.

9. **Construction works** in the cities often continue for indefinite period. This can invite mosquito breeding. Example is construction work in Delhi in 2010, prior to Commonwealth Games and spread of dengue and dengue haemorrhagic fever.

10. Last but not the least comes the question of global warming. Most of those points discussed above are directly or indirectly related to this burning issue. For example, indiscriminate urbanization, industry, insect abating appliances, deforestation are directly adding fuel to global warming. Water supply, migration, agricultural and zoonotic affairs have indirect relationship with global warming, so to say.

Climate Change and Vector-borne Diseases

No prophecy can definitely be made regarding the future effect(s) of global warming, but the future is definitely blake. We may or may not understate cause and effect of global warming, willingly or unwillingly, knowingly and unknowingly but something is happening and something is going to happen probably in very near future. What would be possibilities related to arthropods and arthropod borne diseases? Will any arthropod species be extinct due to this global warming? The possibility is very very rare. These arthropods have occupied this earth long before human beings. They may be very tiny, but very hard and much more adaptive than human beings. In contrast, there is every possibility that a zoophilic species may be converted into an anthropophilic species as a result of metamorphic changes in ecology. Example of *A. subpictus* may be cited. In this primarily zoophilic mosquito species in a recent study anthropophilic index of an indoor resting population is found to be as high as 41.0 per cent in a village in West Bengal, indicating a considerable extent of man–vector contact (Chatterjee and Chandra, 2000). As a consequence, it is possible that a non-vector or a secondary vector species may become the primary vector. *Armigeres subalbatus* is a non-vector species in India. There is every possibility that it may act as a vector of dengue and/or chikungunya, as a result of its prolific presence and human contact, which may enhance due to global warming.

It is possible that *P. falciparum* parasite would be resistant to all antimalarials available and *P. vivax* the benign tertian malaria may follow this path. Complicated vivax malaria with Acute Respiratory Distress Syndrome (ARDS) has already been reported (Price *et al.*, 2007).

A peculiar situation has been observed in some viral diseases, such as DHF, chikungunya, Japanese encephalitis, etc. The disease appears in epidemic form, then it disappears for several years to reappear with more severity. As vector breeding would be unabated due to global warming, such condition will gradually disappear and epidemics would be prolonged and devastating.

Many a disease have particular transmission season. In Kolkata, for example, transmission season of dengue/DHF/chikungunya extends from monsoon to post monsoon season. This is also true for falciparum malaria, in which peak is obtained in November (Hati *et al.*, 2002). No such transmission season will probably exist in those future days. Intense transmission may continue throughout the year as abundant vectors would reign perennially.

Invasion of the vector may occur in new areas. *Aedes aegypti* would possibly invade rural areas while *Aedes albopictus* in urban area posing danger. Some such evidences have recently been documented, which point out that *Aedes aegypti* is now constituting 10 per cent of all *Aedes* mosquitoes in rural areas, whereas in urban areas, like Kolkata, *Aedes albopictus* constitutes 10 per cent of all *Aedes* mosquitoes (Naskar, 2010). The result of A226V mutation in the genome of additional vector, *Aedes albopictus* increases severity of the disease and more infectivity for vectors. Virus would get ample chance of such mutation in the changed condition.

These are assumptions, but supported by some background data, which are difficult to ignore.

References

Chatterjee, S.N. and Chandra, G. 2000. Role of *Anopheles subpictus* as a primary vector of malaria in an area in India. *Jpn. J. Trop. Med. Hyg.*, 28: 177–181.

Fox, C. 2008. *Mosquito Net*, i2i Publishing, Manchester, UK. Pp.253.

Hati, A.K. 2010. *Medical Entomology*(3rd edn.). Allied Book Agency, Kolkata. Pp.238.

Hati, A.K., Chaudhuri, P., Purkayastha, S. *et al.* 2002. A study of transmission dynamics of vivax and falciparum malaria in an endemic area in Kolkata. *Calcutta Statistical Association Bulletin*, 53: 133–143.

Hati, A.K. 1999. *Malaria–diagnosis, treatment and control*. National Book Agency Pvt. Ltd., Kolkata. Pp. 80.

Hati, A.K. 1986. *Studies on Culex vishnui and other related vectors of Japanese encephalitis in rural West Bengal*, Department of Medical Entomology, Calcutta School of Tropical Medicine, Kolkata. Pp.151.

Hati, A.K., Nandy, A. and Chowdhury, A.B. 1985. An epidemic outbreak of Kala–azar in a district in West Bengal, India, WHAO/VBC/85.911.Pp.9.

Nagpal, B.N. and Sharma, V.P. 1995. *Indian Anophelines*. Oxford and IBH Publishing Co. Pvt. Ltd., New Delhi, Mumbai, Kolkata. Pp. 416.

Naskar, R. 2010. *In search of Aedes albopictus*. M.Sc. Dissertation, Bethune College, Kolkata. Pp.24.

Price, L., Planche, T. and Rayner C. 2007. Acute respiratory distress syndrome in *Plasmodium vivax* malaria: Case report and review of the literature. *Trans. Roy. Soc. Trop. Med. Hyg.*, 101: 655–659.

Yoshikawa, M.J., Tang, C. S. and Nishibuchi, M. 2009. Incidence of chikungunya fever in Singapore, public health measures and transnational movements of people. *Emerging Infect. Dis.*, 15: 1243–1249.

Climate Change: Man and Environment (2012)   *Pages* **263–276**
Editor: **Goutam Kumar Saha**
Published by: **DAYA PUBLISHING HOUSE, NEW DELHI**

23

Birds in a Warming World: A Review

Subhendu Mazumdar[1] and Goutam Kr. Saha[2]

Introduction

The impact of climate change on the avifaunal community throughout the globe has been best studied. Birds are considered as good indicators of environmental change (Chambers, 2008), as they are easy to identify, placed high in the food chain, thus they often reflect the events and processes further down the chain, relatively well–studied and are often with pre–existing baseline data. Therefore, changes in bird populations have been proposed to be considered as an indicator to study the impact of climate change (Robinson *et al.*, 2005). Evidences indicate that the climate change influences the distribution, phenologic patterns as well as breeding success of many bird species (Crick, 2004). Many bird species are extending their geographical ranges northwards (Thomas and Lennon, 1999), while some migratory birds arrive earlier on their breeding grounds (Mason, 1995), start breeding earlier (Crick and Sparks, 1999; Dunn and Winkler, 1999) or even migrate to their wintering grounds later (Gatter, 1992).

Altitudinal Shift and Distribution Pattern

Several studies indicate that birds, as well as other flora and fauna, are shifting their ranges towards the pole or to higher altitudes in tropical mountains in response to climate change (Parmesan and Yohe, 2003). Root and Schneider (1993; 2003) found strong statistical correlations between six large scale environmental factors, mainly

1 Shibpur Dinobundhoo College, Howrah. E-mail: subhendumazumdar@gmail.com

2 University of Calcutta, Kolkata. E-mail: gkszoo@rediffmail.com

climatic variables and the distribution and abundance of a majority of 148 wintering land birds of North America. Generally, range shifts are likely to have the greatest impacts on the species occupying tundra, arctic and montane habitats.

The extents of these habitats are likely to be seriously reduced under warmer conditions (IPCC, 2001), with a consequent effect on the population of birds (and other wildlife) inhabiting those habitats. European birds are expected to undergo range shifts causing a net decrease in neotropical migrant bird species in every region of USA, including boundary shifts of more than 1,000 km (Huntley *et al.,* 2006). It is important to note that range contractions are expected to be more frequent than range expansions (Huntley *et al.,* 2006). Studies indicate that future range shifts due to global warming are expected to 29 to 30 per cent net decline in the number of bird species in the eastern Midwest and Great Lakes regions respectively (Price and Glick, 2002). Besides, their new climatically suitable ranges may also be unsuitable for their survival due to fragmented landscapes caused by unscientific human land use. Moreover, presently most of the areas with rich avifaunal diversity have been notified as protected areas. Shift in distribution range, as a consequence of global warming, may force these birds to live into some unprotected areas (Böhning–Gaese and Lemoine, 2004) leading to an even greater challenge for conservation. In Europe, the endangered Spanish imperial eagle, currently found mainly in natural reserves and parks, is expected to lose its entire current range (Huntley *et al.,* 2006). Island and mountain birds may perish with the alteration of climate pattern and rise in temperature, as they are confined to increasingly smaller patches of habitat. With high global warming scenario, it has been projected that there will be 99 per cent shrink in the Scottish highland habitat of the Capercaillie (*Tetrao urogallus*), the world's largest grouse, by 2050 virtually eliminating UK habitat for this bird (Berry *et al.,* 2001; Pounds *et al.,* 1999, 2006).

Range shifts and changes in distribution pattern will finally lead to a change in the species composition of the natural communities as because the response of each constituent species to climate change will be unique. Hence, birds living in such altered natural communities may often come into contact with different prey species, competitors, parasites and predators (Root and Hughes, 2005). Study of an entire ecological community in Mexico including 1,179 bird species, predicted greater than 40 per cent turnover in some local ecological communities by 2055 suggesting severe 'ecological perturbations', as many species are supposed to disappear or displaced by invader species (Peterson *et al.,* 2002).

Alteration and Loss of Bird Habitat

Global warming and climate change will lead to a change in the vegetation pattern and cause subsequent alteration of habitat. Changes in vegetation pattern will be most extreme for the Tundra areas. An overall 40–57 per cent loss of current tundra habitat has been estimated which, in turn, will have adverse impacts on the population of arctic birds. Shrink of tundra habitat will particularly affect birds like geese and sandpipers as they breed exclusively in these regions (Zöckler and Lysenko, 2001). Birds in the antarctic region are affected as well. Population dynamics of several species of penguins inhabiting Antarctic Peninsula are seriously affected with warming and disappearance of the sea ice cover (Ainley *et al.*, 2001). Several studies indicate that crucial bird habitats, especially for the wetland birds, are reducing with the shrinking and shifting of wetlands as a result of global warming. Populations of Siberian crane (*Grus leucogeranus*) consist of 3,000 individuals throughout the globe and have already shown its vulnerabilities to climate change. This bird was once a regular winter visitor in India, but no more visits its wintering grounds in this region (Wormworth and Mallon, 2006).

Sea level rise will also impact a number of coastal ecosystems. Permanent inundation of coastal zones will pose a serious threat to the estuarine mudflats and salt marshes (Pethick and Crooks, 2000) with severe implications for both wintering and breeding waterfowl and wader species. When the sea level rises, the marsh vegetation tends to move upward and inland. However, ridges along the coastline often prevent this movement leading to coastal squeeze and loss of marsh area (Boere, 2006). Estuarine mudflats also represent important feeding sites for wintering waders. The shape of these estuaries may also change (Austin and Rehfisch, 2003) with the increasing sea–level, which will further influence the nature of the sediment thereby altering the species composition and density of the invertebrate populations on which the waders feed. With the advancement of sea, the sediment will become sandier leading to the decline in the densities of waders that prefer muddy estuaries, like redshank and dunlin (Austin and Rehfisch, 2003). However, both density and number of the species that prefer sandy estuaries, such as oystercatcher, are predicted to increase.

Mangroves are highly adapted trees that thrive in the intertidal waters of the tropical region and are treasure house of rich avian diversity. Encroaching seawater may also threaten the survival of several mangrove species (Spalding *et al.*, 2001).

Increase in Disease Prevalence in Avian Communities

A number of authors have raised the possibility that global warming might have played a crucial role in the recent range expansions and outbreaks of vector-borne diseases (Bouma *et al.*, 1994, 1996; Epstein *et al.*, 1998; Epstein, 2000; Kovats *et al.*, 2003). Climate change has the potential to alter the exposure of several bird populations to vector-borne diseases by changing the distribution of geoclimatic conditions that are suitable for the vectors and disease pathogens. An increase in global temperature will result in an expansion of warm temperature regimes into higher altitudes and latitudes. Any associated changes in rainfall in tropical and subtropical zones will also render habitats more or less suitable for vectors. These changes would make birds in the temperate environments more receptive to many tropical vector-borne diseases while having less negative effects on tropical environments. Besides, extreme climatic events are also reported to have major effects on the transmission rates of vector-borne diseases and the climate change is expected to increase the frequency of such events.

There are increasing evidences that warmer temperatures are thought to favour the spread of various infectious diseases and parasites and can influence the dynamics of bird populations through increased mortality and thus can limit population growth in some cases (Newton, 1998; Tompkins *et al.*, 2000). Rate of reproduction of pathogens or the distribution of vectors carrying these pathogens are altered with the change in climate parameters (Faustino *et al.*, 2004). Besides, reduced precipitation and greater drought is expected in some regions and this could compel many species of birds to accumulate around limited water resources, increasing the risk of exposure to the potential pathogens and their quick transmission among the closely associated individuals (BTO, 2002). According to Epstein (2001), warm winters and spring droughts led to the spread of West Nile Virus through avian populations in Europe and North America. Benning *et al.* (2002) suggested that the increase and invasion of mosquitoes carrying malarial pathogens due to temperature rise might lead to the extinction of several species of Hawaii's honeycreepers. However, the actual extent of the impact of climate change on bird population could not be forecasted as relatively little is known about epidemiology of many bird–parasites.

Change in Migration Phenology

Migratory birds may particularly be vulnerable to the impact of climate change as they use separate habitats, often geographically distant, as their wintering and breeding grounds and changes in any one of the sites during their life cycle may have serious impacts on their survival and population dynamics. Climate induced changes in habitat are likely to have impacts on staging, stopover ecology and fuelling in migratory birds as well (Robinson *et al.*, 2005). This is the area of study for which there is the strongest evidence of the impact of climate change on birds.

Spring Migration

The timing of arrival on breeding grounds is very much crucial for the migratory birds, especially as it has to be synchronized with the availability of prey (Robinson *et al.*, 2005). With the changing climate, time of arrival of migratory avifauna to the breeding site (spring migration) is changing as well. As a consequence of this 'phenological mismatch', many migratory species are unable to breed successfully because their arrival time no longer coincides with peak food availability (Parmesan and Galbraith, 2004; Visser *et al.*, 2004). In Netherlands, this mismatch has led to the decline of up to 90 per cent in some populations of pied flycatchers over the past two decades. Observations made at several point locations, as well as over broad areas like English counties show general trends towards earlier spring arrivals for many species (Bradley *et al.*, 1999; Jenkins and Watson, 2000; Spark, 1999; Sparks and Mason, 2001; Hüppop and Hüppop, 2003). One analysis of 64 studies revealed that birds had advanced timing for such spring migration arrival to their respective breeding site at an average rate of 6.6 days per decade (Root and Hughes, 2005). Similar advances in arrival dates of spring migrants have been seen in Europe (Lehikoinen *et al.*, 2004) and North America (Bradley *et al.*, 1999). Changes in the timing of arrival seem to be greater among short–distance migrants, than long–distance migrants (Hubalek, 2003). In Europe, some birds (which normally used to migrate) have even stopped migrating altogether (Lehikoinen *et al.*, 2004). Long–distance migratory birds are facing a greater climate change threat than the resident birds, as they are often unable to adjust their migratory journeys with the peak food availability. The arrival dates of the long–distance migrants are relatively unchanged and the time of peak insect availability has advanced. As a consequence, many long–distance migratory birds are nesting more quickly after their spring arrival in Europe. Many other long–distance migrants are unable to nest and breed sufficiently quickly to match nestlings' food demands with peak insect supplies (Both *et al.*, 2006). Some species of long–distance migrants in Europe and North America are already showing signs of decline. Of 119 long–distance migrants studied in Europe, 54 per cent have already shown a sustained, often severe, decline from 1970 to 2000, with climate change implicated as a major contributing factor (Sanderson *et al.*, 2006).

Winter Migration

Studies of migratory birds indicating later departure from their breeding grounds and earlier arrival on the wintering grounds are scanty and more scattered (Bairlein and Winkel, 2001; Sparks and Mason, 2004). Sparks and Mason (2001) observed that nine species displayed trends towards progressively later departures for the wintering grounds since the 1950s, whereas only one species was leaving the UK at a significantly earlier date than before. Studies on the autumn passage of willow warblers (*Phylloscopus* trochilus) at Dungeness Bird Observatory in Kent (Marchant and Wernham, 2003) revealed that departure dates were on average 5–10 days later over the period 1994–2000 than they had been between 1962–1968 (Robinson *et al.*, 2005). There are few evidences suggesting that short–distance winter migrants in UK, such as merlin (*Falco columbarius*) and hen harrier (*Circus cyaneus*), are appearing on the wintering grounds earlier (Sparks and Mason, 2004). However, this trend is less consistent across species as there are many other short–distance migrants displaying no advancement at all.

Effect on Migratory Routes and Stopover Sites

Climate induced changes in habitat are likely to have impacts on staging, stopover ecology and fuelling in migratory birds. Migratory fuelling generally occurs before crossing a large ecological barrier, such as a large expanse of desert, water and high mountains. For example, the garden warbler (*Sylvia borin*) weighs 16–18g in the breeding and wintering seasons and increases its body mass to 37g before crossing the Sahara desert by depositing large amount of fat within the body tissues (Bairlein and Winkel, 2001). Appropriate time and amount of deposition of fat is crucial for successful completion of their migratory journeys. Therefore, any mismatch in this timing may even threaten their survival.

Stopover sites are of special significance for successful migration of birds as they use them to rest and feed during their migratory journeys. Besides, weather conditions *en route* are also known to affect an individual's ability to migrate. Changes in wind direction and velocity could also have a major impact on the physical condition of the migrants (Piersma and Jukema, 1990; Butler *et al.*, 1997). Therefore, loss or destruction of stopover sites and changes in several other climatic parameters both in the stopover sites as well as in the migratory pathway may compromise the ability of the migrants to complete their migratory journeys (Robinson *et al.*, 2005).

Impact on Breeding Phenology

Except a few opportunistic breeders, many bird species use temperature as direct cue to time their breeding (Visser *et al.*, 2002, 2004). The changing climatic conditions may also lead to the alteration in the availability of food resources, which, in turn, affects the breeding success of both migratory and resident species of avifauna. Studies on the El Nino reveal that periodic warming alters oceanic currents along the west coast of the America, leading to a crash in fish abundance and catastrophic breeding failure or even adult mortality among seabirds in the Pacific (Barber and Chavez, 1983, 1986). A similar study between 1950 and 2000 on fulmars (*Fulmarus glacialis*) in Orkney, located off the north coast in Scotland, revealed that hatching and fledging success were related to the North Atlantic Oscillations (NAO), which potentially influence the abundance of their crustacean and fish food supplies (Thompson and Ollason, 2001). In north Scotland, internationally important populations of sea birds have suffered massive breeding failures, which appear to be due to the unavailability of food as a result of the collapse of an entire food web. The warmer seawaters caused a serious decline in the number of plankton subsequently leading to a crash in the fish population, which in turn affected the sea birds (Crick, 2004).

Analyzing the breeding data from the British Trust for Ornithology's Nest Record Scheme (NRS) it has been found that 51 species show a trend towards laying earlier (Crick *et al.*, 1997, 2003). Dunn and Winkler (1999) analysed 3450 nest records from the North American nest record schemes between 1959 and 1991 and found that the mean lay date of tree swallow (*Tachycineta bicolor*) had shifted an average of 9 days earlier which is mainly attributed to the change in air temperature (Wormworth and Mallon, 2006). Further analysis of nesting data for 65 species of birds in the northern latitudes of the United Kingdom over the period 1971–1995, revealed that laying dates were significantly advanced, on average by 8.8 days, in 20 species including waders, resident and migrant insectivores, corvids and seed–eaters (Crick *et al.*, 1997). Crick and Sparks (1999) found that laying dates were significantly related to either spring temperatures or to spring rainfall. They also calculated that by the year 1995, 53 per cent of species in their study showed a significant advancement in their laying date and they predicted that the average laying dates will be earlier for 75 per cent of species in UK by the year 2080. Hence, there are good evidences to indicate that change in egg–laying date are driven by changes in the climate. Although earlier laying appears to be advantageous as it permits a longer breeding season, but climate change actually lead to a lack of synchrony between the peak nestling food requirements and peak prey availability. This phenological dysfunction could potentially result in reduced breeding success as many new–born offspring might be deprived of sufficient nourishment due to poor prey availability, which in turn might give rise to weaker individuals and, in worse cases, might even lead to a population crash (Robinson *et al.*, 2005). However, it is interesting to note that ambient temperatures during the laying period have been found to positively influence the clutch size of a number of passerine species, including the migratory pied flycatcher (*Ficedula hypoleuca*) and common redstart (*Phoenicurus phoenicurus*) (Järvinen, 1989; Winkel and Hudde, 1997) and increased clutch sizes have also been reported to be positively correlated to the warmer temperatures and earlier laying dates (Winkel and Hudde, 1997).

Few studies suggested that adverse climatic conditions, especially during the onset of the breeding season, might also influence adult birds. Wet summers are associated with poor breeding success for migratory raptors. It has been documented that breeding performance of peregrine falcon (*Falco peregrinus*) and merlin (*Falco columbarius*) (Leech and Crick, 2007), that hunt other bird species on the wing, are adversely affected as heavy rain reduces their visibility, as well as, minimize the activity levels of their avian prey species (Ratcliffe, 1993).

Impact Leading to Population Change

The size of all populations is essentially controlled, at a fundamental level, through the balance of several demographic rates, *i.e.* natality, mortality and migration (Newton, 1998). Significant impact of climate change on both productivity and survival of many bird species have been well documented and the relative magnitudes of these impacts determine the overall change in population size. Number of studies indicate that the impact of local weather, mostly temperature and precipitation variables, largely influence the population size of many species of avifauna. Extreme weather events can have catastrophic effects on bird populations (Stenseth *et al.*, 2002). Increased storm frequencies, as a consequence of climate change, affect the productivity of species that nest in low–lying coastal areas, particularly on tropical islands (Robinson *et al.*, 2005). It may also affect the ability of birds to complete their migratory journeys. Severe storms can also have a direct impact on an individual's survival. Lens and Dhondt (1992) reported the death of 62 per cent of their study population of crested tit (*Parus cristatus*) after a severe storm in Belgium. The effect of tornadoes and hurricanes may be similarly devastating (Wauer and Wunderle, 1992) and the seabirds are particularly vulnerable to those severe adverse conditions, which may result in the mortality of large number of individuals. Migratory species passing through hurricane belts, in areas such as the Caribbean, often face serious threat due to adverse climatic conditions during their migratory journeys (Gauthreaux and Belser, 1999). With the changing climatic conditions, the intensity and frequencies of these storms are supposed to increase making the scenario even worse for those seabirds and migratory species. Extreme climates not only lead to the increased mortality of individuals, but they also influence the abundance and availability of the food supply. For example, extended periods of frost or snow cover may prevent ground–feeding birds from foraging, leading to increased rates of mortality and hence a reduction in population size. Survival rates of many passerines that spend the winter in Britain are known to be related to winter weather conditions (Robinson *et al.*, 2004). About 65 days gap between the date of the first robin sighting and the first date of bare ground at the snow measuring station has been reported, which is 18 days longer than in 1981. As a result, the birds are forced to wait longer for the snow to melt before they can feed facing a greater risk of starvation (Inouye *et al.*, 2000). In UK, annual census datasets of grey heron (*Ardea cinerea*) since 1928 show significant weather driven population crashes due to poor survival in severe winters (Marchant *et al.*, 2004). Climate induced shift in their distribution is leading to a decline in the numbers of many species of Britain's wader populations (Rehfisch *et al.*, 2004; Austin and Rehfisch, 2005).

Similarly, climatic conditions may also influence prey availability for seabirds, which not only affects their productivity, but may also impact on their survival. Population of emperor penguins (*Aptenodytes forsteri*) in Terre Adélie, Antarctica decreased during a relatively warm period (Barbraud and Wiemerskirch, 2001) as the increased sea surface temperatures led to poor krill (*Euphausia superba*) production, resulting in reduced populations of fish and squid (which prey on the krill) and hence poorer foraging success for the penguins. Once survival recovered, population of these penguins stabilized at a new lower level. Similar 94 per cent decline in rockhopper penguins (*Eudyptes chrysocome*) population on Campbell Island since the early 1940s and 90 per cent decline in the numbers of sooty shearwaters (*Puffinus griseus*) off the west coast of North America in the non-breeding season since 1987 were reported to be associated with rises in sea–surface temperature and subsequent declines in the availability of their prey species (Veit *et al.,* 1997; Roemmich and McGowan, 1995). On the contrary, the low temperatures and high rainfall in winter also adversely affect the foraging success of many migratory shorebirds (Insley *et al.,* 1997).

Remarks

It is evident that the global warming and climate change is seriously affecting many aspects of avifauna and there are many reasons to be concerned about the plight of several species of birds induced by climate change. Birds play an important role in nature, pollinating plants, dispersing seeds and eating insects. Several species of warblers are thought to be one of the major bio–control agents in parts of the Great Lakes and Rocky Mountain regions and are responsible for eating up to 84 per cent of spruce budworm larvae, mountain pine beetles, other pests possibly controlling insect outbreaks (Crawford and Jennings, 1989) and play a vital role in the protection of the commercially and ecologically important forests in these regions. Some birds are critical to the reproduction of plants. The long–distance migrants, for example rufous hummingbird (*Selasphorus rufus*), is the primary pollinator for the wild blueberry in southeast Alaska and helps benefit the entire ecosystem as many other wildlife species in the region depend on the blueberry for food (Calder, 1993). Besides playing vital ecological role, they also serve as one of the major indicator of the ecosystem functioning. Species that are unable to respond to climate change through phenotypic and/or genotypic adaptability may be most vulnerable to climate change and species with poor dispersal ability may not be able to adapt with the changing climate parameters. Rapid changes in the environment are likely to favour generalist/ adaptable species and be detrimental to highly specialized species adapted to a specific ecological niche. Survival of birds with small population will be threatened if the climate becomes more variable, as predicted. From the review of several literatures it is evident that several bird populations may be drastically reduced by extreme climatic events and an increase in the frequency of such events may altogether affect their viability. Climate change may lead to the alteration in the bird habitats by changing the time and abundance of their food supplies and other resources leading to the altitudinal shift in distribution and abundance, as well as, change in the migration and breeding phenology of many species. Besides, inundation of most of the low–lying and coastal areas due to rise of sea–level will lead to the loss of many natural habitats which in turn will lead to the unavailability of suitable habitat for a good number of coastal species. With the decline and change in the rainfall pattern, the floral composition of several region of the globe will be altered, which subsequently, will alter the faunal community dependent on it. Rainfall will also influence the crop pattern and agricultural biodiversity. Large–scale changes in agriculture, including crop types and pesticide usage (in response to changing pest populations), will also threaten many bird species that live in such agricultural lands. Species range expansion or contraction in response to climate change, may lead to changes in the level of competition and predation (Harrington *et al.*, 1999). However, information on the impact of climate change on Indian birds is very scanty. More ornithological researches, involving several aspects of birds found in this Indian subcontinent, are essential to understand the impact of climate change on the population dynamics, distribution and survival of individual avifaunal species in this region. Long-term monitoring schemes and databases on population and nests of several bird species can help to identify several new and unforeseen impacts of climate change. Another important area that needs to be addressed is the potential detrimental impacts of the invasive and alien species on the survival of the native flora and fauna under changed environmental conditions.

Moreover, the responses of avifaunal species to global warming are early signals of the wider threat not only to the wildlife and ecosystems, but also to the very existence of human civilization. The shift in climate regime has already started to influence the epidemiology of several vector-borne diseases in human beings as well. India is surrounded by oceans on its three sides. An intensification of tropical and sub-tropical cyclones, sea level rise and subsequent flooding in coastal regions will not only lead to habitat unavailability and population decline of many coastal birds, but will affect millions of people living in coastal villages as well. In India, it is estimated that about seven million people would be displaced with one–meter sea level rise (Asthana, 1995). More than 1,000 people died and there was huge property loss as severe cyclone hit Andhra Pradesh in 1996 (Parikh and Parikh, 2002). Therefore, priority need to be given to the researches on climate change, so that proper adaptation strategy for future could be designed well in advance based on a robust scientific database. The impact and consequences of global warming and climate change is alarming and it is time to act both at local to global level to save many species from the imminent extinction in near future.

Acknowledgement

Sincere thanks are due to the Head, Department of Zoology, University of Calcutta and the Principal, Shibpur Dinobundhoo Institution (College) for their encouragement and support.

References

Ainley, D., Wilson, P. and Fraser, W.R. 2001. Effects of climate change on Antarctic sea ice and penguins. In: *Impacts of climate change on wildlife*, (Green, R.E., Harley, M., Spalding, M. and Zöckler, C. eds.) Royal Society for the Protection of Birds,UK. Pp. 26–27.

Asthana, V. 1995. *Climate Change in Asia*, Asian Development Bank.

Austin, G. and Rehfisch, M.M. 2003. The likely impact of sea level rise on waders (Charadrii) wintering on estuaries. *Journal of Nature Conservation*, 11: 43–58.

Austin, G. and Rehfisch, M.M. 2005. Shifting non-breeding distributions of migratory fauna in relation to climate change. *Global Change Biology*, 11: 31–38.

Bairlein, F. and Winkel, W. 2001. Birds and climate change. In: *Climate of the 21st century: changes and risks* (Lozan, J.L., Grassl, H. and Hupfer, P. eds.). Hamburg, Germany. Pp. 278–282.

Barber, R.T. and Chavez, F.P. 1983. Biological consequences of El Nino. *Science*, 222: 1203–1210.

Barber, R.T. and Chavez, F.P. 1986. Ocean variability in relation to living resources during the 1982–83 El Nino. *Nature*, 319: 279–285.

Barbraud, C. and Weimerskirch, H. 2001. Emperor penguins and climate change. *Nature*, 411: 183–186.

Benning T.L., LaPointe D., Atkinson, C.T. *et al.* 2002. Interactions of climate change with biological invasions and land use in the Hawaiian Modelling the fate of endemic birds using geographic information system. *Proc. Natl. Acad. Sci.USA*, 99(22): 142–146.

Berry, P.M., Vanhinsberg, D., Viles, H.A. *et al.* 2001. Impacts on terrestrial environments. In: *Climate change and Nature Conservation in Britain and Ireland: Modelling natural resource responses to climate change (the MONARCH project).* (Harrison, P.A., Berry, P.M. and Dawson, T.P. eds.). UKCIP Technical report, Oxford. Pp. 43–150.

Boere, G.C., Galbraith, C.A. and Stroud, D.A. (eds.). 2006. *Waterbirds around the world*, The Stationery Office, Edinburgh, UK. Pp. 960.

Böhning–Gaese, K. and Lemoine, N. 2004. Importance of Climate Change for the Ranges, Communities and Conservation of Birds. In: *Birds and Climate Change. Advances in Ecological Research*, (Møller, A., Berthold, P. and Fiedler, W. eds.). Elsevier Academic Press, 35: 211.

Both, C., Bouwhuis, S., Lessells, C.M. *et al.* 2006. Climate change and population declines in a long–distance migratory bird. *Nature*, 441: 81.

Bouma, M.J., Sondorp, H.E. and Van der Kaay, H.J. 1994. Health and climate change. *Lancet*, 343: 302.

Bouma, M. J., Dye, C. and Van der Kaay, H.J. 1996. Falciparum malaria and climate change in the Northwest Frontier Province of Pakistan. *Am. J. Trop. Med. Hyg.*, 55: 131–137.

Bradley, N.L., Leopold, A.C., Ross, J. *et al.* 1999. Phenological changes reflect climate changes in Wisconsin. *Proc. Natl. Acad. Sci. USA.*, 96: 9701–9704.

British Trust for Ornithology (BTO). 2002. *The Effect of Climate Change on Birds*, Information pages by David Leech. Available at: http://www.bto.org/research/advice/ecc/index.htm.

Butler, R., Williams, W.T.D., Warnock, N. *et al.* 1997. Wind assistance: a requirement for migration of shorebirds? *Auk*, 114:456–466.

Calder, W.A. 1993. Rufous Hummingbird (*Selasphorus rufus*). In: *Birds of North America* (Poole, A. and Gill, F. eds.). The Academy of Natural Sciences, Philadelphia, PA, and The American Ornithologists Union, Washington DC.

Chambers, S.A. 2008. *Birds as Environmental Indicators: Review of Literature, Parks Victoria Technical Series No. 55*, Parks Victoria, Melbourne.

Crawford, H.S. and Jennings, D.T. 1989. Predation of birds on spruce budworm *Choristoneura fumiferana*: functional, numerical, and total responses. *Ecology*, 70: 152–163.

Crick, H.Q.P. 2004. The impact of climate change on birds. *Ibis*, 146 (Suppl.1): 48–56.

Crick, H.Q.P. 2006. Migratory Wildlife in Changing Climate. In: *Migratory Species and Climate Change–Impacts of a changing environment on wild animals*, (Frisch H. ed.). UNEP/CMS, Bonn. Pp. 42–43.

Crick, H.Q.P., Dudley, C., Glue, D.E. *et al*. 1997. UK birds are laying eggs earlier. *Nature*, 388: 526.

Crick, H.Q.P. and Sparks, T.H. 1999. Climate change related to egg–laying trends. *Nature*, 399: 423– 424.

Crick, H.Q.P., Baillie, S.R. and Leech, D.I. 2003. The UK Nest Record Scheme: its value for science and conservation. *Bird Study*, 50: 254–270.

Dunn, P.O. and Winkler, D.W. 1999. Climate change has affected the breeding date of tree swallows throughout North America. *Proceedings of the Royal Society of London*, 266: 2487–2490.

Epstein, P.R., Diaz, H.F., Elias, S. *et al*. 1998. Biological and physical signs of climate change: focus on mosquito–borne diseases. *Bull. Am. Meteorol. Soc.*, 79: 409–417.

Epstein, P.R. 2000. Is global warming harmful to health? *Sci. Am.*, 283: 50–57.

Epstein, P.R. 2001. West Nile virus and the climate. *Journal of Urban Health*, 78: 367–371.

Faustino, C.R., Jennelle, C.S., Connolly, V. *et al*. 2004. *Mycoplasma gallisepticum* infection dynamics in a house finch population: seasonal variation in survival, encounter and transmission rate. *J. Anim. Ecol.*, 73: 651–669.

Gatter, W. 1992. Zugzeiten und Zugmuster im Herbst: Einfluß des Treibhauseffekts auf den Vogelzug? *Journal für Ornithologie*, 133: 427–436.

Gauthreaux, S.A. Jr. and Belser, C.G. 1999. Bird Migration in the region of the Gulf of Mexico. *The Ostrich*. In: *Proceedings of the International Ornithological Congress XXII*, (Adams, N.J. and Slotow, R.H. eds.), BirdLife South Africa, Johannesburg.

Harrington, R., Woiwod, I.P. and Sparks, T.H. 1999. Climate change and trophic interactions. *Trends Ecol. Evol.*, 14: 146–150.

Hubalek, Z. 2003. Spring migration of birds in relation to North Atlantic Oscillation. *Folia Zool.*, 52: 287–298.

Huntley B., Collingham Y.C., Green, R.E. *et al*. 2006. Potential impacts of climate change upon geographical distributions of birds. *Ibis*, 148: 8.

Hüppop, O. and Hüppop, K. 2003. North Atlantic Oscillation and timing of spring migration in birds. *Procedings Royal Society London*, 270: 233–240.

Inouye, D.W., Barr, B., Armitage, K.B. *et al.* 2000. Climate change is affecting altitudinal migrants and hibernating species. *Proc. Natl. Acad. Sci., USA*, 97(4): 1630–1633.

Insley, H., Peach, W., Swann, B. *et al.* 1997. Survival rates of redshank *Tringa totanus* wintering on the Moray Firth. *Bird Study*, 44: 277–289.

Intergovernmental Panel on Climate Change (IPCC). 2001. *Climate Change: Impacts, Adaptation, and Vulnerability* (McCarthy, J.J., Canziani, O.F., Leary, N.A. *et al.* eds.). Cambridge University Press, Cambridge, UK. Pp. 1031.

Järvinen, A. 1989. Patterns and causes of long-term variation in reproductive traits of the Pied Flycatcher *Ficedula hypoleuca* in Finnish Lapland. *Ornis Fenn.*, 66: 24–31.

Järvinen, A. 1996. Correlation between egg size and clutch size in the Pied Flycatcher *Ficedula hypoleuca* in cold and warm summers. *Ibis*, 138: 620–623.

Jenkins, D. and Watson, A. 2000. Dates of first arrival and song of birds during 1974–99 in mid–Deeside, Scotland. *Bird Study*, 47: 249–251.

Kovats, R.S., Bouma, M.J., Hajat, S. *et al.* 2003. El Nino and health. *Lancet*, 362: 1481–1489.

Leech, D.I. and Crick, H.Q.P. 2007. Influence of climate change on the abundance, distribution and phenology of woodland bird species in temperate regions. *Ibis*, 149: 128–145.

Lehikoinen, E., Sparks, T. H. and Žalakevicius, M. 2004. Arrival and departure dates. In. *Birds and Climate Change. Advances in Ecological Research*, (Møller, A.P., Fiedler, W. and Berthold, P. eds.). Elsevier Academic Press, 35: 1–31.

Lens, L. and Dhondt, A.A. 1992. The effect of a severe storm on a population of Crested Tits *Parus cristatus* in Belgium. *Bird Study*, 39: 1–33.

Marchant, J.H. and Wernham, C.V. 2003. *Phenology of autumn passage of Willow Warblers at Dungeness, 1960–2000. BTO Research Report No. 323*, British Trust for Ornithology, Thetford. Pp.34.

Marchant, J.H., Freeman, S.N., Crick, H.Q.P. *et al.* 2004. The BTO Heronries Census of England and Wales 1928–1998: new indices and a comparison of analytical methods. *Ibis*, 146: 323–334.

Mason, C. F. 1995. Long-term trends in the arrival dates of spring migrants. *Bird Study*, 42: 182–189.

Newton, I. 1998. *Population Limitation in Birds*, Academic Press, London. Pp. 597.

Parikh, J.K. and Parikh, K. 2002. Climate Change:India's Perceptions, Positions, Policies and Possibilities. *Climate Change and Development*, OECD. 1–30.

Parmesan, C. and Yohe, G. 2003. A globally coherent fingerprint of climate change impacts across natural systems. *Nature*, 421: 37–42.

Peterson, A.T., Ortega–Huerta, M.A., Bartley, J. *et al.* 2002. Future projections for Mexican faunas under global climate change scenarios. *Nature*, 416: 626–629.

Parmesan, C. and Galbraith, H. 2004. *Observed impacts of global climate change in the U.S*, Pew Center on Global Climate Change. Arlington, VA. Pp.67.

Pethick, J.S. and Crooks, S. 2000. Development of a coastal vulnerability index: a geomorphological perspective. *Environmental Conservation*, 27: 359–367.

Piersma, T. and Jukema, J. 1990. Budgeting the flight of a long–distance migrant: changes in nutrient reserve levels of bar–tailed godwits at successive spring staging sites. *Ardea*, 78: 315–337.

Pounds, J.A., Fogden, M.P.L. and Campbell, J.H. 1999. Biological response to climate change on a tropical mountain. *Nature*, 398: 611–615.

Pounds, J.A., Bustamante, M.R., Coloma, L.A. *et al.* 2006. Wide spread amphibian extinctions from epidemic disease driven by global warming. *Nature*, 439: 161–167.

Price, J. and Glick, P. 2002. *A birdwatcher's guide to global warming*, National Wildlife Federation and American Bird Conservancy. Pp. 34.

Ratcliffe, D.A. 1993. *The Peregrine Falcon* (2nd ed.). London. Pp. 416.

Rehfisch, M.M., Austin, G.E., Freeman, S.N. *et al.* 2004. The possible impact of climate change on the future distributions and numbers of waders on Britain's non-estuarine coast. *Ibis*, 146(Suppl.1): S70–S81.

Robinson, R.A., Green, R.A., Baillie, S.R. *et al.* 2004. Demographic mechanisms of the population decline of the Song Thrush *Turdus philomelos* in Britain. *J. Anim. Ecol.*, 73: 670–682.

Robinson, R., Learmonth, A., Hutson, M., *et al.* 2005. *Climate change and migratory species. BTO Research Report*, Department for Environment, Food and Rural Affairs (Defra), London. Pp. 414.

Roemmich, D. and McGowan, J.A. 1995. Climatic warming and the decline of zooplankton in the California Current. *Science*, 267: 1324–1326.

Root, T.L. and Schneider, S.H. 1993. Can large–scale climatic models be linked with multiscale ecological studies? *Conserv. Biol.*, 7: 256–270.

Root, T.L., Price, J.T., Hall, K.R. *et al.* 2003. Fingerprints of global warming on wild animals and plants. *Nature*, 421(6918): 57–60.

Root, T.L and Hughes, L. 2005. Present and future phenological changes in wild plants and animals. In: *Climate Change and Biodiversity* (Lovejoy, T.E. and Hannah, L. eds.). Yale University Press, New Haven and London. Pp. 61–69.

Sanderson F.J., Donald P.F., Pain D.J. *et al.* 2006. Long-term population declines in Afro–Palearctic migrant birds. *Biological Conservation*, 131: 93.

Spalding, M., Grady, S. and Zöckler, C. 2001. Changes in tropical regions. In: *Impacts of climate change on wildlife* (Green, R.E., Harley, M., Spalding, M. and Zöckler, C. eds.), Royal Society for the Protection of Birds, UK. Pp. 28–29.

Sparks, T.H. 1999. Phenology and the changing pattern of bird migration in Britain. *Int. J. Biometeorol.*, 42: 134–138.

Sparks, T.H. and Mason, C.F. 2001. Dates of arrivals and departures of spring migrants taken from the *Essex Bird Reports* 1950–1998. *Essex Bird Report*, 1999: 154–164.

Sparks, T.H. and Mason, C.F. 2004. Can we detect change in the phenology of winter migrant birds in the UK? *Ibis*, 146(Suppl.1): 58–61.

Stenseth, N.C., Mysterud, A., Ottersen, G. *et al.* 2002. Ecological effects of climate fluctuations. *Science*, 297: 1292–1296.

Thompson, P.M. and Ollason, J.C. 2001. Lagged effects of ocean climate change on fulmar population dynamics. *Nature*, 413: 417–420.

Tompkins, D.M., Greenman, J.V., Robertson, P.A. *et al.* 2000. The role of shared parasites in the exclusion of wildlife hosts: *Heterakis gallinarum* in the ring–necked pheasant and the grey partridge. *J. Anim. Ecol.*, 69: 829–840.

Viet, R.A., McGowan, J.A., Ainley, D.G. *et al.* 1997. Apex marine predator declines ninety per cent in association with changing oceanic climate. *Global Change Biology*, 3: 23–28.

Visser, M.E., Silverin, B., Lambrechts, M.M. *et al.* 2002. No evidence for tree phenology as a cue for the timing of reproduction in tits *Parus* spp. *Avian Science*, 2: 77–86.

Visser, M.E., Both, C. and Lambrechts, M.M. 2004. Global climate change leads to mistimed avian reproduction. *Advances in Ecological Research*, 35: 89–110.

Wauer, R.H. and Wunderle, J.M. 1992. The effect of hurricane Hugo on bird populations on St Croix, United States Virgin Islands. *Wilson Bull.*, 104: 656–673.

Winkel, W. and Hudde, H. 1997. Long-term trends in reproductive traits of tits (*Parus major, P. caeruleus*) and Pied Flycatcher *Ficedula hypoleuca. Journal of Avian Biology*, 28: 187–190.

Wormworth, J. and Mallon, K. 2006. *Bird Species and Climate Change: The Global Status Report (Version 1.0)*, WWF, Australia. Pp. 75.

Zöckler, C. and Lysenko, I. 2001. Waterbirds on the edge: climate change impact on Arctic breeding waterbirds. In: *Impacts of climate change on wildlife* (Green, R.E., Harley, M., Spalding, M. and Zöckler, C. eds.). Royal Society for the Protection of Birds, UK. Pp. 20–25.

Climate Change: Man and Environment (2012)   *Pages* 277–283
Editor: **Goutam Kumar Saha**
Published by: **DAYA PUBLISHING HOUSE, NEW DELHI**

24

Impact of Micro and Macro Level Climate Change on Animal Behaviour

☆ *Pulak Lahiri**

Introduction

Climate, weather and environment are interconnected and interdependent. As concepts, they are multi dimensional. Climate is a region's long-term atmospheric condition typically over decades; while weather is an area's short term atmospheric conditions, typically over hours or days (Tyler Miller, 2005). The term environment has somewhat different connotation especially from the perspective of animal behaviour. It implies the kind of habitat in which an animal lives and responds. The animal experiences here the totality of all external factors, both biotic and physical. In this sense it is logical to describe environment as the animal's microhabitat or microclimate, since the environment *per se* is modified by the animal's own behavioural choices, more so by its own presence (Willmer *et al.*, 2000; Lahiri, 2006a).

The behaviour of an animal depends upon climate (*e.g.* annual cycles like migration, hibernation etc.), weather (seeking shades or burrows depending upon high temperature and tide) and its microclimate (top canopy, liter layer etc. where it lives). Change in any component of these systems is bound to affect an animal's behaviour. Gross change is quite perceptible, while subtle change brought about by minor change of climate is difficult to notice, until it snow balls into a crisis. The near extinction of many species of birds caused by exposure to organochlorine (OC)

* Formerly of Department of Zoology, University of Calcutta, Kolkata. E-mail: pulaklahiri1@rediffmail.com

insecticides causing shell thinning of eggs is a glaring example of such crisis (Lahiri, 2006b). Humans have always been intrigued by animal behaviour and the interest has stemmed from both survival need and curiosity about natural world. Behaviour is centrally important for any species. The survival of a species depends on its individual member's ability to obtain food and shelter (foraging behaviour), to find appropriate mate and reproduce (reproductive and mating behaviour), to raise offspring and to protect themselves from predators (parental care and anti–predator behaviour). All such behaviour and strategies ensure survival, and survival in turn ensures evolutionary success. An individual of a species must therefore adopt set of behaviour, some of which are programmed and others are learned.

When a behaviour pattern gets stabilized (through the process of evolution), it repeats itself in a kind of stereotyped pattern, independent of immediate control and learning (*e.g.* Fixed Action Pattern or FAP).

The prevailing climate plays an important role in shaping such behaviour; therefore, any change in climatic conditions is bound to influence the behaviour adopted by an animal. An important evidence of evolution is the fossil remains of a species. Unfortunately, behaviour itself does not leave behind any fossil remains. However, many of the fossils of animals indicate kind or change in behaviour subsequent to climate change. Take for example, the evolution of horse. The fossils of horse lineage in North America indicate that primitive horses were browsers, eating leaves and succulent plants. With climate in North America becoming drier during Miocene, grasslands replaced riparian forests forcing the horses to be fast running and to adopt ability to chew tough grasses. These behavioural changes are quite evidenced in the change of their foot and tooth structures (Simpson, 1945, 1949; Stanley, 1993). The evolution of horse tribe involved progression through eight genera and took some last 60 million years (Dobzhansky, 1951). Indeed evolution is a slow process and so also the corresponding changes in behaviour. However, a strong selection pressure may sometimes act on a population and if generation time is short, there are opportunities to follow changes from one generation to the next. Consider for example, Kettlewell's (1965) well known study on industrial melanism, a kind of mimicry, *i.e.* change in color in the peppered moth (*Biston betularia*) brought by local climate change due to industrial revolution in Britain. The preponderant natural population of these moths is light coloured. They are active during night and rest on the lichen infested bark of trees during day remaining camouflaged with lichens, thereby evading predation by the birds. Pollution, as a fallout of industrial growth, killed the very sensitive lichens, exposing the light coloured moths to their predators; as a result the less predominant darker moths were more likely to survive and reproduce, therefore their frequency in the population increased. Interestingly, as air became cleaner later due to subsequent anti–pollution measures, lichens resurfaced and so also the light coloured moths!

Another example of recent local weather change affecting population and behaviour of animals may be cited from the well coordinated study of Grant (1981, 1986) on two species of finches living in the Daphne island of Galapagos. *Geospiza magnirostris* is a bigger bird with relatively large and strong beak enabling it to crack open easily the large and tough fruits like caltrops. The ground finch *G. fortis* on the other hand, is smaller with smaller beak and is a seedeater. In 1977–78, a drought prevailed, availability of small seeds become scare and the ground finches failed to breed, their population suffered almost 85 per cent decline. With hard seed and fruits remaining relatively common, *G. magnirostris* with larger and stronger beaks was favoured and the frequency of these birds increased substantially within a short time. The situation changed in 1982 when a spate of torrential rain and storm following El Nino, swept Galapagos, the weather eventually leading to a huge excess of small seeded plants. Over next few generations, frequency of smaller bird population increased. Such effects do not only show the population dynamics pattern as affected by weather conditions, but also the underlying adaptive mechanisms that include behavioural strategies under duress. Indeed studies in behavioural ecology show both the ecological realm of the species under study and the way natural selection might shape the observed behaviour even during periodic climate change.

Behaviour can be divided either into broad descriptive categories (*e.g.* courtship, nesting, sleeping, feeding, etc.) or into more restricted units *e.g.* specific patterns shown during various phases of courtship (Dricamer *et al.,* 2002). A chain of events spans these aspects, triggering one action after another. The physiological and developmental state of the animal is very important in stimulating a specific behavioural action. If any of the physiological or development step is affected by some unwarranted agent or event, the normal physiological or developmental cycles become disturbed, upsetting the chain of expected behaviour pattern.

In recent past many abnormalities or aberrations in behaviour have been noticed, generally caused by change in animal's microclimate. One such effect as mentioned earlier is the episode of eggshell thinning and related problems in birds and other animals caused by the exposure to organochlorine insecticides. It should be mentioned here that Rachel Carson (1962) in her famous book, 'Silent Spring' first described how wide spread use of DDT had caused unusual behaviour like silencing of "robins, catbirds, doves, jays, wrens and scores of bird voices". Ultimately, DDT and other organochlorines were banned for agricultural practices in USA and other countries.

An important aspect of physiology directing animal behaviour is its precise regulation by endocrine system. Over years, a finely tuned system of complex interaction between behaviour, hormones and external cues have developed in many animals. Of these, special mention must be made about gonadal hormones *viz.,* androgen, progesterone and estrogen. Interestingly over species barrier the structure of these steroid hormones remain largely conserved unlike many peptide hormones. Therefore, if the functions of these hormones become affected by some xenobiotics or other chemicals that mimic such hormone action, their effects do not remain restricted to a particular species. Besides controlling aggressive, territorial or mating behaviour, steroid hormones have marked organizational effects during critical period in behavioural development. Therefore, such effects produce relatively permanent change in the organism's nervous and other tissues. Thyroid and adrenal hormones also have organizational effects on behaviour. Hypothyroid animal exhibit characteristics of cretinism, slower growth, delayed sexual maturation and retarded development of nervous system. As a result their actions are slower and occur with great difficulty. Non availability of iodine in the microclimate strongly affects metamorphosis, and/

or retention of larval characters in many amphibians (*e.g.* Axolotol larva). Environmental disruption that may affect iodine content of the soil is bound to affect such susceptible animals (Tyler Miller, 2005).

Since hormones are molecular messengers, precise functioning of the endocrine system via modification of rates and directions of various cellular activities is of fundamental importance to behaviour. A number of endocrine disrupting chemicals (EDCs) like dioxins, polychlorinated biphenyl (PCB), phenolics, pthalates and of course DDT, lindane and other chlorinated insecticides affect microclimate of animals (Lahiri, 2006b). Exposure of animals to such xenobiotics early in life may lead to profound and irreversible damage to behaviourally relevant neural circuits, gender development and morphology. The well known effect of Tributyl tin (TBT) used as anti fouling paint for oil containers on dog whelk (*Nucella* sp.) populations in Britain may be cited here. TBT induced development of penis–like organ in the females causing phenotypic sex reversal in these barnacles ("imposex"); as a result the population sex ratio became highly upset (Turk and Turk, 1988). Similar effects have also been found in fish, alligator and bird exposed to such chemicals (Krimsky, 2001). How unsuspecting chemicals might affect key physiological functions like sex differentiation reflected in the behaviour effect, is sometimes beyond imagination (Bretveld *et al.*, 2006)! The normal nesting behaviour in birds has been adversely affected by DDT and other organochlorines. Similarly migratory behaviour in Atlantic salmons having high residues of pesticides is adversely affected (Guilette *et al.*, 1995). Below is an account of some altered development and reproductive effects caused by hormone–disrupting chemicals (EDCs) in vertebrates (Colborn *et al.*, 1987; Gray *et al.*, 2001; Guillete, *et al.*, 1995; Bretveld *et al.*, 2006).

☆ Sex reversal – Imposex, female phallus development, abnormal steroid genesis, spermatogenesis and behaviour (salmon, red–eared slider turtle); ovotestis, hermaphrodites (frogs, toads).

☆ Sex reversal – Abnormal penis development, steroid genesis and gene expression (alligator).

☆ Abnormal gonadal differentiation and altered reproductive behaviour (egrets and other birds).

☆ Abnormal genital development, steroid genesis and gene expression (grizzly bear and other mammals).

☆ Even humans are not spared from adverse effects of EDCs (Krimsky, 2000, 2001; Swan *et al.*, 2005).

While the above effects generally indicate microclimate alteration affecting fine tuning of precise physiological processes in a particular habitat, projected effects of global climate change on animals is far wide and obviously of much larger scale. There is no doubt that climate change is real and happening now. The average global surface temperature has increased by 0.8°C in the past century and 0.6°C in past three decades (Anon, 2001) largely because of human activities. If this trend continues, the change could result in an increase in the number of days falling outside of temperature thresholds set by basic physiological tolerance for each species and especially for sensitive species. Amphibians are very sensitive to climate change. Their populations are declining at an alarming rate even in protected wildlife reserves and parks (Tyler Miller, 2005). The eggs of amphibians have no protective shells to block ultra violet radiation or pollution. Increase in UV caused by reduction of stratospheric ozone can harm young embryos in shallow ponds (Tyller Miller, 2005). In three of five major reptilian lineages, sex is determined by nest and ambient temperature, since they lack heteromorphic sex chromosome. However, it is just not temperature *per se* at any particular moment, but cumulative effects of temperature through a critical phase that matter (Dugatkin, 2004). In the field, the sex ratio of hatchlings is largely affected by location of nests and the thermal microclimate of the nest interior, which in turn depends on the ambient temperature condition. A two degree overall rise in ambient temperature will affect temperature dependent sex determination (TSD) in reptiles. Similarly old world leaf eating monkeys like colobine will be harder hit by a moderate 2°C rise in global temperature. Already there are signs of subtle behavioral change indicating stress effects in the colobine monkeys. Some of the observed behavioural changes believed to be induced by global warming are indicated below:

1. Marmots ending their hibernation about three weeks earlier now, compared to thirty years ago.
2. Many fish species are moving northward in search of cooler water.
3. A fruit fly gene normally associated with hot, dry conditions has spread through populations living in traditionally cooler southern regions.
4. Grizzly bears are increasingly moving into areas long been considered as polar bear habitat. The latter are already imperiled by ice loss in the Arctic.

Such alteration of behaviour is quite indicative of subtle climate change. It seems that the climate change effect is no more remaining restricted to a particular habitat of an animal, rather it is becoming far wide and on global scale. It is time to wake up. It also needs a knee–jerk reaction for immediate preventive action!

Remarks

Climate, weather and environment are interconnected. Together and individually they affect animal behaviour. An animal remains fine tuned to its environment, more so to its microhabitat or microclimate. On daily or seasonal basis the animal responds appropriately to environmental cues for foraging, predation, anti–predation, reproduction, migration, etc. The survival strategies reflected in their behaviour are product of selection and therefore evolution. Failure in adaptive behaviour during cataclysmic climate change (*e.g.* glaciation) in the past has resulted in extinction of many species. In recent times a different kind of climate change caused due to anthropogenic activities has become a matter of grave concern. The animals are not adapted to the abruptness or kind of climate change in action and therefore they are not programmed to respond appropriately. This is leading to mismatching of behaviour, which is quite detrimental for the animal's well being and survival. Studies of wildlife indicate associations between hormone–disrupting chemicals in the environment and declining population, thinning eggshells, morphologic abnormalities, impaired viability of offspring and neurological disorders including cognitive and neuro–behavioural effects. An area of grave concern is the EDC–induced sex ratio skew, which has adversely affected behaviour and breeding patterns in wildlife. The projected behavioural effects of global climate change as sequels of global warming are of wide significance – the red alert area include migration, hibernation, reproduction and foraging. Behaviour is a sentinel of well being of an animal and of its immediate surrounding. Behaviour also becomes the first casualty of induced weather or climate change. Since climate change is happening, it is high time to recognize the threat of climate change at both macro and microhabitat level and seek preventive measures for survival and well being of animals.

References

Anon. 2001. *Climate Change 2001: Synthesis Report.* Integrated Panel on Climate Change (IPCC), Cambridge University Press, Cambridge.

Bretveld, R.W., Thomas, C.M.G., Scheepers, P.T.G. *et al.* 2006. Pesticide exposure: The hormone function of the female reproductive system disrupted? *Repro.Biol. Endocrinol.*, 4: 30–44.

Carson, R. 1962. *Silent Spring.* Houghton Mitten, Boston.

Colborn, T., Fry, D.M., Toone, C.K. *et al.* 1987. Sex ratio skew and breeding patterns of gulls: demographic and toxicological considerations. *Studies in Avian Biol.*, 10: 26–43.

Dobzhansky, T. 1951. *Genetics and the Origin of Species* (3rd ed.). Oxford and IBH Publishing Company, Calcutta.

Dricamer, L.C., Vessey, S.H. and Jacob, E.M. 2002. *Animal Behaviour* (5th ed.). Mcgraw Hill.

Dugatkin, L.A. 2004. *Principles of Animal Behaviour.* WW Norton and Company, NY.

Grant, P.R. 1981. Speciation and the Adaptation of Darwin's finches. *Amer Sci.*, 69: 653– 663.

Grant, P.R. 1986. *Ecology and Evolution of Darwinian Finches.* Princeton Univ. Press, Princeton, NJ.

Gray, L.E. Jr., Ostby, J., Furr, J. *et al.* 2001. Effects of environmental antiandrogens on reproductive development in experimental animals. *Human Repro. Update.*, 7(3): 248– 264.

Guillette, L.J. Jr., Crain, D., Roony, A.A. *et al.* 1995. Organisational versus Activation: The role of endocrine–disrupting contaminants (EDCs) during embryonic development in wild life. *Env. Hlth. Persp.*, 103(Suppl.7): 157–165.

Kettlewell, H.B.D. 1965. Insect Survival and Selection for Pattern. *Science*, 148: 1290–1296.

Krimsky, S. 2000. *Hormonal Chaos.* Johns Hopkins Univ. Press, Baltimore, Md.

Krimsky, S. 2001. Hormone Disruptor: A clue to understanding the environmental causes. *Environment*, 43(5): 23–31.

Lahiri, P. 2006a. Environment Is. In: *Environmental Change and its Impact*. (Mitra, M.K., Sarkar, P. and Biswas, S. eds.) Classique Books, Kolkata. Pp. 11–18.

Lahiri, P. 2006b. Lindane, an endocrine disruptor causes reproductive failure and fetotoxicity. In: *Environmental Awareness and Wildlife Conservation* (Basu, R.C., Khan, R.A. and Alfred, J.R.B. eds.). Zoological Survey of India, Kolkata. Pp. 241–249.

Simpson, G.G. 1945. *Tempo and Mode in evolution*. Columbia Univ. Press, NY.

Simpson, G.G. 1949. *The Meaning of Evolution*. Yale Univ. Press, New Haven, CT.

Stanley, S.M. 1993. *Earth and Life through Time* (2nd ed.). W.H. Freeman, NY.

Swan, S.H., Main K.M., Liu, F. *et al.* 2005. Decrease in anogenital distance among male infants with prenatal phthalate exposure. *Env. Hlth Persp.*, 113(8): 1056–1061.

Turk, J. and Turk, A. 1988. *Environmental Science* (4th ed). Saunders College Publishing, NY.

Tyler Miller, G. Jr. 2005. *Essentials of Ecology*. Thomson, Singapore.

Willmer, P., Stone, G. and Johnson, I. 2000. *Environmental Physiology of Animals*. Blackwell Science, London.

Climate Change: Man and Environment (2012)   *Pages* **284–290**
Editor: **Goutam Kumar Saha**
Published by: **DAYA PUBLISHING HOUSE, NEW DELHI**

25

Climate Change and Biodiversity: Indian Initiatives

☆ *Ramakrishna**

Introduction

India's 3,166,414 sq km area shows a notable diversity of habitats due to significant variations in rainfall, altitude, topography and latitude that cause major seasonal changes in vegetation. They include wide range of eco–zones like desert, high mountains, highlands, tropical and temperate forests, swamplands, plains, grasslands, riverine areas, as well as, island archipelago. It hosts four biodiversity hotspots: the Himalayas, the Western Ghats, the hilly ranges that straddle the India–Myanmar border and the Nicobar group of islands. These hotspots have numerous endemic species. Because of its size and range of latitude, topography and climate, India is home to a great diversity of ecoregions, ranging from permanent ice and snow covered Himalayas to tropical rainforests.

India is endowed with an immense variety of natural resources in its rich animal (91,200 species) and plant heritage (44.447 species), besides over 250 breeds of farm animals. According to world biogeographic classification, India represents two of the major realms (Palaearctic and Indo–Malayan) and three biomes (Tropical Humid Forests, Tropical Dry/Deciduous Forests and Warm Deserts/Semi–Deserts). These include 12 biogeographic regions. However, the Wildlife Institute of India has proposed a modified classification–which divides the country into 10 biogeographic regions and 20 biogeographic zones. The rich vegetation wealth and diversity in the country is enormous due to the variety of climatic and altitudinal variations coupled

* Zoological Survey of India, Kolkata. E-mail: ramakrishna.zsi@gmail.com

with diverse ecological habitats. Champion and Seth (1968) recognized 16 major forest types comprising 221 minor types, of these, tropical moist deciduous forms the major percentage (37 per cent), followed by tropical dry deciduous (28.6 per cent), tropical wet evergreen (8 per cent), tropical thorn forest (2.6 per cent) and others are of minor values. Recent Forest Cover Assessment by Forest Survey of India (2009) indicates forest and tree cover of the country is 78.37 million hectares (mha) amounting to a dense forest cover of 8.35 mha (2.54 per cent of geographic area), medium dense forest cover of 31.90 mha (9.71 per cent) and open forest of 28.84 mha (8.77 per cent), thus totaling 21.02 per cent; while the non forest cover constitute a major portion (77.72 per cent) with scrub forest depicting only 1.26 per cent of the total area.

Forests are an important reservoir of biodiversity. Ancient and frontier forests, because of their long standing and relatively lower levels of human disturbance, are typically richer in biodiversity than other natural or semi–natural forests. Loss of forest will inevitably result in reduction of biodiversity as a direct result of loss of habitat. There are many anthropocentric reasons to preserve biodiversity (direct use value, option value, etc.) but the principal such reason considered here is the indirect use value in the form of ecosystem services. A loss in biodiversity affects the stability of an ecosystem resulting in a reduction of its resistance to disruption of the food web (by loss of the weak interaction effect), resistance to species invasion and resilience to global environmental change. Ecosystem diversity supplies important ecosystem services, *viz.,* carbon sequestration, plant pollination, pest control by natural predators, watershed protection and erosion control, maintenance of soil fertility and pasture regeneration and breakdown of waste and pollutants, etc. Additionally, ecosystem diversity can contribute to development by encouraging nature–based tourism. Thus, ecosystem services improve local well being by providing clean water and productive agricultural systems.

India is a large developing country with nearly 700 million rural population directly depending on climate–sensitive sectors (agriculture, forests and fisheries) and natural resources (such as water, biodiversity, mangroves, coastal zones, grasslands) for their subsistence and livelihoods (Sathaye *et al.,* 2006). Further, the adaptive capacity of dry land farmers, forest dwellers, fisherfolk and nomadic shepherds is very low (Ravindranath and Sathaye, 2002). Climate change is likely to impact all the natural ecosystems as well as socio–economic systems as shown by the National Communications Report of India to the UNFCCC (2004). The latest high resolution climate change scenarios and projections for India, based on Regional Climate Modeling (RCM) system, known as PRECIS developed by Hadley Center and applied for India using IPCC scenarios A2 and B212 shows the following:

☆ An annual mean surface temperature rise by the end of century, ranging from 3–5° C, with warming more pronounced in the northern parts of India.

☆ A 20 per cent rise in all India summer monsoon rainfall and further rise in rainfall is projected over all states except Punjab, Rajasthan and Tamil Nadu, which show a slight decrease.

☆ Extremes in maximum and minimum temperatures are also expected to increase.

☆ Similarly extreme precipitation also shows substantial increases, particularly over the west coast of India and west central India.

Expected impacts due to climate change in India projected by India's Initial National Communications to the United Nations Framework Convention on Climate Change is as below.

Water Resources

The hydrological cycle is likely to be altered and the severity of droughts and intensity of floods in various parts of India is likely to increase. Further, a general reduction in the quantity of available run–off is predicted. Water yield (which is a function of precipitation, total surface run–off, evapo–transpiration and soil properties), is projected to increase in the Himalayan region in 2030s by 5–20 per cent. However, water yields are likely to be variable across the northeastern region, Western Ghats and coastal region. In some places of these regions, water yield is projected to increase and in some places it is projected to decrease. Moderate to extreme drought severity is projected in 2030s for the Himalayan region, as compared to the other regions. All the regions are likely to experience flooding which are exceeding existing magnitudes by 10 per cent to 30 per cent.

Agriculture

Simulations using dynamic crop models indicate a decrease in yield of crops as temperature increases in different parts of India. However, this is offset by an increase in CO_2 at moderate rise in temperature and at higher warming; negative impact on crop productivity is projected due to reduced crop durations. According to the latest report, irrigated rice in all the regions are likely to gain in yields marginally due to warming as compared to the rain fed crop as the irrigated rice tends to benefit from 'CO_2 fertilization effect'. Maize and sorghum are projected to have reduced yields in all the regions. The coconut productivity is projected to rise in the western coast and reduce in the eastern coastal region. Observations indicate a reduction in apple production in the Himalayan region, which is likely to continue in the future.

Forests

Climate impact assessments using BIOME–3 model and climate projections for the year 2085 show 77 per cent and 68 per cent of the forested grids in India are likely to experience shift in forest types. Indications show a shift towards wetter forest types in the northeastern region and drier forest types in the northwestern region in the absence of human influence. Increasing atmospheric CO_2 concentration and climate warming could also result in a doubling of net primary productivity under the A2 scenario and nearly 70 per cent increase (Ravindranath *et al.*, 2005). Change is projected for 8 per cent, 18 per cent, 56 per cent and 30 per cent of the vegetation grids an increase in net primary productivity by 23 per cent, 20 per cent, 57 per cent and 31 per cent is projected in Western Ghats, northeastern region, Himalayan region and the coastal region respectively.

Coastal Zone

Simulation models show an increase in frequencies of tropical cyclones in the Bay of Bengal; particularly intense events are projected during the post–monsoon period. Sea level rise is projected to displace populations in coastal zones, increase flooding in low–lying coastal areas, loss of crop yields from inundation and salinization.

Human Health

Malaria and dengue incidents are likely to persist in many states and new regions may become malaria–prone and the duration of the malaria transmission windows is likely to widen in northern and western states and shorten in southern states. Malaria is projected to spread in new areas in Jammu and Kashmir in the Himalayan region. In the northeastern region opportunities for transmission is likely to increase for a longer period. In the Western Ghats, no change is observed between in 2030s and the trends observed in 1970s. However, in the coastal region, especially in the eastern coast, a marked decrease in number of months is projected in which that the malaria transmission window would be open.

Desertification

Globally, about 1900 million hectares of land are affected by land degradation, of which 500 million hectares each are in Africa and the Asia–Pacific and 300 million hectares in Latin America. Climate change leading to warming and water stress could further exacerbate land degradation, leading to desertification. The United Nations Convention to Combat Desertification (UNCCD) aims to address the problem of land degradation, which is linked to climate change. It is important to note that the climate–sensitive sectors (forests, agriculture, coastal zones) and the natural resources (groundwater, soil, biodiversity, etc.) are already under stress due to socio–economic pressures. Climate change is likely to exacerbate the degradation of resources and socio–economic pressures. Thus, countries such as India with a large population dependent on climate–sensitive sectors and low adaptive capacity have to develop and implement adaptation strategies (Sathaye *et al.*, 2006).

India's National Action Plan on Climate Change (NAPCC)

On June 30, 2008, Prime Minister released India's first National Action Plan on Climate Change (NAPCC) outlining existing and future policies and programmes addressing climate mitigation and adaptation. The plan identifies eight core "national missions" emphasizing the overriding priority of maintaining high economic growth rates to raise living standards. The plan "identifies measures that promote our development objectives while also yielding co–benefits for addressing climate change effectively." It says these national measures would be more successful with assistance from developed countries and pledges that India's per capita greenhouse gas emissions "will at no point exceed that of developed countries even as we pursue our development objectives."

National Solar Mission

The NAPCC aims to promote the development and use of solar energy for power generation and other uses with the ultimate objective of making solar competitive with fossil–based energy options.

National Mission for Enhanced Energy Efficiency

Current initiatives are expected to yield savings of 10,000 MW by 2012. The plan includes specific goals for increasing use of solar thermal technologies in urban areas, industry and commercial establishments; a goal of increasing production of photovoltaic to 1000 MW/year; and a goal of deploying at least 1000 MW of solar thermal power generation.

National Mission on Sustainable Habitat

To promote energy efficiency as a core component of urban planning, the plan calls for: Extending the existing Energy Conservation Building Code; a greater emphasis on urban waste management and recycling, including power production from waste; strengthening the enforcement of fuel economy standards in the automobiles and using pricing measures to encourage the purchase of efficient vehicles; and incentives for the use of public transportation.

National Water Mission

With water scarcity projected to worsen as a result of climate change, the plan sets a goal of a 20 per cent improvement in water use efficiency through pricing and other measures.

National Mission for Sustaining the Himalayan Ecosystem

The plan aims to conserve biodiversity, forest cover and other ecological values in the Himalayan region, where glaciers that are a major source of India's water supply are projected to recede as a result of global warming.

National Mission for a "Green India"

Goals include the afforestation of 6 million hectares of degraded forestlands and expanding forest cover from 23 per cent to 33 per cent of India's territory.

National Mission for Sustainable Agriculture

The plan aims to support climate adaptation in agriculture through the development of climate–resilient crops, expansion of weather insurance mechanisms and agricultural practices.

National Mission on Strategic Knowledge for Climate Change

To gain a better understanding of climate science, impacts and challenges, the plan envisions a new Climate Science Research Fund, improved climate modelling, and increased international collaboration. It also encourages private sector initiatives to develop adaptation and mitigation technologies through venture capital funds.

This is reflected in the document prepared by CBD (22nd May. 2008) that Biodiversity related impacts by anthropogenic climate changes are:

1. Species' geographic ranges are shifting towards higher latitudes and elevations.

2. Changes in timing of life cycles (phenology) of natural events such as flowering, reproduction and leaf unfolding, which has the bearing on the life cycle of many animals dependent on such changes.

3. Interactions between species vary resulting in mismatches between the peak of resource demands by reproducing animals and the peak of resource availability. This is causing population declines in many species and may indicate limits to natural adaptation.

4. Changes in photosynthetic rates, carbon uptake and productivity in response to CO_2 "fertilization" and nitrogen deposition, thus affecting gross primary production.

5. Structural and functional changes in ecosystems are resulting in substantial changes in species abundance and composition. These have impacts on livelihood and traditional knowledge including, for example, changing the timing of hunting and fishing and traditional sustainable use activities, as well as impacting upon traditional migration routes for people.

6. Climate change will interact with other pressures acting on natural systems, most notably land use and land use change, invasive alien species and disturbance by fire.

7. Changes and shifts in the distribution of marine biodiversity resulting from climate change could have serious implications for fisheries, affecting the socio–economic fabric of society.

Remarks

There is currently considerable evidence to suggest that climate change has already affected the abundance, distributions and population dynamics of a wide range of animal and plant species. The prime priority should be oriented towards understanding the impact of climate change on the population dynamics. Breeding biology of several migratory species, especially birds, will offer many clues of long and short–term monitoring of species. Warmer temperatures may result in altitudinal extension of grass and heather moorland at the expense of the lichen–dominated montane habitat on which many bird species breed. This is evident from the studies on relationships between survival rates, breeding success, abundance, distribution, food availability, phenology of bird species and a range of climatic variables. Climate change may lead to shifts in the rate of reproduction of pathogens or in the distribution of vectors carrying these pathogens. In fact, the warm winters and spring droughts may have contributed to the spread of West Nile virus through avian populations in Europe and North America and that the incidence and distribution of this pathogen was likely to increase if predicted long-term trends towards a general amelioration of the climate in these areas are accurate.

Besides above, impact of climate especially the temperature variation on sex determination in selected species, hatching behaviour of marine turtles, thermo regulation in cold blooded animals, altitudinal migration of selected species, rate of extinctions in amphibians due to an emerging disease called chytridiomycosis and global climate change are thought to be the biggest threats to amphibians.

Reference

Champion, H.G. and Seth, S.K. 1968. *Revised forest types of India*. Govt.of India Publications, New Delhi.

Ravindranath, N.H. and Sathaye, J. 2002. *Climate Change and Developing Countries*. Kluwer Academic Publishers, Dordrecht, Netherlands.

Ravindranath, N.H., Joshi, N.V., Sukumar, R. *et al*. 2005. Impact of climate change on forests in India. *Curr. Sci.*, 90: 354–361.

Sathaye, J., Shukla, P.R. and Ravindranath, N.H. 2006. Climate change, sustainable development and India: Global and National concerns, *Curr. Sci.*, 90(3): 315–325.

UNFCCC. 2004. India's Initial National Communications to the United Nations Framework Convention on Climate Change, Ministry of Environment and Forests, New Delhi. dated 16th Nov. 2010 A "4X4" assessment of the impact of climate change on key sectors and regions of India in 2030s. www.moef.nic.in

Index

www.ingramcontent.com/pod-product-compliance
Lightning Source LLC
Chambersburg PA
CBHW050510190326
41458CB00005B/1495